NUREG-1815, Vol. 2

Environmental Impact Statement for an Early Site Permit (ESP) at the Exelon ESP Site

Final Report

Appendices A through K

Manuscript Completed: July 2006
Date Published: July 2006

Division of New Reactor Licensing
Office of Nuclear Reactor Regulation
U.S. Nuclear Regulatory Commission
Washington, DC 20555-0001

Abstract

This environmental impact statement (EIS) has been prepared in response to an application submitted to the U.S. Nuclear Regulatory Commission (NRC) by Exelon Generation Company, LLC (Exelon) for an early site permit (ESP). The proposed action requested in Exelon's application is for the NRC to (1) approve a site within the existing Clinton Power Station (CPS) boundaries as suitable for the construction and operation of a new nuclear power generating facility and (2) issue an ESP for the proposed site identified as the Exelon ESP site located adjacent to the CPS. In its application, Exelon proposes a plan for redressing the environmental effects of certain site-preparation and construction activities, i.e., those activities allowed by Title 10 of the Code of Federal Regulations (CFR) 50.10(e)(1), performed by an ESP holder under 10 CFR 52.25. In accordance with the plan, the site would be redressed if the NRC issues the requested ESP (containing the site redress plan), the ESP holder performs these site-preparation and construction activities, the ESP is not referenced in an application for a construction permit or combined operating license, and no alternative use is found for the site. This EIS includes the NRC staff's analysis that considers and weighs the environmental impacts of constructing and operating a new nuclear unit at the Exelon ESP site or at alternative sites, and mitigation measures available for reducing or avoiding adverse impacts. It also includes the staff's recommendation to the Commission regarding the proposed action.

The staff's recommendation to the Commission related to the environmental aspects of the proposed action is that the ESP should be issued. The staff's evaluation of the site safety and emergency preparedness aspects of the proposed action have been addressed in the staff's final safety evaluation report dated February 17, 2006.

This recommendation is based on (1) the application, including the Environmental Report (ER), submitted by Exelon; (2) consultation with Federal, State, Tribal, and local agencies; (3) the staff's independent review; (4) the staff's consideration of comments related to the environmental review that were received during the public scoping process and on the draft EIS; and (5) the assessments summarized in this EIS, including the potential mitigation measures identified in the ER and this EIS. In addition, in making its recommendation, the staff determined that there are no environmentally preferable or obviously superior sites. Finally, the staff has concluded that the site-preparation and construction activities allowed by 10 CFR 50.10(e)(1) requested by Exelon in its application would not result in any significant adverse environmental impact that cannot be redressed.

Contents

Contents

Contents

Contents

Contents

Contents

Contents

Contents

Figures

Tables

Tables

Tables

Executive Summary

On September 25, 2003, the U.S. Nuclear Regulatory Commission (NRC) received an application from Exelon Generation Company, LLC (Exelon) for an early site permit (ESP) for a location identified as the Exelon ESP site, adjacent to the Clinton Power Station (CPS), Unit 1. The Exelon ESP site is located in DeWitt County, Illinois, approximately 10 km (6 mi) east of the City of Clinton. An ESP is a Commission approval of a location for siting one or more nuclear power facilities and is a separate action from the filing of an application for a construction permit (CP) or combined CP and operating license (combined license or COL) for such a facility. An ESP application may refer to a reactor's or reactors' characteristics or plant parameter envelope (PPE), which is a set of postulated design parameters that bound the characteristics of a reactor or reactors that might be built at a selected site; alternatively, an ESP application may refer to a detailed reactor design. The ESP is not a license to build a nuclear power plant; rather, the application for an ESP initiates a process undertaken to assess whether a proposed site is suitable should Exelon decide to pursue a CP or COL.

Section 102 of the National Environmental Policy Act of 1969 (NEPA) (42 USC 4321 et seq.) directs that an environmental impact statement (EIS) be prepared for major Federal actions that significantly affect the quality of the human environment. The NRC has implemented Section 102 of NEPA in Part 51 of Title 10 of the Code of Federal Regulations (CFR). The NRC regulations related to ESPs are delineated in Subpart A of 10 CFR Part 52. As set forth in 10 CFR 52.18, the Commission has determined that an EIS will be prepared during the review of an application for an ESP. The purpose of Exelon's requested action, issuance of the ESP, is for the NRC to determine whether the Exelon ESP site is suitable for a new nuclear unit by resolving certain safety and environmental issues before Exelon incurs the substantial additional time and expense of designing and seeking approval to construct such a facility at the site. Part 52 of Title 10 describes the ESP as a "partial construction permit." An applicant for a CP or COL for a nuclear power plant or plants to be located at the site for which an ESP was issued can reference the ESP, thus reducing the review of siting issues at that stage of the licensing process. However, a CP or COL to construct and operate a nuclear power plant is a major Federal action and will require an EIS be issued in accordance with 10 CFR Part 51.

Three primary issues – site safety, environmental impacts, and emergency planning – must be addressed in the ESP application. Likewise, in its review of the application, the NRC assesses Exelon's proposal in relation to these issues and determines if the application meets the requirements of the Atomic Energy Act and the NRC regulations. This EIS addresses the potential environmental impacts resulting from the construction and operation of a new nuclear unit at the Exelon ESP site.

In its application, Exelon requested authorization to perform certain site-preparation activities after the ESP is issued. The application, therefore, includes a site redress plan that specifies how Exelon would stabilize and restore the site to its pre-construction condition (or conditions consistent with an alternative use) in the event a nuclear power plant is not constructed on the approved site. Pursuant to 10 CFR 52.17(a)(2), Exelon did not address the benefits of the proposed action (e.g., the need for power). In accordance with 10 CFR 52.18, the EIS is focused on the environmental effects of construction and operation of a reactor, or reactors, that have characteristics that fall within the postulated site parameters.

Upon acceptance of the Exelon ESP application, the NRC began the environmental review process described in 10 CFR Part 51 by publishing in the *Federal Register* a Notice of Intent (68 FR 66130) to prepare an EIS and conduct scoping. The staff held a public scoping meeting in Clinton, Illinois, on December 18, 2003, and visited the Exelon ESP site in March 2004. Subsequent to the scoping meeting and the site visit and in accordance with NEPA and 10 CFR Part 51, the staff determined and evaluated the potential environmental impacts of constructing and operating a new nuclear unit at the Exelon ESP site. Included in this EIS are (1) the results of the NRC staff's analyses, which consider and weigh the environmental effects of the proposed action (issuance of the ESP) and of constructing and operating a new nuclear unit at the ESP site, (2) mitigation measures for reducing or avoiding adverse effects, (3) the environmental impacts of alternatives to the proposed action, and (4) the staff's recommendation regarding the proposed action.

During the course of preparing this EIS, the staff reviewed the application (through revision 4), including the Environmental Report (ER) submitted by Exelon, consulted with Federal, State, Tribal, and local agencies, and followed the guidance set forth in review standard RS-002, *Processing Applications for Early Site Permits*, to conduct an independent review of the issues. The review standard draws from the previously published NUREG-0800, *Standard Review Plans for the Review of Safety Analysis for Nuclear Power Plants*, and NUREG-1555, *Environmental Standard Review Plan (ESRP)*. In addition, the staff considered the public comments related to the environmental review received during the scoping process. These comments are provided in Appendix D of this EIS.

Following the practice the staff used in of NUREG-1437, *Generic Environmental Impact Statement for License Renewal of Nuclear Plants*, and in the supplemental license renewal EISs, environmental issues are evaluated using the three-level standard of significance – SMALL, MODERATE, or LARGE – developed by NRC using guidelines from the Council on Environmental Quality. Table B-1 of 10 CFR Part 51, Subpart A, Appendix B, provides the following definitions of the three significance levels:

SMALL – Environmental effects are not detectable or are so minor that they will neither destabilize nor noticeably alter any important attribute of the resource.

MODERATE – Environmental effects are sufficient to alter noticeably, but not to destabilize, important attributes of the resource.

LARGE – Environmental effects are clearly noticeable and are sufficient to destabilize important attributes of the resource.

Mitigation measures were considered for each environmental issue and are discussed in the appropriate sections.

The results of this evaluation were documented in a draft EIS issued for public comment in February 2005. During the comment period, the staff conducted a public meeting on April 19, 2005, near the Exelon ESP site to describe the results of the NRC environmental review, answer questions, and provide members of the public with information to assist them in formulating comments on the draft EIS. After the comment period closed, the staff considered and dispositioned all the comments received. These comments are addressed in Appendix E of this EIS.

The staff's recommendation to the Commission related to the environmental aspects of the proposed action is that the ESP should be issued. The staff's evaluation of the site safety and emergency preparedness aspects of the proposed action have been addressed in the staff's final safety evaluation report, published May 1, 2006.

This recommendation is based on (1) the application, including the ER submitted by Exelon; (2) consultation with other Federal, State, Tribal, and local agencies; (3) the staff's independent review; (4) the staff's consideration of public comments related to the environmental review that were received during the review process; and (5) the assessments summarized in the EIS, including the potential mitigation measures identified in the ER and this EIS. In addition, in making its recommendation to the Commission, the staff has determined that there are no environmentally preferable or obviously superior sites. Finally, the staff has concluded that the site-preparation and construction activities allowed by 10 CFR 50.10(e)(1) would not result in any significant adverse environmental impact that cannot be redressed.

Abbreviations/Acronyms

ABWR	Advanced Boiling Water Reactor
ac	acre(s)
ACE	U.S. Army Corps of Engineers
ACR-700	Advanced Canada Deuterium Uranium Reactor
ADAMS	Agencywide Document Access and Management System
AEC	U.S. Atomic Energy Commission
ALARA	as low as is reasonably achievable
AmerGen	AmerGen Energy Company, LLC
ANSI	American National Standards Institute
AP1000	Advanced Pressurized Water Reactor
APE	area of potential effect
AQCR	Air Quality Control Region
AQI	Air Quality Index
ATWS	anticipated transient without scram
BEA	Bureau of Economic Analysis
BEIR	Biological Effects of Ionizing Radiation
BLS	U.S. Bureau of Labor Statistics
BOW	Bureau of Economic Analysis
Bq	becquerel
Btu	British thermal unit(s)
BWR	boiling water reactor
°C	Celsius
CANDU	Canada Deuterium Uranium
CARB	
CEQ	Council on Environmental Quality
CFR	Code of Federal Regulations
cfs	cubic feet per second
Ci	curie(s)
cm	centimeter(s)
CNWRA	Center for Nuclear Waste Regulatory Analysis
CO	carbon monoxide
COL	combined license
CP	construction permit
CPS	Clinton Power Station
CWA	Clean Water Act of 1977 (also known as the Federal Water Pollution Control Act)
DBA	design basis accident
DEIS	draft environmental impact statement
DHS	Department of Homeland Security
DO	dissolved oxygen

Abbreviations/Acronyms

DOE	U.S. Department of Energy
DOT	U.S. Department of Transportation
DU	depleted uranium
EAB	exclusion area boundary
ECL	effluent concentration limits
EGC	Exelon Generation Company
EIA	Energy Information Administration
EIS	environmental impact statement
EFL	extremely low frequency
EMF	electromagnetic field
EPA	U.S. Environmental Protection Agency
ESBWR	Economic Simplified Boiling Water Reactor
ESRP	Environmental Standard Review Plan
ER	Environmental Report
ERA	Environmental Resource Associates
ESP	early site permit
Exelon	Exelon Generation Company, LLC
°F	Fahrenheit
FBI	Federal Bureau of Investigation
FEMA	Federal Emergency Management Agency
FERC	Federal Energy Regulatory Commission
FR	Federal Register
fps	feet per second
ft	foot/feet
FWPCA	Federal Water Pollution Control Act (also known as the Clean Water Act of 1977)
FWS	U.S. Fish and Wildlife Service
FY	fiscal year
gal	gallon(s)
GEIS	generic environmental impact statement
GEn&SIS	Geographical, Environmental and Siting Information
GIS	geographic information system
gpm	gallons per minute
GT-MHR	Gas Turbine-Modular Helium Reactor
ha	hectare(s)
hr	hour(s)
HRCQ	highway route controlled quantity

I	interstate
IAC	Illinois Administration Code
IAEA	International Atomic Energy Agency
ICRP	International Commission on Radiation Protection
IDNR	Illinois Department of Natural Resources
IDOCEO	Illinois Department of Commerce and Economic Opportunity
IDOR	Illinois Department of Revenue
IDOT	Illinois Department of Transportation
IEEE	Institute of Electrical and Electronics
IEPA	Illinois Environmental Protection Agency
IHPA	Illinois State Historic Preservation Agency
in.	inch(es)
INEEL	Idaho National Engineering and Environmental Laboratory
INHS	Illinois Natural History Survey
IOC	Illinois Office of the Controller
IPC	Illinois Power Company
IRIS	International Reactor Innovative and Secure
ISA	Illinois Stewardship Alliance
ISGS	Illinois State Geological Survey
ISU	Illinois State University
J	Joules
kg	kilogram(s)
km	kilometers)
kV	kilovolt(s)
kWh	kilowatt hour(s)
L	liter(s)
lb	pound(s)
L/d	liters per day
LLRWPAA	Low-Level Radioactive Waste Policy Amendments
LOCA	loss-of-coolant accident
LOS	level-of-service
LPZ	low population zone
LR	License Renewal
LWR	light water reactor
m	meter(s)
m^3	cubic meter(s)
m/s	meter(s) per second
m^3/d	cubic meter(s) per day

Abbreviations/Acronyms

m³/s	cubic meter(s) per second
m³/yr	cubic meter(s) per year
MEI	maximally exposed individual
mgd	million gallons per day
mg/L	milligrams per liter
mGy	milligray(s)
mi	mile(s)
mL	milliliter(s)
mph	miles per hour
mrad	millirad(s)
mrem	millirem(s)
MSA	Metropolitan Statistical Area
MSDS	Material Safety Data Sheet
MSL	mean sea level
mSv	millisievert(s)
MT	metric ton(s) (or tonne[s])
MTU	metric ton(s) uranium
MW	megawatt(s)
MWd/MTU	megawatt days per metric ton of uranium
MW(e)	megawatt(s) electric
MW(t)	megawatt(s) thermal
MWh	megawatt hour(s)
ng/J	nanogram per Joule
NAGPRA	Native Graves Protection and Repatriation Act
NAS	National Academy of Science
NCI	National Cancer Institute
NCDC	National Climate Data Center
NCRP	National Council on Radiation Protection and Measurements
NEIS	Nuclear Energy Information Service
NEPA	National Environmental Policy Act of 1969
NGO	non-governmental organization
NHPA	National Historic Preservation Act of 1966
NIEHS	National Institute of Environmental Health Sciences
NIST	National Institute of Standards
N	north
NE	northeast
NNE	north northeast
NOx	nitrogen oxide(s)
NOI	notice of intent
NOT	notice of termination
NPDES	National Pollutant Discharge Elimination System

NRC	U.S. Nuclear Regulatory Commission
NUREG	Nuclear Regulation
NWFR	Mississippi River National Wildlife and Fish Refuge
ODCM	Offsite Dose Calculation Manual
ORNL	Oak Ridge National Laboratory
OSHA	Occupational Safety and Health Administration
PARs	Publicly Available Records
PBMR	Pebble Bed Modular Reactor
PGDP	Portsmouth Gaseous Diffusion Plant
pH	potential of hydrogen
PM	particulate matter
PM_{10}	particulate matter with a diameter of fewer than 10 micrometers
PNNL	Pacific Northwest National Laboratory
PPE	plant parameter envelope
PPWMP	Pollution Prevention and Waste Minimization Program
PV	photovoltaic
PVC	polyvinyl chloride
PWR	pressurized water reactor
FRCIC	reactor core isolation cooling
RCRA	Resource Conservation and Recovery Act of 1976
REMP	radiological environmental monitoring program
REPS	Renewable Energy Portfolio Standard
RI	radio interference
rms	root mean square
ROI	region of interest
RPHP	Radiation and Public Health Project
RSICC	Radiation Safety Information Computational Center
RTO	Regional Transmission Operator
Ryr-1	per reactor year
s	second(s)
scf	standard cubic feet
SE	southeast
SEIS	supplemental environmental impact statement
SER	safety evaluation report
SFP	spent fuel pool
SHPO	State Historic Preservation Officer
SNF	spent nuclear fuel
SOx	sulfur oxide(s)

Abbreviations/Acronyms

SPCC	Spill Prevention Control and Countermeasure
Sr-90	strontium-90
SR	State Route
SRS	Savannah River Site
SSAR	site safety analysis report
SW	southwest
SWPPP	stormwater pollution prevention plans
SWR	Service Water Reservoir
SWU	separative work units
TEDE	total effective dose equivalent
TIF	tax increment financing (districts)
TLD	thermoluminescent dosimeter
TSP	total suspended particulates
TVI	television interference
U_3O_8	yellowcake
UF_6	uranium hexafluoride
UFSAR	Updated Final Safety Analysis Report
UHS	ultimate heat sink
UO_2	uranium oxide
U.S.	United States
USCB	U.S. Census Bureau
USDA	U.S. Department of Agriculture
USGS	United States Geological Survey
WCR	Waste Confidence Rule
yr	year(s)
Y-9	yttrium

Appendix A

Contributors to the Environmental Impact Statement

Appendix A

Contributors to the Environmental Impact Statement

The overall responsibility for the preparation of this environmental impact statement was assigned to the Office of Nuclear Reactor Regulation, U.S. Nuclear Regulatory Commission (NRC). The statement was prepared by members of the Offices of Nuclear Reactor Regulation with assistance from other NRC organizations and Pacific Northwest National Laboratory.

Name	Affiliation	Function or Expertise
NUCLEAR REGULATORY COMMISSION		
Thomas Kenyon	Nuclear Reactor Regulation	Project Manager, Socioeconomics, Environmental Justice
John Tappert	Nuclear Reactor Regulation	Section Chief
M. Christopher Nolan	Nuclear Reactor Regulation	Branch Chief
James Wilson	Nuclear Reactor Regulation	Backup Project Manager, Ecology, Land Use
Jennifer Davis	Nuclear Reactor Regulation	Project Management Support, Cultural Resources, Air Quality
Harriet Nash	Nuclear Reactor Regulation	Project Management Support
Laura Quinn	Nuclear Reactor Regulation	Project Management Support
Andrew Kugler	Nuclear Reactor Regulation	Project Management, Section Chief, Water Quality & Use
Mark Notich	Nuclear Reactor Regulation	Socioeconomic, Environmental Justice, Air Quality, Alternative Energy Sources
Michael Masnik	Nuclear Reactor Regulation	Cultural Resources, Water Quality and Use, Transmission System, Ecology
Richard Emch	Nuclear Reactor Regulation	Radiological Impacts, Severe Accidents, DBAs
Barry Zalcman	Nuclear Reactor Regulation	Alternative Energy Sources
Charles Hinson	Nuclear Reactor Regulation	Radiological Impacts
Steve Klementowicz	Nuclear Reactor Regulation	Radiological Impacts
Jay Lee	Nuclear Reactor Regulation	Severe Accidents, Design Basis Accidents
Robert Palla	Nuclear Reactor Regulation	Severe Accidents
Andrew Barto	Nuclear Materials Safety and Safeguards	Spent Fuel Transportation
James Park	Nuclear Materials Safety and Safeguards	Fuel Cycle Impacts

Name	Affiliation	Function or Expertise
PACIFIC NORTHWEST NATIONAL LABORATORY[a]		
Eva Eckert Hickey		Task Leader
Kimberly Leigh		Deputy Task Leader
Amanda Stegen		Deputy Task Leader
Beverly Miller		Deputy Task Leader
James V. Ramsdell		Air Quality, Alternatives
Dennis Strenge		Severe Accidents
John Jaksch		Socioeconomics, Environmental Justice, Alternatives
Susan Southard		Aquatic Ecology
James Becker		Terrestrial Ecology
Gregory Stoetzel		Radiation Protection
Philip Daling		Transportation
Natesan Mahasenan		Transportation
Darby Stapp		Cultural Resources
Dave Anderson		Land Use, Related Federal Programs
Doug Elliott		Geographical Information System Support
Lance Vail		Water Quality, Use, Hydrology
Christopher Cook		Lake Thermal Processes
James Weber		Technical Editing
Barbara Wilson		Technical Editing
Lila Andor		Document Production
Jean Cheyney		Document Production
Zontziry Johnson		Document Production
Debra Schulz		Document Production
Susan Tackett		Document Production
Rose Urbina		Document Production

(a) Pacific Northwest National Laboratory is operated for the U.S. Department of Energy by Battelle Memorial Institute.

Appendix B

Organizations Contacted

Appendix B

Organizations Contacted

During the course of the staff's independent review of potential environmental impacts from siting one new nuclear unit at the Exelon ESP site, the following Federal, State, regional, Tribal and local organizations were contacted:

Advisory Council on Historic Preservation, Washington, D.C., Director, Don Klima

Brady Weaver Real Estate, Clinton, Illinois, General Manager, Camill Tedrick

Chicago Ecological Field Service Office, U.S. Fish and Wildlife Service, Barrington, Illinois, John Rogner

Chief Deputy to the LaSalle County Treasurer, Ottawa, Illinois, Gary Kleinhans

City of Clinton, Illinois, Administrative Assistant, Tim Followell

Clinton Unit School, District #15, Clinton, Illinois, Superintendent, Roger Little

Community and Economic Development Director, Monticello, Illinois, Mary Jo Hetrick

Cooperative Extension Service, University of Illinois, Clinton, Illinois, Argriculturalist, Pat Toohill

Dean Enrollment Services, Richland Community College, Decatur, Illinois, Nancy Cooper

Delaware Nation, NAGPRA Office, Anadarko, Oklahoma, Phyllis Wahahrockah-Tasi

Delaware Tribe of Indians, Bartlesville, Oklahoma, Brice Obermeyer

Delaware Tribe of Western Oklahoma, Anadardo, Oklahoma, Honorable Lawrence F. Snake

DeWitt County Board Administrative Assistant, Clinton, Illinois, Dee Dee Rentmeister

DeWitt County Board Chairman, Clinton, Illinois, H. Duane Harris.

DeWitt County Board Land Use Chairman and Harp Township Highway Commissioner, Clinton, Illinois, Terry Ferguson

DeWitt County Highway Department, Clinton, Illinois, Craig Fink

DeWitt County Planning and Zoning, Clinton, Illinois, Sherrie Brown

Appendix B

DeWitt County Treasurer, Christy Long, and Supervisor of Assessments, Clinton, Illinois, Sandy Moody

Director of Public Works and Construction, Clinton, Illinois, Steve Lobb

Director of Public Works, Mt. Pulaski, Illinois, Michael Partridge

Eastern Delaware Tribe, Bartllesville, Oklahoma, Honorable Dee Ketchum

Economic Development Director, Clinton, Illinois, Stephen Vandiver

Executive Director Dewitt County Human Resource Center, Clinton, Illinois, Cheryl Lietz

Illinois Department of Natural Resources, Springfield, Illinois, Mike Garthaus

Illinois Department of Transportation, Bridge Maintenance for DeWitt County, Paris, Illinois, Kevin Woods

Illinois State Historic Preservation Officer (SHPO), Springfield, Illinois

Illinois Governor's Office, Springfield, Illinois

Illinois Power Company, Decatur, Illinois

Illinois Historic Preservation Agency, Springfield, Illinois, Maynard Crossland

Kickapoo of Oklahoma Business Committee, McCloud, Oklahoma, Honorable Kendall Scott

Kickapoo Tribe of Texas, Miami, Oklahoma, Honorable Raul Garza, Jr.

Kickapoo Kansas Tribal Council, Horton, Kansas, Honorable Carol Anske

LaSalle County Treasurer's Office, Ottawa, Illinois, Chief Deputy to the LaSalle County Treasurer, Gary Kleinhans

Mayor of Monticello, Illinois, Bill Mitze

Mayor of Clinton, Illinois, Roger Cyrulik

Monticello Chamber of Commerce, Monticello, Illinois, Executive Director, Sue Gorton

Monticello Community Unit School District 23, Monticello, Illinois, School Superintendent,
Lawrence McNabb

Mt. Pulaski, Illinois, Mayor, William Glaze

National Oceanic and Atmospheric Administration Fisheries, Gloucester, Massachusetts,
Patricia Kurkul

Ogle County Treasurer, Oregon, Illinois, John Coffman

Peoria Tribe of Indians of Oklahoma, Miami, Oklahoma, Honorable John P. Froman

Rock Island Ecological Field Service Office, U.S. Fish and Wildlife Service, Rock Island, Illinois,
Richard Nelson

Robbins, Schwartz, Nicholas, Lifton and Taylor, LTD, Chicago, Illinois, Counsel, Frederic Lane

Sandi Thayer Real Estate, Clinton, Illinois, Sandi Thayer

Superintendent of City Services, Monticello, Illinois, Floyd Allsop

Town of Monticello, Illinois, Bill Mitze, Mayor

U.S. Ecological Survey, Urbana, Illinois, Gary Johnson

Appendix C

Chronology of NRC Staff Environmental Review Correspondence Related to Exelon Generation Company, LLC's (Exelon's) Application for an Early Site Permit (ESP) at the Exelon ESP Site in Clinton, Illinois

Appendix C

Chronology of NRC Staff Environmental Review Correspondence Related to Exelon Generation Company, LLC's (Exelon's) Application for an Early Site Permit (ESP) at the Exelon ESP Site in Clinton, Illinois

This appendix contains a chronological listing of correspondence between the U.S. Nuclear Regulatory Commission (NRC) and Exelon Generation Company, LLC (Exelon) and other correspondence related to the NRC staff's environmental review, under 10 CFR Part 51, of Exelon's application for an early site permit at the Exelon ESP site in Clinton, Illinois. All documents, with the exception of those containing proprietary information, have been placed in the Commission's Public Document Room, at One White Flint North, 11555 Rockville Pike (first floor), Rockville, MD, and are available electronically from the Public Electronic Reading Room found on the Internet at the following web address: http://www.nrc.gov/reading-rm.html. From this site, the public can gain access to the NRC's Agencywide Document Access and Management System (ADAMS), which provides text and image files of NRC's public documents in the Publicly Available Records (PARS) component of ADAMS. The ADAMS accession numbers or *Federal Register* citation for each document are included below.

February 28, 2003	NRC meeting notice announcing a public meeting in Clinton, Illinois on March 20, 2003, to discuss the review process for Exelon's ESP application for the Clinton site (Accession No. ML030580509).
March 3, 2003	*Federal Register* Notice of public pre-application ESP meeting for the Exelon site in Clinton, Illinois (68 FR 10052).
April 3, 2003	Summary of public pre-application ESP meeting held in Clinton, Illinois to discuss the ESP review process (Accession No. ML030910535).
June 17, 2003	NRC staff letter to Mr. Tom Rudasill, Vespasian Warner Public Library, regarding the maintenance of reference material for public access related to the Exelon ESP review (Accession No. ML031640019).
June 26, 2003	Response from Mr. Tom Rudasill, Vespasian Warner Public Library, regarding the maintenance of reference material for public access related to the Exelon ESP review (Accession No. 032450430).

Appendix C

September 25, 2003   Letter from Ms. Marilyn C. Kray, Exelon, to NRC submitting the
                      application for an ESP at the Exelon site in Clinton, Illinois (Accession
                      No. ML032721594).

October 15, 2003   NRC Press Release No. 03-133 announcing the availability of the ESP
                      application for the Exelon site in Clinton, Illinois (Accession
                      No. ML032880335).

October 24, 2003   *Federal Register* Notice of receipt and availability of the application for an
                      ESP at the Exelon site in Clinton, Illinois (68 FR 61020).

October 27, 2003   NRC staff letter to Ms. Marilyn Kray, Exelon, regarding the receipt and
                      availability of the application for an ESP at the Exelon site in Clinton,
                      Illinois (Accession No. ML032930051).

October 27, 2003   Summary of September 24, 2003, tele-conference with Exelon to discuss
                      the scheduling of the staff's technical review of Exelon's ESP application
                      (Accession No. ML033000489).

October 30, 2003   *Federal Register* Notice of acceptance of the application for an ESP at
                      the Exelon site in Clinton, Illinois (68 FR 61835).

November 19, 2003   NRC staff letter to Ms. Marilyn Kray, Exelon, forwarding the *Federal
                      Register* notice of intent to prepare an environmental impact statement
                      and conduct scoping process for an ESP at the Exelon site in Clinton,
                      Illinois (Accession No. ML033250261).

November 25, 2003   *Federal Register* Notice of Intent to prepare an environmental impact
                      statement and conduct scoping process for an ESP at the Exelon site in
                      Clinton, Illinois (68 FR 66130).

December 3, 2003   NRC meeting notice announcing a public meeting in Clinton, Illinois on
                      December 18, 2003, to discuss the environmental scoping process for the
                      application for an ESP at the Exelon site in Clinton, Illinois (Accession No.
                      ML033380526).

December 8, 2003   NRC Press Release No. 03-160, "NRC Announces Hearing on Early Site
                      Permit for Clinton Site; Opportunity to Request Participation" (Accession
                      No. ML033420171)

December 12, 2003   *Federal Register* Notice of Hearing and Opportunity to Petition for Leave to Intervene regarding an ESP at the Exelon site in Clinton, Illinois (68 FR 69426).

December 15, 2003   NRC Press Release No. III-03-076, "NRC to Hold Public Meeting December 18 on Environmental Scoping Process for Clinton Early Site Permit" (Accession No. ML033490522).

December 16, 2003   NRC staff letter to Clinton Area Local Public Officials providing notification of receipt and review of the Exelon ESP application (Accession No. ML033421293).

December 18, 2003   NRC staff letter to Mr. Don Klima, Director, Advisory Council on Historic Preservation, regarding the ESP review for the Exelon site in Clinton, Illinois (Accession No. ML033520358).

December 22, 2003   Letter from Ms. Phyllis Wahahrockah-Tasi, NAGPRA Director, Delaware Nation, providing comments related to the ESP review for the Exelon site in Clinton, Illinois (Accession No. ML040080737).

December 23, 2003   NRC staff letter to Mr. Maynard Crossland, Director, Illinois Historic Preservation Agency, regarding the ESP review for the Exelon site in Clinton, Illinois (Accession No. ML033630476).

December 30, 2003   NRC staff letter to the Honorable Kendall Scott, Chairman, Kickapoo of Oklahoma Business Committee, regarding the ESP review for the Exelon site in Clinton, Illinois (Accession No. ML033650531).

December 30, 2003   NRC staff letter to the Honorable Raul Garza, Jr., Chairman, Kickapoo Traditional Tribe of Texas, regarding the ESP review for the Exelon site in Clinton, Illinois (Accession No. ML033650530).

December 30, 2003   NRC staff letter to the Honorable Carol Anske, Chairperson, Kickapoo of Kansas Tribal Council, regarding the ESP review for the at the Exelon site in Clinton, Illinois (Accession No. ML033650527).

December 30, 2003   NRC staff letter to the Honorable Lawrence F. Snake, President, Delaware Tribe of Western Oklahoma, regarding the ESP review for the Exelon site in Clinton, Illinois (Accession No. ML033650456).

Appendix C

December 30, 2003   NRC staff letter to the Honorable John P. Froman, Chief, Peoria Tribe of
                     Indians of Oklahoma, regarding the ESP review for the Exelon site in
                     Clinton, Illinois (Accession No. ML033650305).

December 30, 2003   NRC staff letter to the Honorable Dee Ketchum, Chief, Eastern Delaware
                     Tribe, regarding the ESP review for the Exelon site in Clinton, Illinois
                     (Accession No. ML033650325).

January 13, 2004   Letter from the Honorable John P. Froman, Chief, Peoria Tribe of Indians
                     of Oklahoma, providing comments related to the ESP review for the
                     Exelon site in Clinton, Illinois (Accession No. ML040230461).

January 13, 2004   Letter from the Mr. Brice Obermeyer, NAGPRA Director, Delaware Tribe
                     of Indians of Oklahoma, providing comments related to the ESP review
                     for the Exelon site in Clinton, Illinois (Accession No. ML040480535).

January 21, 2004   NRC summary of public scoping meeting to support review of the ESP
                     application for the Exelon site in Clinton, Illinois (Accession Nos.
                     ML040330445 [Package], ML040330375 [Meeting Summary], and
                     ML040230643 [Meeting Handouts and Transcript]).

February 13, 2004   NRC staff e-mail to Mr. Bill Maher, Exelon, forwarding the proposed
                     agenda for alternate site visits for the Exelon ESP site audit review
                     (Accession No. ML041830102).

February 18, 2004   NRC staff e-mail to Mr. Bill Maher, Exelon, forwarding agenda items for
                     the Clinton site audit (Accession No. ML041830095).

February 24, 2004   NRC staff e-mail to Mr. Bill Maher, Exelon, regarding an additional
                     question concerning spent fuel storage for the Clinton site audit
                     (Accession No. ML041820385).

February 25, 2004   NRC staff e-mail to Mr. Bill Maher, Exelon, forwarding additional agenda
                     items for the Clinton site audit (Accession No. ML041830104).

February 26, 2004   NRC staff e-mail to Mr. Bill Maher, Exelon, forwarding discussion items
                     on worker dose for the Clinton site audit (Accession No. ML041830124).

March 17, 2004   NRC staff letter to Ms. Patricia Kurkul, Regional Administrator, NOAA
                     Fisheries, regarding the ESP review for the Exelon site in Clinton, Illinois
                     (Accession No. ML040770284).

March 17, 2004	NRC staff letter to Mr. John Rogner, Field Supervisor, Chicago Ecological Field Services Office, U.S. Fish and Wildlife Service, regarding the ESP review for the Clinton Power Station site (Accession No. ML040770948).
March 17, 2004	NRC staff letter to Mr. Richard Nelson, Field Supervisor, Rock Island Ecological Field Services Office, U.S. Fish and Wildlife Service, regarding the ESP review for the Exelon site in Clinton, Illinois (Accession No. ML040770896).
April 6, 2004	NRC staff letter to Ms. Marilyn Kray, Exelon, regarding the revised date for transmitting environmental requests for additional information regarding the ESP review for the Exelon site in Clinton, Illinois (Accession No. ML040920584).
April 6, 2004	Letter from Mr. Richard Nelson, Field Supervisor, Rock Island Ecological Field Services Office, U.S. Fish and Wildlife Service, to NRC, providing a response to a letter requesting a list of species in the vicinity of the Exelon ESP site, Ogle, Grundy, LaSalle, and Rock Island Counties (Accession No. ML041180181).
April 8, 2004	NRC summary of site visits to alternative sites for the Exelon ESP site in Clinton, Illinois (Accession No. ML041000222).
April 12, 2004	Letter from Mr. John D. Rogner, Field Supervisor, Chicago Ecological Field Services Office, U.S. Fish and Wildlife Service, to NRC, providing a response to a letter requesting a list of species regarding alternate sites in Will and Lake Counties, Illinois (Accession No. ML041200545).
April 15, 2004	NRC staff letter to Ms. Marilyn C. Kray, Exelon, submitting a request for additional information regarding the ESP review for the Exelon site in Clinton, Illinois (Accession No. ML040930400).
May 11, 2004	NRC staff letter to Ms. Marilyn C. Kray, Exelon, submitting a request for additional information regarding the environmental portion of the ESP review for the Exelon site in Clinton, Illinois (Accession No. ML041330188).
May 18, 2004	NRC staff e-mail to Mr. Bill Maher, Exelon, forwarding clarification items regarding the Clinton site audit (Accession No. ML041830135).

Appendix C

May 26, 2004	Note to file: Docketing of references obtained during the site audit conducted in support of the environmental review of the Exelon ESP site in Clinton, Illinois (Accession Nos. ML041470397 [Note] ML041200352 [Package of References]).
May 26, 2004	NRC Summary of staff audit to support review of the Exelon ESP site in Clinton, Illinois (Accession No. ML041560266).
June 18, 2004	NRC staff e-mail to Mr. Bill Maher, Exelon, forwarding files for lake modeling for the Exelon ESP Clinton site review (Accession No. ML041830154).
June 22, 2004	NRC staff letter to Ms. Marilyn C. Kray, Exelon, requesting comments on the early site permit template (Accession No. ML041400206).
July 9, 2004	NRC staff letter to Ms. Marilyn C. Kray, Exelon, forwarding the environmental scoping summary report associated with the ESP review for the Exelon site in Clinton, Illinois (Accession No. ML041950214 [Letter], ML041950227 [Report]).
July 23, 2004	Letter from Ms. Marilyn C. Kray, Exelon, forwarding responses to NRC staff's requests for additional information for the Exelon ESP site in Clinton, Illinois (Accession No. ML042180079).
August 23, 2004	NRC staff letter to Ms. Marilyn C. Kray, Exelon, requesting additional information regarding the Exelon site in Clinton, Illinois (Accession No. ML042370551).
September 7, 2004	Note to file: Availability of Geographical Information Systems (GIS) files concerning the environmental review of the Exelon ESP site in Clinton, Illinois (Accession No. ML042510446).
September 17, 2004	Letter from Ms. Marilyn C. Kray, Exelon, delay in responding to requests for additional information regarding the Exelon ESP site in Clinton, Illinois (Accession No. ML042730435).
September 23, 2004	Letter from Ms. Marilyn C. Kray, Exelon, forwarding response to requests for additional information regarding the Exelon ESP site in Clinton, Illinois (Accession No. ML042730012).

October 6, 2004	NRC Telecommunication summary to clarify responses to NRC requests for additional information regarding the Exelon ESP site in Clinton, Illinois (Accession No. ML042800504).
November 15, 2004	NRC staff letter to Ms. Marilyn C. Kray, Exelon, forwarding request for additional information regarding the environmental portion of the Exelon ESP site in Clinton, Illinois (Accession No. ML043210579).
November 15, 2004	NRC staff letter to Ms. Marilyn C. Kray, Exelon, providing revised schedule for the environmental review of the Exelon ESP site in Clinton, Illinois (Accession No. ML043090029).
November 16, 2004	Letter from Ms. Marilyn C. Kray, Exelon, forwarding corrections/clarifications to the Exelon ESP Application Environmental Report for the Exelon ESP site in Clinton, Illinois (Accession No. ML043290006).
November 18, 2004	E-mail from Mr. Bill Maher, Exelon, regarding ER corrections for the Exelon ESP Clinton site review (Accession No. ML043410062).
March 2, 2005	NRC staff letter to Ms. Marilyn C. Kray, Exelon, forwarding Notice of Availability of the Draft Environmental Impact Statement (DEIS) for an Early Site Permit (ESP) at the Exelon ESP Site (TAC NO. MC1125) (Accession No. ML050620302).
March 2, 2005	NRC staff letter to US Environmental Protection Agency, forwarding Draft Environmental Impact Statement for An Early Site Permit (ESP) at the Exelon ESP Site (TAC MC1125) (Accession No. ML050620431).
March 3, 2005	NRC staff letter to Ms. Marilyn C. Kray, Exelon, forwarding Notice of Change of Location for Public Meeting on the Draft Environmental Impact Statement (DEIS) for an Early Site Permit (ESP) at the Exelon ESP Site (TAC No. MC1125) (Accession No. ML050920004).ML050980379
March 8, 2005	NRC Press Release No. 05-044, "NRC Seeks Public Input On Clinton Early Site Permit Application; Meeting To Be Held April 19" (Accession No. ML050670134).
March 29, 2005	NRC staff e-mails to Mr. Bill Maher, Exelon, for the public hearing record for the Exelon ESP site (Accession No. ML051010274).

Appendix C

April 5, 2005	NRC meeting notice announcing a public meeting in Clinton, Illinois on April 19, 2005, to Receive Comments on the Draft Environmental Impact Statement for the Exelon Early Site Permit Application (Accession No. ML050950238).
April 7, 2005	NRC staff letter to Mr. Richard Nelson, Field Supervisor, Rock Island Ecological Field Services Office, U.S. Fish and Wildlife Service, regarding Biological Assessment for an Early Site Permit (ESP) for the Exelon ESP Site and a Request for Informal Consultation (Accession No. ML050980127).
April 8, 2005	NRC staff letter to Ms. Marilyn C. Kray, Exelon, forwarding Notice of Change of Location for Public Meeting on the Draft Environmental Impact Statement (DEIS) for an Early Site Permit (ESP) at the Exelon ESP Site (TAC No. MC1125) (Accession No. ML050980379).
April 11, 2005	Letter from Anne E. Haaker, State of Illinois Historic Preservation Agency, regarding Clinton Early Site Permit (Accession No. ML0514404280).
April 12, 2005	NRC Press Release No. III-05-013, "NRC Staff to Hold Public Meeting April 19 in Clinton, Ill. for Comments on Proposed Nuclear Plant Early Site Permit" (Accession No. ML051020302).
April 19, 2005	NRC summary of public meeting regarding review of the ESP application for the Exelon site in Clinton, Illinois (Accession Nos. ML051580549 [Package], ML051580393 [Meeting Summary], ML051590238 [Presentation Slides] and ML051590198 [Meeting Transcript] and ML051300569 [Comments and Information Provided]).
April 19, 2005	Letter from Representative Naomi D. Jakobsson, State of Illinois, 103rd District, Regarding Clinton Early Site Permit (Accession No. ML051440368).
May 17, 2005	Letter from Michael T. Chezik on behalf of U.S. Dept. of the Interior, Office of the Secretary, on Environmental Impact Statement (DEIS), NUREG-1815, for an Early Site Permit at the Exelon site in Clinton, Illinois (Accession No. ML051460042).

May 19, 2005   Letter from Mr. Richard Nelson, Field Supervisor, Rock Island Ecological Field Services Office, U.S. Fish and Wildlife Service, submitting comments regarding Biological Assessment for an early site permit for the Exelon ESP site (Accession No. ML051600132).

May 24, 2005   Letter from Marilyn C. Kray on behalf of Exelon Nuclear on Review of Draft Environmental Impact Statement for Clinton ESP Site (Accession No. ML051540317).

June 3, 2005   Letter from Ms. Marilyn C. Kray, Exelon, Transmittal of Early Site Permit (ESP) Application for the Clinton ESP Site, Submittal of Revision 1 to Exelon Generation Company's Early Site Permit, Environmental Report (Accession Nos. ML051640428 [transmittal letter] ML0515640426 [package]).

June 7, 2005   Summary of public meeting to support the Environmental Review for the Exelon Early Site Permit Application with Attachments and Mailing List (Accession No. ML051580393).

June 30, 2005   Letter from Ms. Marilyn C. Kray, Exelon, Submission of Reviewer's Aid for Revision 1 to Exelon Generation Company's Early Site Permit, Environmental Report (Accession Nos. ML05190121 [Cover Letter] ML0519201270 [Reviewer's Aid]).

August 16, 2005   NRC press release No. 05-111, "NRC revises schedule for reviewing existing early site permit applications" (Accession No. ML052280400).

October 11, 2005   Summary of September 13, 2005 Telephone Conference Held with the Illinois Environmental Protection Agency (EPA) Regarding the Review of the Exelon Early Site Permit (ESP) (Accession No. ML052860253).

October 11, 2005   NRC staff letter to Ms. Marilyn C. Kray, Exelon, Transmitting a Request for Additional Information (RAI) Regarding the Environmental Portion of the Early Site Permit Application for the Exelon Generation Company Site (TAC No. MC1125) (Accession No. ML052860325).

October 11, 2005   Summary of September 19, 2005 Telephone Conference Call with the Illinois Department of Natural Resources (IDNR) Regarding the Review of the Exelon Early Site Permit (ESP) (Accession No. ML052860274).

Appendix C

October 12, 2005	NRC staff letter to Mr. Kenneth Barr, Branch Chief, Rock Island District, U.S. Army Corps of Engineers, Regarding Clinton Early Site Permit Review (TAC No. MC1125) (Accession No. ML052910123).
November 23, 2005	Letter from Ms. Marilyn C. Kray, Exelon, Submitting Revision 1 to Exelon Generation Company's Early Site Permit Application (Accession Nos. ML053420057 [Submittal Letter] ML053420053 [Revision 1 Package]).
December 13, 2005	Response from Ms. Marilyn C. Kray, Exelon, to Request for Additional Information (RAI) - Exelon Early Site Permit (ESP) Application for the Clinton ESP Site (Accession No. ML053540218).
January 10, 2006	Letter from Ms. Marilyn C. Kray, Exelon, Submitting Revision 2 to Exelon Generation Company's Early Site Permit Application (Accession Nos. ML06040042 [Submittal Letter], ML06040043 [Revision 2 Package]).
March 3, 2006	Letter from Ms. Marilyn C. Kray, Exelon, Submitting Revision 3 to Exelon Generation Company's Early Site Permit Application (Accession Nos. ML060950517 [Submittal Letter], ML060950511 [Revision 3, Package]).
April 14, 2006	Letter from Ms. Marilyn C. Kray, Exelon, Submitting Revision 4 to Exelon Generation Company's Early Site Permit Application (Accession Nos. ML061100261 [Submittal Letter], ML061100260 [Revision 4, Package]).
June 30, 2006	NRC letter to Mr. Bruce Yurdin, Illinois Environmental Protection Agency, Transmitting a Summary of Discussions Regarding Compliance with Section 401 of the Federal Water Pollution Control Act Concerning Exelon Generation Company's Application for an Early Site Permit (ESP) at the Exelon ESP Site (Accession No. ML061790097).

Appendix D

Scoping Meeting Comments and Responses

Appendix D

Scoping Meeting Comments and Responses

On November 25, 2003, the U.S. Nuclear Regulatory Commission (NRC) published a Notice of Intent in the *Federal Register* (68 FR 66130) to notify the public of the staff's intent to prepare an environmental impact statement (EIS) to support the early site permit (ESP) application for the Exelon Generation Company, LLC (Exelon) ESP site. This EIS has been prepared in accordance with the National Environmental Policy Act of 1969 (NEPA), Council on Environmental Quality guidelines, and Title 10 of the Code of Federal Regulations (CFR) Parts 51 and 52. As outlined by NEPA, the NRC initiated the scoping process with the issuance of the *Federal Register* Notice. The NRC invited the applicant; Federal, Tribal, State, and local government agencies; local organizations; and individuals to participate in the scoping process by providing oral comments at the scheduled public meeting and/or submitting written suggestions and comments no later than January 9, 2004.

The scoping process included a public scoping meeting, which was held at the Vespasian Warner Public Library in Clinton, Illinois, on December 18, 2003. Approximately 100 members of the public attended the meeting. This session began with NRC staff members providing a brief overview of the ESP process and the NEPA process. Following the NRC's prepared statements, the meeting was open for public comments. Thirty-seven attendees provided either oral comments or written statements that were recorded and transcribed by a certified court reporter. The transcript of the meeting can be found as an attachment to the scoping meeting Summary, which was issued on January 21, 2004. The meeting summary is available electronically for public inspection in the NRC Public Document Room or from the Publicly Available Records (PARS) component of NRC's document system (ADAMS) under accession number ML040330445. ADAMS is accessible from the NRC Web site at http://www.nrc.gov/reading-rm/adams.html (the Public Electronic Reading Room). Note: the URL is case-sensitive. Additional comments received later are also available.

The scoping process provides an opportunity for public participants to identify issues to be addressed in the EIS and highlight public concerns and issues. The Notice of Intent identified the following objectives of the scoping process:

- Define the proposed action which is to be the subject of the EIS

- Determine the scope of the EIS and identify significant issues to be analyzed in depth

- Identify and eliminate from detailed study those issues that are peripheral or that are not significant

- Identify any environmental assessments and other EISs that are being prepared or will be prepared that are related to, but not part of, the scope of the EIS being considered

- Identify other environmental review and consultation requirements related to the proposed action

- Indicate the relationship between the timing of the preparation of the environmental analyses and the Commission's tentative planning and decision-making schedule

- Identify any cooperating agencies and, as appropriate, allocate assignments for preparation and schedules for completing the EIS to the NRC and any cooperating agencies

- Describe how the EIS will be prepared and include any contractor assistance to be used.

At the conclusion of the scoping period, the NRC staff and its contractor reviewed the transcripts and all written material received and identified individual comments. Twelve letters and nine e-mail messages containing comments were received during the scoping period. All comments and suggestions received orally during the scoping meeting or in writing were considered. Each set of comments from a given commenter was given a unique alpha identifier (commenter ID letter), allowing each set of comments from a commenter to be traced back to the transcript, letter, or e-mail in which the comments were submitted.

Table D-1 identifies the individuals providing comments and the commenter ID letter associated with each person's set(s) of comments. The commenter ID letter is preceded by EGCESP (short for Exelon Generation Company Early Site Permit). For oral comments, the individuals are listed in the order in which they spoke at the public meeting. Accession numbers indicate the location of the written comments in ADAMS.

Comments were consolidated and categorized according to the topic within the proposed EIS or according to the general topic if outside the scope of the EIS. Comments with similar specific objectives were combined to capture the common essential issues that had been raised in the source comments. Once comments were grouped according to subject area, the staff and contractor determined the appropriate action for the comment. The staff made a determination on each comment that it was one of the following:

- A comment that was actually a question and introduced no new information

Table D-1. Individuals Providing Comments During Scoping Comment Period

Commenter ID	Commenter	Affiliation (if stated)	Comment Source (ADAMS Accession #)
EGCESP-01	Shannon Fisk	Environmental Law and Policy Center	12/18/03 Scoping Meeting Transcript (ML040330445)
EGCESP-02	Steve Davenport	Farmer	12/18/03 Scoping Meeting Transcript (ML040330445)
EGCESP-03	Sandy Moody	DeWitt County	12/18/03 Scoping Meeting Transcript (ML040330445)
EGCESP-04	Kathleen Frick	Citizens Advisory Panel	12/18/03 Scoping Meeting Transcript (ML040330445)
EGCESP-05	Mr. Frank		12/18/03 Scoping Meeting Transcript (ML040330445)
EGCESP-06	Oscar Shirani	Q-A Consultants	12/18/03 Scoping Meeting Transcript (ML040330445)
EGCESP-07	Kevin Calna		12/18/03 Scoping Meeting Transcript (ML040330445)
EGCESP-08	Kim Gaff	Clinton Resident	12/18/03 Scoping Meeting Transcript (ML040330445)
EGCESP-09	Gregg Brown	No New Nukes	12/18/03 Scoping Meeting Transcript (ML040330445)
EGCESP-10	Mayor Cyrulik	Mayor of Clinton	12/18/03 Scoping Meeting Transcript (ML040330445)
EGCESP-11	Bryan Hickman	City of Clinton	12/18/03 Scoping Meeting Transcript (ML040330445)
EGCESP-12	Terry Ferguson	DeWitt County Board	12/18/03 Scoping Meeting Transcript (ML040330445)
EGCESP-13	Bob Bement	Exelon	12/18/03 Scoping Meeting Transcript (ML040330445)
EGCESP-14	Carolyn Treadway	No New Nukes	12/18/03 Scoping Meeting Transcript (ML040330445)
EGCESP-15	Pat Allison	Clinton School District	12/18/03 Scoping Meeting Transcript (ML040330445)
EGCESP-16	Roger Little	Clinton School District	12/18/03 Scoping Meeting Transcript (ML040330445)
EGCESP-17	Steve Vandiver	Economic Development Director	12/18/03 Scoping Meeting Transcript (ML040330445)
EGCESP-18	Ken Bjelland	DeWitt County Economic Development Committee	12/18/03 Scoping Meeting Transcript (ML040330445)

Table D-1. (contd)

Commenter ID	Commenter	Affiliation (if stated)	Comment Source (ADAMS Accession #)
EGCESP-19	Corey Conn	Board of Nuclear Energy Information Service	12/18/03 Scoping Meeting Transcript (ML040330445)
EGCESP-20	Ruth Ann Lowers	Board of Education	12/18/03 Scoping Meeting Transcript (ML04033445)
EGCESP-21	Ted Lowers	Clinton Businessman	12/18/03 Scoping Meeting Transcript (ML040330445)
EGCESP-22	Harold Weinberg	Clinton Resident	12/18/03 Scoping Meeting Transcript (ML040330445)
EGCESP-23	Robert Adcocit	Welding Inspector	12/18/03 Scoping Meeting Transcript (ML040330445)
EGCESP-24	C. Lee Baker	Past President of Intervenor of ILP Development	12/18/03 Scoping Meeting Transcript (ML040330445)
EGCESP-25	Phil Huckleberry	Illinois Green Party	12/18/03 Scoping Meeting Transcript (ML040330445)
EGCESP-26	Geoff Ower	Illinois State University Chapter of the Student Environmental Action Coalition	12/18/03 Scoping Meeting Transcript (ML040330445)
EGCESP-27	Elizabeth Burns	Illinois Stewardship Alliance	12/18/03 Scoping Meeting Transcript (ML040330445)
EGCESP-28	Karen Lowery	Citizen	12/18/03 Scoping Meeting Transcript (ML040330445)
EGCESP-29	John Workman	IBEW 146	12/18/03 Scoping Meeting Transcript (ML040330445)
EGCESP-30	Dick Baldwin	Clinton Resident	12/18/03 Scoping Meeting Transcript (ML040330445)
EGCESP-31	Monte Campbell	Clinton Resident	12/18/03 Scoping Meeting Transcript (ML040330445)
EGCESP-32	Richard Douglas	Clinton Resident	12/18/03 Scoping Meeting Transcript (ML040330445)
EGCESP-33	Matt Reeder	Illinois Green Party	12/18/03 Scoping Meeting Transcript (ML040330445)
EGCESP-34	Dr. Samuel Galusky	No New Nukes	12/18/03 Scoping Meeting Transcript (ML040330445)
EGCESP-35	Rachel Goad	Student Peace Action Network	12/18/03 Scoping Meeting Transcript (ML040330445)
EGCESP-36	Given Harper	Professor, Illinois Wesleyan University	12/18/03 Scoping Meeting Transcript (ML040330445)
EGCESP-37	Robert Bishop	Nuclear Energy Institute	12/18/03 Scoping Meeting Transcript (ML040330445)
EGCESP-38	Phyllis Wahahrochah-Tasi	Delaware Nation NAGPRA Office	Letter (ML0400807370)

Table D-1. (contd)

Commenter ID	Commenter	Affiliation (if stated)	Comment Source (ADAMS Accession #)
EGCESP-39	Patricia Arbunkle		Letter (ML0402304550)
EGCESP-40	Julie Gowen		Letter (ML0402304570)
EGCESP-41	Gregg Brown	No New Nukes	Letter (ML0402304580)
EGCESP-42	Shannon Fisk	Environmental Law and Policy Center	Letter (ML0402304600)
EGCESP-43	John Froman	Peoria Tribe of Indians of Oklahoma	Letter (ML0402304610)
EGCESP-44	Donald Deiker	Resident of Clinton	E-mail (ML0402304640)
EGCESP-45	Kevin Murphy		E-mail (ML0402304660)
EGCESP-46	Robb Hoover		E-mail (ML0402304680)
EGCESP-47	Dan Moriarity		E-mail (ML0402304710)
EGCESP-48	Ryan Doyle		E-mail (ML0402304730)
EGCESP-49	Roy and Carolyn Treadway		E-mail (ML0402304750)
EGCESP-50	Brooke Barber		E-mail (ML0402304810)
EGCESP-51	Paul Gunter	Nuclear Information and Resource Service	E-mail (ML0402304870)
EGCESP-52	Tina L. Prudhomme	IBEW Local 51	Letter (ML0402304910)
EGCESP-53	Kevin Heiden		Letter (ML0402304950)
EGCESP-54	Unknown		Letter (ML0402304990)
EGCESP-55	Donald Gruber	Clinton Community Schools	Letter (ML0402305160)
EGCESP-56	Dale Holtzscher		E-mail (ML0403308330)
EGCESP-57	Brice Obermeyer	NAGPRA Director, Delaware Tribe of Indians	Letter (ML0404805350)
EGCESP-58	Helen PavLak	Clinton Junior High School	Letter (ML0411900600)

- A comment that was either related to support or opposition of early site permitting in general (or specifically the Exelon ESP) or that made a general statement about the early site permit process. In addition, it provided no new information and did not pertain to 10 CFR Part 52.

- A comment about an environmental issue that
  - provided new information that would require evaluation during the review, or
  - provided no new information

- A comment that was outside the scope of the ESP, which included, but was not limited to
  - a comment regarding the need for, or cost of, power
  - a comment regarding alternative energy sources
  - a comment on the safety of the existing units.

The comments that are considered in the evaluation of environmental impacts in this EIS are summarized in the following pages. All comments received during scoping are included in the meeting summary (ML040330445). For reference, the unique identifier for each comment (commenter ID letter listed in Table D-1 plus the comment number) is provided. The responses provided here have been updated to provide the appropriate section in the EIS where the subject is addressed.

Preparation of the EIS took into account all the relevant issues raised during the scoping process. The draft EIS was made available for public comment. The comment period for the draft EIS offered a second opportunity for the applicant; interested Federal, Tribal, State, and local government agencies; local organizations; and members of the public to provide input to the NRC's environmental review process. The comments received on the draft EIS were considered in the preparation of the final EIS. Those comments and the staff's responses are provided in Appendix E of this EIS. This final EIS, along with the staff's Safety Evaluation Report (SER), will provide much of the basis for the NRC's decision on whether to grant the Exelon ESP.

D.1   Comments and Responses

This section summarizes the in-scope comments and suggestions received as part of the scoping process, and discusses their disposition. Parenthetical numbers after each comment refer to the commenter's ID letter and the comment number. Comments can be tracked to the commenter and the source document through the ID letter and comment number listed in Table D-1.

Comments are grouped by the following categories:

D.1.1 Comments Concerning National Environmental Policy Act Compliance
D.1.2 Comments Concerning Land Use
D.1.3 Comments Concerning Air Quality
D.1.4 Comments Concerning Surface Water Use and Quality
D.1.5 Comments Concerning Aquatic Ecology
D.1.6 Comments Concerning Terrestrial Ecology
D.1.7 Comments Concerning Socioeconomic Issues
D.1.8 Comments Concerning Cultural Resources
D.1.9 Comments Concerning Human Health Issues
D.1.10 Comments Concerning the Uranium Fuel Cycle and Waste Management Issues
D.1.11 Comments Concerning Postulated Accidents
D.1.12 Comments Concerning Alternatives and Alternative Sites

D.1.1 Comments Concerning National Environmental Policy Act Compliance

Comment: Maybe you can tell me what to read out there but how big is your environment that you're looking at? Is it southern United States? Northern United States? Southern DeWitt County? DeWitt County? I don't know how big your environment is that you're looking at (EGCESP-S-03-1).

Response: *The area of review by the NRC in this EIS depends upon the environmental resource being reviewed. For example, the northern transmission line runs toward Bloomington, Illinois, for 37 km (23 mi). The southern transmission lines run south through DeWitt County for 13 km (8 mi). The NRC's assessment of the environmental impacts associated with the 50 km (31 mi) of transmission lines is discussed in Sections 4.1.2 and 5.1.2 of this EIS.*

Comment: On that transmission line, you're talking about the owners of the plant, those transmission lines?...So it's transmission lines of that power (EGCESP-S-03-2).

Response: *The transmission lines are not owned by Exelon. They are owned and maintained by Illinois Power Company. Exelon has an agreement to use the Illinois Power transmission lines. The environmental review included the environmental impacts associated with the transmission lines. This is discussed in Sections 4.1.2 and 5.1.2 of this EIS.*

D.1.2 Comments Concerning Land Use

Comment: But from 14,750 head of cattle diminishing to 750 from the time that the Illinois Power Plant was starting to go and land being purchased. We lost that much in agriculture. And today that is still, and this isn't my figures, this comes from the Extension Office and people where we had to get in order to testify before the Nuclear Regulatory Commission (EGCESP-S-24-1).

Response: *The impacts on land use resulting from construction and operation of the proposed facility is discussed in Sections 4.1.1 and 5.1.1 of this EIS.*

D.1.3 Comments Concerning Air Quality

Comment: Nuclear power makes global warming worse. "Whether nuclear can beat coal does not matter because neither of them can beat other options that are free of carbon dioxide," such as wind and solar power (EGCESP-S-09-13).

Comment: Nuclear power is clean. It does not emit greenhouse gases, sulfur dioxide or nitrogen oxide (EGCESP-S-13-5).

Comment: It is the only large-scale, emission-free electricity source that can be readily expanded. Nuclear power plants avoid the emission of sulphur dioxide and nitrogen oxides…the major greenhouse gas, carbon dioxide (EGCESP-S-37-10). Note: This comment was provided in writing and is in addition to the comments taken from the transcript.

Comment: It does produce emissions into our air and water - coal plants are used to create the energy needed in the uranium enrichment process, and so they do pollute contrary to popular belief (EGCESP-S-47-4).

Comment: It seems that the nuclear industry is not held to clean up any facilities after they are built. And of course, safety is another key reason why the proposed plant should not be constructed. Any nuclear facility has the ability to leak out contaminants into the air and water, even through openings as small as 1/16 of an inch. And as it happens, the first Clinton reactor did not have a clean safety record-and now to build another?? (EGCESP-S-50-3).

Comment: There will be drifting of some solid materials from the plume associated with the cooling towers. These "salts" or minerals will deposit on downwind areas and could have an impact on residential and agricultural activities. The impact of this deposition should be evaluated for nearby areas (EGCESP-S-56-1).

Response: *This information was considered in the staff's evaluation of air quality impacts in the EIS. The results of the analysis are presented in Sections 4.2 and 5.2 of this EIS.*

D.1.4 Comments Concerning Surface Water Use and Quality

Comment: My question is the lake capacity adequate now for the second unit? Do you've got enough water already? (EGCESP-S-05-1).

Response: *The NRC evaluated the impacts of the additional direct and indirect evaporative losses of a wet cooling tower for the early site permit unit. The results of the assessment are provided in Sections 4.3 and 5.3 of the EIS.*

Comment: I would imagine part of your environmental impact would have to be measuring the temperature fluctuation of the Clinton Lake in means of the cooling capability. What input does that have on the final design submittal for the cooling aspect of it? That it would be acceptable to use a lake or would it be necessary the design to have a cooling tower? (EGCESP-S-07-1).

Comment: It is presumed that Clinton Lake will be used as a cooling lake for the second nuclear power plant. What affects will this additional heated water have on the fish and other organisms inhabiting the lake? (EGCESP-S-36-2).

Response: *The impact from any cooling system using the parameters identified in the plant parameter envelope (PPE) was reviewed in accordance with the Environmental Standard Review Plan (NUREG-1555) and discussed in the EIS in Sections 5.3 and 5.4.2. At this time, Exelon has indicated that a closed cooling system employing a cooling tower will be used and not a once-through cooling system. Therefore, the staff did not consider once-through cooling. If the applicant were subsequently to decide that they were interested in once-through designs, it would be required to revise its application. The particular cooling system ultimately chosen by the applicant will have to fall within the PPE submitted by the applicant or, if it does not, that portion of the review will need to be reassessed at the combined license stage. The environmental impacts of a cooling tower and any temperature fluctuations it would have on Clinton Lake are assessed in Section 5.3 of this EIS.*

Comment: The water quality impacts, Clinton Lake, which serves as a cooling source for Clinton 1 is formed by damming up Salt Creek in the north fork of Salt Creek. Salt Creek itself is part of a much larger watershed being part of the head waters of the Sangiman River. The waters of this creek pass through numerous small to medium sized communities as they make their way to the Sangiman River and eventually to the Illinois River. The lake itself is used for recreational purposes, boating and swimming and managed by the Illinois Department of Natural Resources. The fisheries of the lake are used by people from throughout Illinois as well as visitors from other states.

According to the National Pollution Discharge Elimination System, NPDES, the permit that is in place for Clinton 1, there is a limit on the temperature change that can occur to the affluent water discharged from the plant (EGCESP-S-27-1).

Comment: Siting a second nuclear plant at the Clinton site could create adverse water supply and quality impacts. First, as acknowledged in Section 5.2 of the Environmental Report, most of the potential designs for a new Clinton 2 nuclear plant would require more water for cooling than would be available in Clinton Lake during drought periods. Second, the additional effluent discharge from the proposed Clinton 2 could increase water temperatures in Clinton Lake, thereby harming aquatic life. These water-related issues must be thoroughly addressed by the NRC in the EIS (EGCESP-S-42-3).

Comment: The EIS for the Clinton nuclear power station is therefore required to address all of the following environmental impacts, including but not limited to: 1. All impacts on the water levels in Clinton Lake arising from increased intake of reactor cooling water for the operation of any proposed new nuclear power units (EGCESP-S-51-1).

Comment: 4. All impacts arising from the increase in the routine discharge of chemicals, heavy metals, cleaning solvents, biocides and radioactive isotopes into Clinton Lake arising from the operation of additional nuclear power units (EGCESP-S-51-4).

Comment: The cooling towers will be discharging stream of about 12,000 gpm into the discharge canal so as to control the concentration of dissolved minerals in the closed cooling water system that runs from the main condenser to the cooling towers. The water in the discharge canal will eventually end up in the lake. The lake has been characterized as a large body of water which has a small inflow and small outflow as compared to lake volume. Such a configuration can lead to a build-up in the lake when a material is constantly being discharged into it. The EIS should review the impact of the cooling tower "blow-down" on the concentration of dissolved solids in the lake and any potential impact on aquatic life in the lake (EGCESP-S-56-2).

Response: *This information was considered in the staff's evaluation of surface water impacts. The results of the analysis are presented in Sections 4.3 and 5.3 of this EIS.*

D.1.5 Comments Concerning Aquatic Ecology

Comment: By adding a second plant to this location, there's a possibility for significant increases in lake temperatures, which will in turn result in significant impacts on a water body that's already listed on the Illinois Environmental Protection Agency's list of impaired waters (EGCESP-S-27-3).

Comment: 2. All impacts on the aquatic environment of Clinton Lake arising from the increase in thermal discharge of reactor cooling water as result of the operation of additional nuclear power units (EGCESP-S-51-2).

Comment: 3. All impacts on Clinton Lake arising from the increased impingement and entrainment of fish, fish spawn, other aquatic life and nutrients arising from the increased reactor cooling water intake for any proposed additional nuclear power units (EGCESP-S-51-3).

Response: *The NRC staff has assessed potential impacts from the cooling system and resulting aquatic and terrestrial impacts during its evaluation of the ESP application. The results of the analysis are presented in Sections 5.3, 5.4.2, and 5.4.3 in this EIS.*

D.1.6 Comments Concerning Terrestrial Ecology

Comment: One of the gauges I like to use to determine a healthy environment is the amount of wildlife there is in the area. It seems each year we have more pheasants, more quail, more deer, excellent fishing. You know, I would have to gauge that as a testimony that, you know, the Clinton Power Station is not being very detrimental to the environment (EGCESP-S-12-2).

Response: *The NRC staff assessed aquatic and terrestrial impacts during its evaluation of the ESP application, and the results of the analysis are presented in Sections 4.4 and 5.4 in this EIS.*

D.1.7 Comments Concerning Socioeconomic Issues

Comment: Of course, jobs, lower real estate taxes that would come with the second unit, of course (EGCESP-S-02-5).

Comment: Last year we paid a little over 10 million dollars in taxes. We contribute thousands of dollars to organizations. There are some recent – we got the opportunity to participate in the Clinton Ultimate Play Space that was drawn up by children from Clinton. And we got to participate financially and some of our workers helped build that. We also participated in the last United Way campaign, increasing our contributions to the county. Over $10,000 to this county, which is one of the three counties we split our money with. And as part of the larger companies, larger nuclear company that we are a part of, the company nuclear employees contributed over a million dollars to United Way. I take great pride in the recent contribution or gifting or donation of the Clinton Lake Marina to the county this past September. We're pleased to have the DeWitt County Board receive the ownership of the marina. The marina is a big part of DeWitt County. Over a million people use the lake annually. And it helps keep revenue coming into this county and we're proud to be a part of gifting that to DeWitt County (EGCESP-S-13-3).

Comment: I also am concerned about funding for schools. Our funding is decreasing and even though I'm going to be retiring in a few years, I would like to see our school system be as good as it has been in the past few years. Also, I'm very interested in economic development. I have seen our people move out. I've seen our unemployment increase tremendously. I would like for us to have a way to increase our economic development again (EGCESP-S-15-1).

Comment: I have found the power plant to be a partner in the education of the children in the community. A lot of the people that work there have children in our schools. And therefore they have concerns as all of us do (EGCESP-S-16-1).

Comment: This plant has meant a lot to this school district obviously financially. That's not all, though. It's been more important than that because it has been a place for people in the community to have a job and raise children and that's our concern (EGCESP-S-16-3).

Comment: And speaking economically, the Clinton Power Station has been a socioeconomic work horse in DeWitt County for over 30 years, for almost 30 years. Through that time it's provided hundreds of jobs for our area. But it's not just the jobs that it's done for our community. There's a tremendous amount of people the plant has brought to us who have become valuable Clinton DeWitt County residents. Several are friends of mine personally.

They are now volunteers, church members and other contributing citizens for the Clinton area. The taxes paid by the plant have improved our schools, making them some of the finest in the state and helped our county services. And although it doesn't sit within the city limits, it continues to help our city tax base. The plant has purchased fire trucks for our city and helps us cultivate a highly qualified fire and emergency personnel with experience not found in municipalities of our size or even larger because of the extra emergency planning for natural disasters for which they train (EGCESP-S-17-1).

Comment: I'm here tonight representing the DeWitt County Economic Development Committee. And the Committee has discussed this and does support the expansion, the second unit and we feel that the problems that we've had with our local economy, with the loss from Revere, the loss of Troll, and the loss of Imperial China, we really need another opportunity to provide some work in the county for our available work force. And we would welcome the second unit if it's sought to be available (EGCESP-S-18-1).

Comment: But it has provided many construction and permanent jobs in DeWitt County and in the surrounding counties. Our power plant has been a good neighbor and has helped, as we've heard, in many community and civic organizations. Myself and the 600 construction electricians that I represent strongly support the construction of Unit 2 and thank you for your time (EGCESP-S-29-2).

Comment: That it is recognized that the better the economy in a area, the more care is given to the environment. The addition of a second unit at Clinton will provide short-term and long term support to the local economy (EGCESP-S-44-2).

Comment: On behalf of all teachers and staff (about 175 people) of the Clinton Community School District, I would like to express our enthusiastic support of a second nuclear power station at Clinton. We are eternally grateful for the economic benefits our district received from Unit One- as well as those enjoyed by the local economy (EGCESP-S-55-1).

Comment: So I know a few things about living in an area where the unemployment rate is very high, where jobs are leaving and not arriving, about going to a school district that's rural and that doesn't seem to have enough money to actually take care of its students. So I really sympathize with a lot of the things that you're dealing with at Clinton and it really sickens me to see the way that the Exelon Corporation is taking care of people by using them. This is the same Exelon Corporation that just last month tried to jack rate hikes through the State Legislature for no particular reason in the process of attempting to buy out Illinois Power. Doesn't seem to be a friend to the taxpayer. Doesn't seem to be a friend to the consumer. This is also the same company that not only near where I lived at the Byron plant but also here, in the process of buying out the plant, human victims of a devaluation scheme that significantly lowered the property tax revenue from the plant before. There is no reason to believe that this wouldn't happen again and again with a new reactor as well (EGCESP-S-25-1).

Comment: We don't need the tax dollars in terms of property taxes. We have a tax structure that needs to be changed significantly any way to support poor and more rural districts and we've known that for decades (EGCESP-S-25-8).

Comment: I also know of socioeconomic problems. And I, as well as anybody else, wants food on my table and I want electricity. But I also want to be healthy (EGCESP-S-28-4).

Comment: And I think the negative consequences of building a new plant completely outweigh new jobs that could be brought in from some other source or some other company that's willing to move in here (EGCESP-S-33-3).

Comment: And I understand anxiety and the difficulty that the community is in, any local community that is in economic distress I can appreciate your concerns (EGCESP-S-19-3).

Comment: Now, we have had the change of a marina. In the beginning the Illinois Power would not have gotten their construction permit unless they presented an analysis of the cost of the recreation plan for Clinton Lake to be executed. And that was one of the last questions and it was 30 days before they were given their construction permit until they did supply that analysis. And they did. So they were responsible then for the recreation on Clinton Lake. What's happened? That's been changed. The plant's been sold to another firm, organization and who ends up then with the liability of the recreation plan for Clinton Lake? You, the DeWitt County people (EGCESP-S-24-3).

Comment: 10. All potential socio-economic impacts from the elevated national security requirements and countermeasures to protect a larger target of terrorism with the expansion of the nuclear power station site including the indefinite and possibly permanent closure of Clinton Lake to public access for sporting, recreation and other means of community economic livelihood (EGCESP-S-51-10).

Comment: Also, how will the recent sale affect the plant to move forward with the new unit (EGCESP-S-58-2).

Response: *These comments discuss socioeconomic issues. The NRC staff assessed the socioeconomic impacts of the proposed action in Sections 4.5 and 5.5 of the EIS, including impacts related to taxes, property values, and recreational use of the lake.*

D.1.8 Comments Concerning Cultural Resources

Comment: Given the location of the proposed project, we request that you conduct a file search in conjunction with the State Office of Historic Preservation and the state's Archaeological Survey. These state agencies will advise you of the potential for archaeological resources, particularly sites of significant cultural interest or sites that contain human remains. Should either of these agencies determine that there are potentially significant archaeological sites in the area and that these sites are related to the tribe's heritage, the Delaware Nation requests that you contact our offices. Together with the SHPO and State Archaeologist we will develop a plan to best protect these archaeological resources. Should either of these agencies recommend an archaeological survey or test excavation of the proposed construction site, we ask that the Delaware Nation be informed of the results of the survey. The Delaware Nation also requests copies of any accompanying site forms or reports. Also, any changes to the above referenced project should be resubmitted to the NAGPRA Director of the Delaware Nation for review. Should this project inadvertently uncover an archaeological site and/or human remains, even after an archaeological survey, we request that you immediately contact the appropriate state agencies, as well as the Delaware Nation. Also, we ask that you halt all construction activities until the tribe and these state agencies are consulted (EGCESP-S-38-1).

Comment: The Peoria Tribe of Indians of Oklahoma is currently unaware of any documentation directly linking Indian Religious Sites to the proposed construction. In the event any items falling under the Native American Graves Protection and Repatriation Act (NAGPRA) are discovered during construction, the Peoria Tribe request notification and further consultation. The Peoria Tribe has no objection to the proposed construction. However, if any human skeletal remains and/or any objects falling under NAGPRA are uncovered during construction, the construction should stop immediately, and the appropriate persons, including state and tribal NAGPRA representatives contacted (EGCESP-S-43-1).

Comment: Our review indicates that this project is located in an area that was not inhabited by the Delaware Tribe. As such, there is little potential for impacting unknown archaeological sites culturally affiliated with the Delaware Tribe and we have no particular objection to the proposal (EGCESP-S-57-1).

Response: *As part of its environmental review of historic and cultural resources, the staff met with the Illinois State Historic Preservation Office (SHPO) and other appropriate information sources. The results of the analysis are presented in Sections 2.9, 4.6, and 5.6 of this EIS. Should an application for a construction permit or combined license be submitted, the staff will take any appropriate action called for as a result of its review of that application.*

D.1.9 Comments Concerning Human Health Issues

Comment: Breast cancer rates in communities within 50 miles of a nuclear reactor increase by an average of 14-40% while the reactor is operating. Areas with more than one reactor have higher cancer rates than single-reactor sites. The increases cannot be attributed to fallout from nuclear weapons tests. Nationally, breast cancer increases by an average of 1% per year in areas without nuclear reactor exposure (Radiation and Public Health Project) (EGCESP-S-09-21).

Comment: Babies born within 50 miles of a reactor have a higher risk of suffering low birth weights or newborn death. While health experts hoped these figures would fall as U.S. neonatal and natal care improved, our country's figure have actually gone up significantly, by 4-8% over expected cases. Thyroid cancer and hypothyroidism rates are also increasing in areas near nuclear reactors. No New Nukes hopes to work with the Radiation and Public Health Project to get current figures for the existing Clinton reactor (EGCESP-S-09-22).

Comment: By analyzing 50 years of U.S. National Cancer Institute data, Dr. Gould showed that "of the 3,000-odd counties in the United States, women living in about 1,300 nuclear counties (located within 100 miles of a reactor) are at the greatest risk of dying of breast cancer." Dr. Gould found similar risks for prostate cancer among men living in nuclear counties (EGCESP-S-09-7).

Comment: The Radiation and Public Health Project (RPHP) Baby Teeth Study is the first to measure radioactivity in the bodies of Americans living near nuclear reactors. It will also help determine whether this radioactivity raises the risk of cancer in children and adults. The study grew out a Jay M. Gould's book "The Enemy Within: The High Cost of Living Near Nuclear Reactors," which found that women living within 100 miles of nuclear reactors are at greatest risk of dying of breast cancer. An earlier study showed that radioactivity in baby teeth rose rapidly due to fallout from atomic bomb tests above the Nevada desert in the 1950s and 1960s, a time when childhood cancer rates were also rising. This information was instrumental in the 1963 ban of above-ground tests by the United States and Soviet Union. The federal government withdrew funding for the study in 1970, and no longer collects information on how much radioactivity is entering our bodies (EGCESP-S-09-10).

Comment: This plant is a danger to our health. And if we allow it to not only stay, but also to grow, it is a danger to our conscience.

Any source of energy that causes tremendous amounts of death and suffering is immoral. End of story.

And this damage is not just a local problem. According to the speaker last Monday night, infant mortality as well as breast cancer rates caused by the plant, are up all the way into Indiana.

These statistics are similar for all of the 11 plants in Illinois, and the 113 in America. This is a lot of death we're talking about.

In order to gauge the severity of nuclear contamination in humans, the Radiation and Public Health Project has put together an experiment to see how much Strontium-90 is in baby teeth. Strontium-90 is produced only by atomic bombs and nuclear reactors, and is chemically similar to calcium. So when the body finds the poison, it uses it as calcium and stores it in teeth and bones.

Earlier studies showed that radioactivity levels were raised in the 1950s and 1960s, and were continued until the government withdrew funding in 1970.

The government no longer does any research on Americans to find out how much radioactivity is entering our bodies.

Well, let me get this straight. The U.S. government allows and even encourages the production of nuclear energy, even though there is solid proof people are dying because of it? We are allowed to live in towns surrounding these plants, but I highly doubt citizens of and around, Braidwood, Byron, Clinton, Dresden, LaSalle County, Limerick, Oyster Creek, Peach Bottom, the Quad Cities, Rock Island and Zion know precisely what they're up against. Do they know why their babies are dying? Probably not. I highly doubt the families who suffer this tremendous loss would just let the perpetrator go on committing the crime if they did (EGCESP-S-09-11).

Comment: Reactors currently in operation cause cancer, heart disease, immune deficiency disorders, fetal deformities, and still births every day. Legal radiation releases harm us. We don't need to add to our radiation burden by building another reactor (EGCESP-S-09-17).

Comment: Most citizens believe that reactors don't routinely release radiation and radioactive particles into the air and water. By the Nuclear Regulatory Commission's (NRC) own calculations, U.S. reactors released 370 curies, or about 1.6 curies per million persons during the 1970-1987 period. ("The Enemy Within") Those living closest to reactors got the highest doses. Because anything released from a nuclear reactor is considered "background radiation" after one year, the NRC can make yearly releases look very small. Unfortunately, some radioactive releases accumulate over time, increasing our health risks in the process (EGCESP-S-09-20).

Comment: We do know that radiation is destructive to persons, to living creatures and to the environment. Why then would we ever possibly risk destruction of our lives and the web of life? Notice I said risk. I didn't say we would. I said we would risk it. Why would we even consider unleashing the power of the atom in ways that allow incomprehensible risks. I say incomprehensible because we have not even yet begun to comprehend those risks or to take them seriously (EGCESP-S-14-2).

Comment: We also know it's not clean because we have evidence that suggest that in DeWitt and Pyatt County that when the Clinton Reactor No. 1 has been running in the '90's as opposed to when it has not been running, the infant mortality rates rise. There's also evidence to suggest that cancer rates rise. A lot of people have spoken saying that they haven't seen any environmental concerns. These are concerns that leap right out in your face. Certainly everyone in the room knows someone who has suffered from cancer, possibly even died from it. You don't know what caused that cancer. Why would you take that risk that cancer might have been somehow related to the operation of a nuclear power plant near you?

That's a risk that isn't going to go away. And we're never going to be able to convincingly prove one way or the other, perhaps, that it was actually nuclear power that did it. So those problems are visible (EGCESP-S-25-5).

Comment: Building a new reactor in Clinton, Illinois would pose a threat to our national food supply. Even during normal operation, nuclear reactors knowingly release radioactive fission products that fall out over surrounding lands. In the case of central Illinois that means agriculture lands. The proposed site for the new reactor is located in the midst of some of the richest agricultural land in the world...One of the radioactive daughter products find its way into our food is strontium-90, which falls onto broad leaves which in turn are consumed by either people or animals. We see greens of all kinds absorb high doses of radioactive particles, as do grasses that are fed to livestock. There are a myriad of ways that radioactive particles enter the food chain. They can also fall out onto fresh water lakes and streams or be released into these water bodies in coolant water (EGCESP-S-26-4).

Comment: I would like to address environmental concerns affecting infant mortality that we've been discussing. The Clinton Nuclear Reactor was off line, shut down during the period of 1996 to 1998. Using State of Illinois Health Department data on infant mortality, and this is defined as deaths in children under one year of age, infant mortality data for calculated for the three years prior to the shut down, 1993 to 1995, the three-year period surrounding the shut down of '96 to '98 and the three years after restart, '99 to '01. Based on the prevailing winds, the following counties were considered downwind of the Clinton Reactor plume. And I might note that it is more than just DeWitt and Pyatt County. These counties include DeWitt, Pyatt, Champaign, Moltry, Douglas, Coles and Vamilia. Two other counties as well in Indiana were considered but I won't be using those in terms of our data discussion this evening. The surrounding counties in the north, south and west are considered up wind. They are Taswell, Christian, Ford, McClain, Megan, Logan, and Sangiman. And every studied county downwind to the Clinton Reactor, infant mortality dramatically decreased during the shut down period from 9.04 deaths per 1,000 live births in the period prior to the restart to 4.6 deaths per 1,000 live births during the period where the reactor was shut down.

During the same period infant mortality rates in the surrounding upwind counties remain statistically unchanged; 8.5 deaths per 1,000 live births down to 8.35 deaths per 1,000 live births. After restart, infant mortality rates soared upwards all of the downwind counties from 4.6 deaths per 1,000 live births to 9.8 deaths per 1,000 live births. But it continued to drop in the upwind counties.

This study strongly suggests the presence of the Clinton Reactor when it is on line is decreasing infant health. Additionally, this study is not alone in its findings. The Radiation Public Health Project studied infant mortality in cancer rates in counties surrounding eight reactors across the country after shut down. In all eight cases, infant deaths and childhood cancers dropped dramatically two years after shut down (EGCESP-S-34-1).

Comment: There is a hidden health cost to nuclear power. The NRC regulation regarding low level radiation releases into the environment need to be re-examined. What will the health costs continued operations of power station be and what will the health cost of a second reactor be? (EGCESP-S-34-2).

Comment: And so that this observation is made in public, I want to point out just one underhanded use of language that the NRC and the nuclear industry uses over and over again to lull concerned citizens in to believe that the NRC is, in fact, safeguarding the public's interest. We are told repeatedly that radiation emissions from a nuclear reactor are far lessor, far less radiation that – exposed to background radiation. What the NRC does not point out is that background radiation includes emissions from radioactive chemicals which occur naturally and those which result in a nuclear effluent process itself, whereas part of the munitions manufacturing or nuclear energy reactors. In fact, emissions release by a nuclear reactor are considered background radiation after one year, whether this one year old particulate is still dangerous or not. NRC guidelines also say that should a second reactor open in Clinton, each reactor would be entitled to count emissions from the plant next door as background radiation. So, the citizens of central Illinois would never know exactly how much radiation is being released from the two plants unless they calculated themselves if they could even find the data necessary for such a calculation given the fact the NRC has stopped publishing its yearly report on radioactive particular emissions from U.S. reactors. What citizens need to realize is the NRC never talks about natural background radiation, which includes emissions from radioactive chemicals which are not man made. The NRC can't talk about natural background radiation because there's nothing natural about their standards of background radiation though they will make it sound like their standards are as safe as living in a basement apartment with a radon remediation system in place (EGCESP-S-34-6).

Comment: In your booklet "Citizen's Guide to US Nuclear Regulator Commission Information" I found two disturbing quotes on page seven. The first, in the section on high-level waste states "The disposal of high-level radioactive waste requires a determination of acceptable health and environmental impacts over thousands of years." Who gave you the right to determine what is "acceptable" harm to inflict on the future? If we can't create something without harming the future, we shouldn't create it at all (EGCESP-S-41-2).

Comment: 6. All impacts on the public health and environment arising out of the increase in routine and accidental radioactive emissions to the air and to the water as the result of the operation of additional nuclear power units. The analysis should consider work by Dr. John Gofman, showing that low-level radiation, at levels considered to be safe for medical use, is a significant contributor to deaths from heart disease and cancer. See Radiation from Medical Procedures in the Pathogenesis of Cancer and Ischemic Heart Disease (Committee for Nuclear Responsibility: 1999) (EGCESP-S-51-6).

Comment: And I want to tell you all that I was this size before the nuclear power plant was built. So, that had no affect on me that I know of (EGCESP-S-02-3).

Comment: Reasons for this include the possible negative impacts on aquatic life and possible increase in the populations of N. fowleri (Naegleria fowleri) (EGCESP-S-27-2).

Comment: In addition, should a significant event occur at the plant or plants and a radioactive release occurs to the lake, the impacts will be far reaching not only to those in the immediate area but to a significant portion of central Illinois. Water supplies and land use will be negatively impacted possibly for decades to come (EGCESP-S-27-4).

Comment: They send the survey that one guy's talking about that checks my quality of life, my animals, my garden. I've never heard of any negative impacts of that (EGCESP-S-31-2).

Comment: According to the NRC's own guidelines, NRC 10 CFR 52.18, Part 100 regarding this ESP scoping meeting, the NRC must evaluate the nature and proximity of human related hazards at the proposed reactor site. Proximity of the current Clinton Reactor No. 1 is a human related hazard that should be sufficiently investigated before any plans for an ESP for a second Clinton Reactor is approved (EGCESP-S-34-9).

Comment: There is clear evidence that nuclear reactors adversely affect public health. As a society we have a moral obligation to our present and future citizens to prevent these hazards if at all reasonably possible (EGCESP-S-40-2).

Comment: I want to tell you that infant mortality rates that they're spouting up here are not only incorrect, what they're telling you is absolutely and totally wrong and I can tell you why. I happen to be the Birth through Three Teacher for the Clinton School District and I work with

84 families right now and 92 babies. I work in concert with the DeWitt Fl County Health Department, which means I have to gather information for them to compile and report through the state. You need to know this. The babies that have died in Clinton have not died as a result of radiation or any other hazard such as that. However, I'd like to tell you what they have died from. We happen to have one of the highest rates of domestic abuse and violence in the state. I also happen to have one of the highest teen pregnancy rates in the state. And we also have a very high unemployment rate. Now, if you know anything about socioeconomic factors, that certainly plays into what has happened to these young babies (EGCESP-S-08-3).

Comment: Second, I have the envelope put out by the Tooth Fairy Project, which is measuring the level of radioactive isotopes strontium in our baby's teeth. Since the government is no longer monitoring the level of radioactivity that is entering our bodies, at least not in an official way, it seems to me that someone has to do it. And the new information on the infant mortality rates downwind of the Clinton facility makes the Tooth Fairy Project Study even more important (EGCESP-S-09-3).

Due to a 60 percent rise in radioactive isotope Strontium-90 in our babies' teeth since the late 1980s, with the counties closest to nuclear reactors having the highest levels, I urge you to avoid using a second nuclear reactor at the Clinton, Illinois facility (EGCESP-S-39-1). Radioactive Sr-90 [Strontium-90] is one of the deadliest elements release by nuclear facilities. The chemical structure of Sr-90 is so similar to that of calcium that the body gets fooled and deposits Sr-90 in the bones and teeth where it remains, continually emitting cancer-causing radiation. Most of the strontium in the baby teeth is transferred to the fetus by the mother during pregnancy. Because we know when and where the baby was born, and where the mother lived while carrying, we can accurately determine when and where radioactivity was absorbed from the environment (EGCESP-S-09-9).

The Radiation and Public Health Project has found a 60 percent rise in radioactive isotope Strontium-90 in our babies' teeth since the late 1980s, with the counties closest to nuclear reactors having the highest levels. It is important to understand that Strontium-90 doesn't occur in nature. It is produced by the fission of either nuclear bombs or nuclear power plants. It is also important to understand that it doesn't take an accident for a nuclear power plant to release radioactive material: That material is released during the routine operation of those facilities. RPHP has found significant elevations in the infant mortality rates of counties downwind of the Clinton facility during the years the plant is operating and reductions of that rate when the plant is shut-down. That data has been previously published in The Pantagraph. Our babies' bodies weren't meant to hold Strontium-90. That was not part of the creator's plan. The NRC must hear from us. Tell them you don't want Strontium-90 in our children's bodies. Tell them that is too high a price (EGCESP-S-41-6).

Response: *The NRC's regulatory limits for radiological protection are set to protect workers and the public from the harmful health effects of radiation on humans. The limits, including effluent release limits, are based on the recommendations of standards-setting organizations. Radiation standards reflect extensive ongoing study by national and international organizations (e.g., the International Commission on Radiological Protection [ICRP], the National Council on Radiation Protection and Measurements, and the National Academy of Sciences) and are conservative to ensure that the public and workers at nuclear power plants are protected. The NRC radiation exposure standards are presented in 10 CFR Part 20, "Standards for Protection Against Radiation," and are based on the recommendations in ICRP Publications 26 and 30. In addition, the U.S. Environmental Protection Agency has established a whole body dose limit of 25 millirem per year (see 40 CFR Part 190). Finally, Appendix I to 10 CFR Part 50 provides dose design objectives for exposure of the public to radioactive effluents from nuclear reactors. Numerous scientifically designed, peer-reviewed studies of personnel exposed to occupational levels of radiation (versus life-threatening accidental doses or medical therapeutic levels) have shown minimal effect to human health, and any effect was from exposures well above the exposure levels of the typical member of the public from normal operation of a nuclear power plant.*

Regarding health effects to populations around nuclear power plants, NRC relies on the studies performed by the National Cancer Institute (NCI). NCI conducted a study in 1990, "Cancer in Populations Living Near Nuclear Facilities," to look at cancer mortality rates around 52 nuclear power plants, 9 U.S. Department of Energy facilities, and 1 former commercial fuel-reprocessing facility. The NCI study concluded from the evidence available that there is no suggestion that nuclear facilities may be linked causally with excess deaths from leukemia or from other cancers in populations living nearby. Additionally, the American Cancer Society has concluded that although reports about cancer case clusters in such communities have raised public concern, studies show that clusters do not occur more often near nuclear plants than they do by chance elsewhere in the population.

Strontium-90 (Sr-90) is produced in roughly 5.8% of nuclear fissions in a reactor's fuel elements and undergoes radioactive decay with a half-life of almost 29 years. Sr-90, and its radioactive decay product yttrium-90 (Y-90), are not harmful unless they are near or inside the body. They are easily shielded if outside the body, resulting in no radiation exposure. The statement is made in one of the comments that the government does not require environmental measurements of Sr-90. On the contrary, NRC licensees perform environmental monitoring for radionuclides in the vicinity of each nuclear reactor. Based on the results of their environmental monitoring program, no elevated levels of radionuclides in the environment attributed to plant operation have been detected. Compared to other radionuclides, both natural and human-made, Sr-90 is not one of the more toxic. For example, naturally occurring thorium-230 is 700 times more radiotoxic for inhalation.

The issue of radioactive effluents and their impacts on human health are assessed in Sections 4.9 and 5.9 of this EIS.

Comment: A particular concern is the potentially pathogenic amoeba, Naegleria fowleri that resides in Clinton Lake. And actually the fact that it does reside in Clinton Lake has been documented in a study published in a scientific journal applied in environmental microbiology. When exposed to warm water this amoeba can become pathogenic and can cause a deadly type of encephalitis in humans. Will the construction of the additional nuclear power plant increase the likelihood of the presence of the deadly form of this amoeba in Clinton Lake? And finally, what affects will this have on the people swimming and skiing in the lake? (EGCESP-S-36-3),

Response: *The NRC assessed human health impacts of the proposed action and presents the results in Section 5.8.1 of this EIS.*

Comment: The Federal Government no longer collects information on how much radioactivity is entering our bones. Yet this information is crucial for determining whether nuclear power plants and weapons facilities are affecting our health and contributing to America's cancer epidemic (EGCESP-S-09-8).

Response: *Measurements of radioactive substances in the body would be misleading and unwarranted. Radioactive substances come from a variety of sources. Interpreting measurements of radioactive materials in people is difficult unless one knows what each individual was exposed to, when the exposure occurred, and by what routes they occurred (ingestion, inhalation, etc.). Also, mitigation must be accounted for, because people may have lived and acquired radionuclides elsewhere than near a nuclear power plant. Finally, substances in the human body are dynamic, not static. This includes radioactive and nonradioactive substances. The dynamic processes include intake of material; uptake to systemic circulation from the gastrointestinal tract, respiratory tract, or skin; translocation throughout the body system; retention over time; and elimination via excretion and radioactive decay.*

Nevertheless, the NRC requires the licensee to perform environmental monitoring for radionuclides in the vicinity of each nuclear reactor to ensure that regulatory limits set to protect workers and public health are maintained. The limits, including effluent release limits, are based on recommendations of standards-setting organizations. Radiation standards reflect extensive ongoing study by national and international organizations (e.g., the International Commission on Radiological Protection, the National Council on Radiation Protection and Measurements, and the National Academy of Sciences) and are conservative to ensure that the public and workers at nuclear power plants are protected. The issue of radioactive effluents and their impact to human health are assessed in Sections 4.9 and 5.9 of this EIS.

Comment: NRC is acting and talking like it's already decided this plant will go through. For real discussion, experts need to present the grave dangers with equal time. Or even more time, since the health of everyone in downstate Illinois is at risk from nuclear plants (EGCESP-S-54-2).

Response: *The decision to issue an ESP has not been made at this time. This EIS has been prepared in accordance with the requirements of 10 CFR 52.18 and 10 CFR Part 51. The evaluation of impacts to human health is discussed in Sections 4.8, 4.9, 5.8, and 5.9 of this EIS.*

D.1.10 Comments Concerning the Uranium Fuel Cycle and Waste Management Issues

Comment: And I'm here because I'm very, very concerned about radioactive nuclear waste from Clinton Power Plant 1 and proposed Clinton Power Plant 2 (EGCESP-S-14-1).

Comment: The fact is that nuclear energy, whether it's unleashed through nuclear bombs or small deadly munitions or a nuclear power plant, all leads to the same end product, which is radioactive nuclear waste. We humans who have made the terrible mistake of creating this waste have absolutely no clue what to do with it now that it exists. No clue where to store it, how to transport it nor how to store it in ways that will keep it for the tens of thousands to millions of years that this radioactivity will remain extraordinary lethal. And who will keep it safe? Who will keep it safe? The radioactivity of the radioactive waste that already exist will need to be cared for far longer than human civilization has even existed. In a nuclear plant, every day routine operation radioactivity is released into our air, water and soil (EGCESP-S-14-3).

Comment: If you had a large medical center with a thousand laboratories using radioactive materials, you would have a combined inventory of about two curies of radiation, I understand from my sources, and in contrast operating a nuclear power reactor will have about 16 billion queries [curies] in its reactor core. This is the equivalent of a long lived radioactivity of at least 1,000 Hiroshima bombs, 1,000 Hiroshima bombs in the size of a reactor like Clinton. Just one pound of plutonium, which is the most toxic known element and remains deadly for 250,000 years. If it was evenly distributed and ingested will kill everybody on the planet, one pound. And yet a thousand megawatt power plant the size of Clinton 1 produces nearly 180 metric tons of radioactivity waste per year, high level radioactive waste. Is all of this waste plutonium? No, it's not. But do we need more high level radioactive waste of any kind? No (EGCESP-S-14-5).

Comment: What is happening to the spent fuel rods and other radioactive waste in Clinton Reactor 1, let alone for Clinton Reactor 2? How full is the storage? How safe is the storage? What's going to happen when the storage here is filled? What's going to happen about transporting it? How and when and where will it be transported? Where will it be kept? Who on earth would want this waste near them or transported through them? And what if there is no safe place? We do not know how to keep this safe for 250,000 years or millions of years (EGCESP-S-14-6).

Comment: There's a discussion about Yucca Mountain being a site. If it is ever approved, it would not open until 2010. And so waste wouldn't even start flowing until then. And in addition, Yucca Mountain doesn't even have enough capacity to hold all the waste that is being produced by plants that are currently operating, much less new plants (EGCESP-S-01-4).

Comment: Neither the industry nor the government knows exactly what to do with nuclear waste. A national waste repository in Yucca Mountain, NV is likely to be held up in court for many years - the state of Nevada does not want the site. Native people are being forced to take some of the waste, again(st) the wishes of the people who live there (EGCESP-S-09-24).

Comment: Nuclear energy is not safe for our environment or to our public health. It creates waste that we currently do not know how to dispose of. Yucca Mountain is definitely not a safe option, the science tells us that, and the transportation to such a location would endanger all the American people that live near the transportation routes. Not to mention the devastating effects that an accident could have on our food supply - as most of the routes to Yucca through the Great Plains are surrounded by farms. Even besides all this, if Yucca was approved, all the space in it is accounted for already. There would be no room for more waste from Clinton, IL that's for sure (EGCESP-S-48-2).

Comment: We have to be careful about the legacy we are leaving to our children's children's children's children. A legacy of lethal radiation relieved [left] to them to tend (EGCESP-S-14-10).

Comment: That is just like the waste that it produces and that also has to be disposed of and put under ground away from man for the next 45,000 years (EGCESP-S-24-5).

Comment: It's also not clean. We know that it's not clean because we have the nuclear waste to deal with (EGCESP-S-25-4).

Comment: This waste that we have that we're developing, we can't comprehend the damage it will do and the way it will have to be stored (EGCESP-S-28-2).

Comment: Nuclear power is dirty. It creates waste that will be horribly dangerous to every single future generation to come (EGCESP-S-47-3).

Comment: High level wastes, some of which would be stored at the Clinton site, are very lethal when exposed directly to human beings. While they may be contained for many years at the site without direct deaths to humans, they cannot be stored there or any where without exposure directly to humans. No place, even the proposed Yucca Mountain area proposed for long-term storage, can be maintained for the thousands of years that some of the nuclear wastes will be lethal to humans. Further, just proximity to a nuclear reactor and wastes may indirectly raise the death rates of persons living nearby. The nuclear wastes at the second (or first) nuclear power plant cannot be made safe. They pose an environmental danger to the population living near the Clinton plant (EGCESP-S-49-2).

Comment: 5. All impacts arising from the additional accumulation of high-level nuclear waste generated and indefinitely stored on-site at Clinton nuclear power station as the result of the operation of additional nuclear power reactors. This discussion is required, given that the Waste Confidence Rule applies only to waste generated by "existing facility licenses." 55 Fed. Reg. 38,474 (September 18, 1990) (EGCESP-S-51-5).

Response: *The safety and environmental effects of long-term storage of spent fuel onsite have been assessed by the NRC, and, as set forth in the Waste Confidence Rule (10 CFR 51.23), the Commission generically determined that such storage could be accomplished without significant environmental impact. In the Waste Confidence Rule, the Commission determined that spent fuel can be stored onsite for at least 30 years beyond the licensed operating life, which may include the term of a renewed license. At or before the end of that period, the fuel would be removed to a permanent repository. In its Statement of Consideration for the 1990 update of the Waste Confidence Rule (55 FR 38472), the Commission addressed the impacts of both license renewal and potential new reactors. Therefore, the current rule can be used in the staff's review of an early site permit application. In its most recent review of the Waste Confidence Rule on December 6, 1999 (64 FR 68005), the Commission reaffirmed the findings in the rule. In addition to the conclusion regarding safe onsite storage of spent fuel, the Commission states in the rule that there is reasonable assurance that at least one geologic repository will be available within the first quarter of the twenty-first century, and sufficient repository capacity for the spent fuel will be available within 30 years beyond the licensed life for operation of any reactor. The NRC staff assessed the environmental impacts of nuclear waste and the results of this analysis are presented in Chapter 6 of this EIS.*

Comment: The production of nuclear waste kills babies, women, men, children. This is not just another left-wing plight. This is a matter of sanity (EGCESP-S-09-12).

Comment: On transportation issues related to spent fuel; as stated at the March 20th, 2003 Pre-Application Early Site Permit Public Meeting, Clinton 1 is already at 60 percent capacity for storage of spent fuel. The management there is considering asking for permission to rerack this spent fuel to allow for more storage space at the site. Assumptions are that a national depository will open in the near future and that this spent fuel will be transported to this site for final storage.

In order to transport this waste, it could be moved by rail and tracks leased to Canadian National. Those tracks not only go through the heart of the City of Clinton, the cars will also be traveling through many more Illinois communities before exiting the state on the way to Yucca Mountain. You heard the railroad go by tonight. Should an incident occur on this route, the immediate community could suffer an extreme radiological event with long term radiation and an inevitable result. No matter what jobs could be generated by building and operating a second nuclear reactor at the Clinton site, it is highly unlikely that the benefits afforded to the people in portions of DeWitt County could counter act such an event. Economic impacts on the citizens of Illinois; much is made of the green benefits of nuclear power. However, in good conscience, we must look at long term generational impacts and cause of nuclear waste on the citizens of Illinois and of this nation. Since all we know is that Exelon wants to have permission to build a second nuclear plant on this site, we can therefore conclude that there will be waste associated with the plant. For reasons stated above, ISA believes this is not in the best of interest of the citizens of Illinois to have to assume the risk of such generation of high level nuclear waste entails (EGCESP-S-27-5).

Response: *The NRC staff assessed the environmental impacts of the uranium fuel cycle, including the impacts of fuel manufacturing, transportation, and the onsite storage and eventual disposal of spent fuel. Results of this analysis are presented in Chapter 6 of this EIS.*

D.1.11 Comments Concerning Postulated Accidents

Comment: Each reactor has the potential to have a catastrophic accident severe enough to destroy for thousands of years all land within 250 miles of the reactor. Industry observers admit that a core meltdown accident has a 50 percent probability of occurring in any decade (EGCESP-S-09-16).

Comment: Each reactor has potential to have a catastrophic accident severe enough to destroy for thousands of years all life within 250 miles and with a fifty percent possibility occurring in any decade, in every decade. This possibility is too high for me (EGCESP-S-14-8).

Comment: A worst case accident resulting in a breach in the containment building at any nuclear reactor here in the United States would be devastating not only to the people of our country but also to the global community as a bloom of deadly radioactive fall out would spread worldwide, just as it did in the Chernobyl tragedy. Clinton, Illinois specifically is not a suitable site for numerous reasons. One of them is its close proximity to Chicago. It is not a smart decision to build a new reactor up wind to a major population center. If the containment building were breached in an accident with winds blowing from the southwest to the northeast, Chicago would be contaminated and destroyed in what would be the worst tragedy in the United States history (EGCESP-S-26-2).

Comment: It doubles the risk of something happening. And there is no guarantee in life, as it has been said. But if there is no guarantee in life and there's always a risk that a catastrophic accident could happen, and that's going to affect us, that's going to affect everybody who lives here (EGCESP-S-33-2).

Response: *The environmental impacts of postulated accidents are discussed, and the results of this analysis are presented in Section 5.10 of this EIS.*

Comment: 7. All impacts on public health and safety arising out of a severe accident, including the impacts of the accident itself, sheltering, evacuation, radiation exposure treatment and reoccupation or relocation of entire communities in the event of an accident at an expanded Clinton site (EGCESP-S-51-7).

Response: *The SER prepared for the early site permit application assesses issues related to emergency planning (see 10 CFR 52.18), including consultation with the Department of Homeland Security/ Federal Emergency Management Agency (DHS/FEMA). In addition, the staff documented in the SER whether the site characteristics are such that adequate security plans and measures can be developed (see 10 CFR 100.21). The environmental impacts of postulated accidents are assessed, and the results of this analysis are presented in Section 5.10 of this EIS.*

Comment: 8. All impacts arising from the simultaneous operation of the existing and aging Clinton power reactor in close proximity to any new proposed advanced reactor design, including the possibility of multiple, simultaneous accidents, whether related (e.g., by fire or natural disaster) or unrelated (EGCESP-S-51-8).

Response: *Existing requirements provide assurance that the probability of simultaneous accidents at multiple units would be substantially less (e.g., over an order of magnitude) than the probability of accidents involving a single unit. For example, 10 CFR Part 50, General Design Criterion 5, "Sharing of structures, systems, and components," requires that structures, systems, and components important to safety not be shared unless it can be shown that such sharing will not significantly impair their ability to perform their safety functions, including, in the event of an accident in one unit, an orderly shutdown and cooldown of the remaining units. Also, a plant- and site-specific probabilistic risk assessment (PRA) will be required prior to operation of any future plant pursuant to 10 CFR 50.34(f)(1)(i). This PRA will determine whether the risk from the as-built units will be low and will account for any inter-unit dependencies. In contrast, the consequences associated with an accident involving multiple units (e.g., a multi-unit core-melt accident) could reasonably be expected to be only marginally greater than with a single-unit event. For example, given the same accident release* *characteristics for both units, the total releases from two reactor cores (and the associated accident consequences) would, as a first-order-of-magnitude approximation, be about twice that for a single unit. The substantially lower frequency of a multiple-unit accident would more than offset the potentially greater consequences of the multiple-unit accident. Thus, the risk associated with multiple, simultaneous accidents would be a negligible contributor to the overall risk from all units on the site. Accordingly, the staff does not plan to address multi-unit accidents as part of the ESP review.*

D.1.12 Comments Concerning Alternatives and Alternative Sites

Comment: Second issue I wanted to address is alternatives. We believe that the NRC is legally required to objectively evaluate alternative sources of energy, especially removable [renewable] energy sources and energy conservation (EGCESP-S-01-6).

Comment: And, in fact, the National Environmental Policy Act specifically requires a consideration of all alternatives, which includes alternative energy sources. Exelon's application relies on 20 year old data to basically dismiss clean energy alternatives as, you know, unreliable and not realistic. But, in fact, renewable energy sources and energy efficiency present a lower cost, safer and environmentally cleaner approach to meeting Illinois' energy needs than nuclear power would. For example, federal studies show that wind power can supply up to 20 percent of the U.S.'s energy needs and energy efficiency efforts can reduce energy demand by 33 percent by 2020. Of course, jobs and economic develop(ment) are at issue, obviously. It's very important to the community. But clean energy alternatives and energy efficiency provides significant job opportunities. For example, wind turbines are considered the cash crop of the 21st Century because they very easily fit in a farm where a farmer can get extra cash from the energy produced by wind turbines. In addition, the opportunities for economic development and energy efficiency technology are great. And we're currently falling behind other countries that invest in that. Therefore, we believe that the NRC should give fair consideration to alternative ways of meeting whatever power to be produced by this proposed second unit (EGCESP-S-01-8).

Comment: Conservation and economical alternative energy sources will one day make nuclear power obsolete. U.S. energy intensity is down 40% from doomsday government and industry projections announced in the 1980's (EGCESP-S-09-15).

Comment: And then I invite you to act with me in every way possible to decrease energy consumption, to develop renewable and safe clean energy and that will allow Clinton 1 and every other plant to be shut down forever (EGCESP-S-14-11).

Comment: This is also the same company that has repeatedly blocked in the last year attempts on the part of the Illinois Legislature to institute renewable energy portfolio standards, which would institute and guarantee that wind power, solar power would be explored, used, power that if you do the research you'll find can be cheaper than nuclear power (EGCESP-S-25-2).

Comment: We don't need the power from nuclear power. We can get it from wind and other renewable energy sources (EGCESP-S-25-7).

Comment: We encourage Exelon to look toward more renewable energy sources (EGCESP-S-27-8).

Comment: And I challenge the Chamber of Commerce, I challenge the DeWitt County Board, I challenge you to bring in industry into this county that is alternative energy, that is healthy industry that will not affect our future children (EGCESP-S-28-5).

Comment: The NRC also sets out in the guidelines for this meeting that it is interested in those facts that demonstrate their obviously superior alternative energy sources for this region. Based on reports and articles in the Environmental Law and Policy Center, the Nuclear Energy Institute's 20th anniversary conference wind, solar, biomass of geothermal energy approaches are far more cost effective than anything nuclear power has to offer. And these alternative energy approaches also would offer an incredible number of jobs for citizens in the region far more quickly than the proposed Clinton Reactor No. 2 can offer and should be seriously considered by those running this meeting that these alternative energy approaches do not produce the intensely hazardous radioactive waste products that nuclear reactors produce every day (EGCESP-S-34-13).

Comment: But large scale generation of electricity does not lend itself to solar generation, to windmills. They all are contributors. So I would suggest to you, from my perspective and having worked in energy policy for quite some time, it's not a question of which. It's a question of all.

I don't think we have the luxury with the population growth, with the demand growth that we see in the future to dismiss out of hand any source. We need everything we can get. They all have their risk, they all have their benefits (EGCESP-S-37-7).

Comment: Instead of a second nuclear reactor at this site which would release radioactive material into the environment, The Environmental Law and Policy Center has developed a plan called "Repowering the Midwest: the Clean Energy Development Plan for the Heartland." Please consider this plan instead of a second nuclear reactor at the Clinton site (EGCESP-S-39-2).

Comment: Without question there are reasonable alternatives even though pursuing them may require conservation, putting up with energy shortages at least in the short-run, and investing in the development of alternative sources of energy (EGCESP-S-40-3).

Comment: The Environmental Law and Policy Center has developed a plan called "Repowering the Midwest: The Clean Energy Development Plan for the Heartland." That plan reduces our use of nuclear power while creating more jobs and making/saving more money than building more nuclear reactors would. Ask the NRC to seriously consider that plan (EGCESP-S-41-7).

Comment: While consideration of whether there is a need for the power from construction and operation of a new Clinton 2 nuclear plant is barred by the NRC, id., the consideration of alternative means of meeting a need for that power is not foreclosed. In fact, the NRC is required to develop and explore, pursuant to Section 102(2)(E) of NEPA, "appropriate alternatives to recommended courses of action in any proposal, which involves unresolved conflicts concerning alternative uses of available resources" 10 CFR 51.45. Energy efficiency and renewable energy resources clearly qualify as "appropriate alternatives" to the siting of the proposed new Clinton 2 nuclear plant and must be rigorously explored and objectively evaluated as part of the EIS. Although Exelon included a discussion of renewable energy resources and energy efficiency in Section 9.2 of its Environmental Report, Exelon nonetheless improperly relied on outdated information to conclude that such alternatives are not feasible. Exelon's discussion relies heavily on the NRC's 1996 Generic Environmental Impact Statement for License Renewal of Nuclear Plants, NUREG-1437, which, in turn, is based on data from the early 1990s regarding the viability of wind power, solar power, and energy efficiency. Technological improvements and market developments since the early 1990s, however, have greatly increased the efficiency and capacity of these alternatives, while at the same time reducing their costs and environmental impacts. The NRC's analysis of renewable energy resource and energy efficiency alternatives must reflect current knowledge and information regarding the economic and technological feasibility of these alternatives, as well as the comparative environmental impacts (EGCESP-S-42-2).

Comment: I urge you to consider the plan put forth by the Environmental Law and Policy Center, 'Repowering the Midwest: the Clean Energy Development Plan for the Heartland.' It outlines ways to reduce our use of nuclear power without sacrificing jobs (EGCESP-S-46-1).

Comment: We need to start using safe energy alternatives such as wind and solar power not dangerous nuclear power (EGCESP-S-48-3).

Comment: Instead put money, time, and investigation into constructing clean energy sources that can create a safe environment, permanent safe jobs, revenue for communities, and save government and tax payer money (EGCESP-S-50-5)

Comment: 1. Whether effects on the environment would be reduced if Exelon alternatively implemented more applications of energy efficiency technologies and energy conservation rather than the development of additional nuclear power capacity at the Clinton site. The Renewable Energy Policy Project has demonstrated that innovative and well-managed efficiency programs would reduce annual increases in electric growth by 61%, substantially reducing demand over a twenty-year period (EGCESP-S-51-12).

Comment: 2. Whether effects on the environment would be reduced if Exelon alternatively implemented use of passive solar, photovoltaic, wind turbines and hybrid renewable energy systems rather than the development of additional nuclear power capacity at the Clinton site (EGCESP-S-51-13).

Comment: 3. Whether effects on the environment would be reduced if Exelon alternatively implemented greater use of natural gas energy rather than the development of additional nuclear power capacity at the Clinton site (EGCESP-S-51-14).

Comment: 4. Whether effects on the environment would be reduced if Exelon alternatively implemented broader applications of the above mentioned resources as distributed power systems rather than increased reliance on an increasingly vulnerable electrical grid system connecting any additional new power capacity at the Clinton site (EGCESP-S-51-15).

Response: *The staff prepared this EIS in accordance with the requirements of 10 CFR 52.18 and 10 CFR 51. As discussed in proposed changes to Part 52 published in the Federal Register on July 3, 2003 (68 FR 40025), consideration of alternative energy sources need not be included in the applicant's ER. In the case of the Exelon application, Exelon did choose to include a consideration of alternative energy sources, and, therefore, the staff assessed energy conservation using current available data. Results of the staff's analysis are discussed in Chapters 8 and 9 of this EIS.*

D.2 References

10 CFR Part 20. Code of Federal Regulations, Title 10, *Energy*, Part 20, "Standards for Protection Against Radiation."

10 CFR Part 50. Code of Federal Regulations, Title 10, *Energy*, Part 50, "Sharing of Structures, Systems, and Components."

10 CFR Part 51. Code of Federal Regulations, Title 10, *Energy*, Part 51, "Environmental Protection Regulations for Domestic Licensing and Related Regulatory Functions."

10 CFR Part 52. Code of Federal Regulations, Title 10, *Energy*, Part 52, "Early Site Permits, Standard Design Certifications, and Combined Licenses for Nuclear Power Plants."

10 CFR Part 100. Code of Federal Regulations, Title 10, Energy, Part 100, "Reactor Site Criteria."

40 CFR Part 190. Code of Federal Regulations, Title 40, *Protection of Environment*, Part 190, "Environmental Radiation Protection Standards for Nuclear Power Operation."

International Commission on Radiological Protection (ICRP). 1977. *Recommendations of the International Commission of Radiological Protection*. ICRP Publication 26, Pergamon Press, New York.

International Commission on Radiological Protection (ICRP). 1979. *Limits for Intakes of Radionuclides for Workers*. ICRP Publication 30, Pergamon Press, New York.

U.S. Nuclear Regulatory Commission (NRC). 2000. *Environmental Standard Review Plan*. NUREG-1555, Vol. 1, NRC, Washington, D.C.

Appendix E

Comments on the Draft Environmental Impact Statement and Responses

Appendix E

Comments on the Draft Environmental
Impact Statement and Responses

This environmental impact statement (EIS) has been prepared in response to an application submitted to the U.S. Nuclear Regulatory Commission (NRC) by Exelon Generation Company, LLC (Exelon) for an early site permit (ESP). The proposed action requested in Exelon's application is for the NRC (1) to approve a site within the existing Clinton Power Station boundaries as suitable for the construction and operation of a new nuclear power-generating facility, (2) to issue an ESP for the proposed site identified as the Exelon ESP site co-located with the existing Clinton Power Station, and (3) to authorize site-preparation activities as described in the site redress plan. This EIS includes the NRC staff's analysis that considers and weighs the environmental impacts of constructing and operating one or more new nuclear units at the Exelon ESP site or at alternative sites, and mitigation measures available for reducing or avoiding adverse impacts. It also includes the staff's recommendation to the Commission regarding the proposed action.

As part of the NRC review of the application, the NRC solicited comments from the public on a draft of this EIS (DEIS). A 75-day comment period began on March 10, 2005, when the NRC issued a Notice of Availability (70 FR 12022) of the DEIS to allow members of the public to comment on the results of the NRC staff's review. On April 19, 2005, a public meeting was held in Clinton, Illinois. At the meeting, the staff described the results of the NRC environmental review, answered questions related to the review, and provided members of the public with information to assist them in formulating their comments.

As part of the process to solicit public comments on the draft EIS, the staff:

- Placed a copy of the draft EIS at the Vespasian Warner Public Library

- Made the draft EIS available in the NRC's Public Document Room in Rockville, Maryland

- Placed a copy of the draft EIS on the NRC website at: www.nrc.gov/reading-rm/doc-collections/nuregs/staff/sr1815/index.html

- Provided a copy of the draft EIS to any member of the public who requested one

- Sent copies of the draft EIS to certain Federal, State, Tribal, and local agencies

- Published a notice of availability of the draft EIS in the *Federal Register* on March 10, 2005 (70 FR 12022)

- Filed the draft EIS with the U.S. Environmental Protection Agency (EPA)

- Announced and held a public meeting on April 19, 2005, in Clinton, Illinois, to describe the results of the environmental review, answer any related questions, and take public comments.

Approximately 300 people attended this meeting and 60 attendees provided oral comments. A certified court reporter recorded these oral comments and prepared written transcripts of the meeting. The transcripts of the public meetings are part of the public record for the proposed project and were used to establish correspondence between comments contained in this volume of the EIS to oral comments received at the public meeting. In addition to the comments received at the public meeting, the NRC received 113 letters and e-mail messages with comments. The comment period closed on May 25, 2005; however, the NRC did, to the degree permitted by the schedule, consider comments submitted after the comment period ended.

The NRC has published a compendium of the transcript and the written comments received during the public comment period in a public record dated June 7, 2005. The comment letters, e-mail messages, and the transcripts of the public meeting are available from the Publicly Available Records component of NRC's Agencywide Documents Access and Management System (ADAMS). ADAMS is accessible at http://www.nrc.gov/reading-rm/adams.html, which provides access through the NRC's Public Electronic Reading Room link. Persons who do not have access to ADAMS or who encounter problems in accessing the documents located in ADAMS, should contact the NRC's Public Document Room reference staff at 1-800-397-4209 or 301-415-4737, or by e-mail at pdr@nrc.gov. The ADAMS accession numbers for the letters and e-mail messages are provided in Table E-1. The NRC staff has reviewed each written comment and the transcript of the public meeting.

E.1 Disposition of Comments

This volume contains all of the comments abstracted from the comment letters and e-mail messages provided to the staff during the comment period as well as the comments from the transcripts.

Each set of comments from a given commenter was given a unique alpha identifier (commenter ID letter), allowing each set of comments from a commenter to be traced back to the transcript, letter, or e-mail in which the comments were submitted.

After the comment period, the staff considered and dispositioned all comments received. To identify each individual comment, the NRC staff reviewed the transcript of the public meeting and each letter and e-mail received related to the draft EIS. As part of the review, the staff identified statements that they believed were related to the proposed action and recorded the statements as comments. Each comment was assigned to a specific subject area, and similar comments were grouped together. Finally, responses were prepared for each comment or group of comments.

For each comment, the staff determined whether a comment:

- Related to the Exelon ESP and discussed a specific environmental impact

- Related to an issue considered outside the scope of this environmental review (emergency response, alternative energy sources, cost of power, need for power, operational safety, safeguards and security related to terrorism)

- Opposed or supported nuclear power

- Opposed or supported the Exelon ESP

- Discussed NRC's ESP process

- Discussed National Environmental Policy Act (NEPA) requirements.

This appendix presents the comments and the NRC responses to them grouped by similar issues as follows:

- Comments Related to the ESP Process

- General Comments in Support of NRC and its ESP Process

- General Comments in Opposition to NRC and its ESP Process

- General Comments in Support of the Applicant and its ESP Application

- Comments Related to Environmental Impacts

- Comments Related to Alternatives and Alternative Sites

- Comments Concerning the Site Redress Plan

- Comments Concerning Editorial Issues

- Comments Concerning Out-of-Scope Issues: Safety, Safeguards and Security, Emergency Preparedness, Cost of Power, and Need for Power

- Comments Concerning NRC's Administrative Process

- Comments in Support of or Opposition to Nuclear Power

When the comments resulted in a change in the text of the draft EIS, the corresponding response refers the reader to the appropriate section of the report where the change was made. Revisions to the text from the draft EIS are indicated by vertical lines beside the text. Table E-1 provides a list of commenters identified by name, affiliation (if given), comment number, and the source of the comment.

Many comments addressed topics and issues that are not part of the environmental review for this proposed action. These comments included questions about the NRC's safety review, general statements of support or opposition to nuclear power, observations regarding national nuclear waste management policies, comments on the NRC regulatory process in general, and comments on NRC regulations. These comments are included, but detailed responses to such comments are not provided because they addressed issues that do not directly relate to the environmental effects of this proposed action and are thus outside the scope of the NEPA review of this proposed action.

Many comments specifically addressed the scope of the environmental review, analyses, and issues contained in the draft EIS, including comments about potential impacts, proposed mitigation, the agency review process, and the public comment period. Detailed responses to each of these comments are provided in this appendix.

Table E-1. Individuals Providing Comments on the Draft Environmental Impact Statement

Commenter ID	Commenter	Affiliation (if stated)	Comment Source and ADAMS Accession #
01	Harry Borrenpohl		E-mail (ML050800063)
02	Susan O'Rourke		E-mail (ML050830288)
03	Durango Mendoza		E-mail (ML050960345)
04	Rich Katz		E-mail (ML051050345)
05	Jeff Semmerling		E-mail (ML051050328)
06	Bernice Barta		E-mail (ML051050338)
07	Scott Ollar		E-mail (ML051160038)
08	Anne Haaker	DeWitt Historic Preservation Agency	Letter (ML051440428)

Table E-1. (contd)

Commenter ID	Commenter	Affiliation (if stated)	Comment Source and ADAMS Accession #
09	Mailie La Zarr		E-mail (ML051160040)
10	Gina Cassidy		E-mail (ML051160039)
11	Marsha Puthoff		E-mail (ML051160035)
12	Rudolf Mortimer		E-mail (ML051440359)
13	E. McCabe		E-mail (ML051440361)
14	Thomas Hieronymus	DeWitt County Farm Bureau	Letter (ML051440360)
15	Sara Stevenson		E-mail (ML051440357)
16	John Veirs		E-mail (ML051440364)
17	Naomi Jakobssen	State Representative	Letter (ML051440368)
18	David Baggott		E-mail (ML051440363)
19	Linda Weber		E-mail (ML051440378)
20	Joy Reese		E-mail (ML051440370)
21	Armine Kotin Mortimer		E-mail (ML051440367)
22	Carol Preston		E-mail (ML051440372)
23	Darcy Gentner	Sierra Club	E-mail (ML051180462)
24	John Schaefer		Letter (ML051440450)
25	Dennis Nelson		E-mail (ML051440374)
26	Philp Nelson	Illinois Farm Bureau	Letter (ML051440383)
27	Katherine Ferguson		Letter (ML051440392)
28	Eric Ferguson		Letter (ML051440400)
29	Terry Ferguson		Letter (ML051440385)
30	Micheal Chezik	U.S. Fish and Wildlife Service	Letter (ML051460042)
31	Dan Hang		Transcript (ML051590198)
32	Phil Huckelberry	Illinois Green Party	Transcript (ML051590198)

Table E-1. (contd)

Commenter ID	Commenter	Affiliation (if stated)	Comment Source and ADAMS Accession #
33	Gary Lambert		Transcript (ML051590198)
34	Cheryl Springwood		Transcript (ML051590198)
35	Amy Butterworth		Transcript (ML051590198)
36	Matt Rader		Transcript (ML051590198)
37	Karen Lowery	Teacher	Transcript (ML051590198)
38	UNKNOWN		Transcript (ML051590198)
39	Rachel Herbener		Transcript (ML051590198)
40	Kathleen Garibaldi		Transcript (ML051590198)
41	Bill Row		Transcript (ML051590198)
42	Terry Ferguson	Resident	Transcript (ML051590198)
43	Roger Massey	Sheriff of DeWitt County	Transcript (ML051590198)
44	Curt Hochbein	Representative of Naomi Jackobssen	Transcript (ML051590198)
45	Steve Vandiver	Economic Development Director for Clinton	Transcript (ML051590198)
46	Carolyn Treadway		Transcript (ML051590198)
47	Bruce Macking		Transcript (ML051590198)
48	Laura Ekem	Resident	Transcript (ML051590198)
49	Sandra Lindberg	Resident	Transcript (ML051590198)
50	Harold Weinberg	Resident	Transcript (ML051590198)
51	Cheryl Lietz	Resident	Transcript (ML051590198)
52	Corey Conn	Resident	Transcript (ML051590198)
53	Ken Bjelland	DeWitt County Economic Development	Transcript (ML051590198)
54	Nan Craig	Resident	Transcript (ML051590198)
55	Michael Duerr		Transcript (ML051590198)

Table E-1. (contd)

Commenter ID	Commenter	Affiliation (if stated)	Comment Source and ADAMS Accession #
56	Delores Pino	Nuclear Energy Information Services (NEIS)	Transcript (ML051590198)
57	Darren Black	Fire Department	Transcript (ML051590198)
58	Roy Treadway	Illinois State University	Transcript (ML051590198)
59	Shannon Fisk	Environmental Law and Policy Center	Transcript (ML051590198)
60	Gregg Brown		Transcript (ML051590198)
61	Kelly Taylor		Transcript (ML051590198)
62	Roger Blomquist		Transcript (ML051590198)
63	Patricia Swarts	Clinton Elks Lodge	Transcript (ML051590198)
64	Delbert Horn		Transcript (ML051590198)
65	Sydney Baiman		Transcript (ML051590198)
66	Michael Stuart		Transcript (ML051590198)
67	Paul Gunter	Nuclear Information and Resource Service	Transcript (ML051590198)
68	Brendan Hoffman		Transcript (ML051590198)
69	Lee Jankowski		Transcript (ML051590198)
70	Craig Pohlod		Transcript (ML051590198)
71	Dennis Nelson	NEIS	Transcript (ML051590198)
72	Dorian Breuer		Transcript (ML051590198)
73	Vic Connor		Transcript (ML051590198)
74	Norris McDonald	African-American Environmentalist Association	Transcript (ML051590198)
75	David Pointer		Transcript (ML051590198)
76	Ross Radel	Student	Transcript (ML051590198)
77	Tracy Radel	Student	Transcript (ML051590198)

Table E-1. (contd)

Commenter ID	Commenter	Affiliation (if stated)	Comment Source and ADAMS Accession #
78	Kevin Austin	Student	Transcript (ML051590198)
79	Alan Bolind	Student	Transcript (ML051590198)
80	George Gore		Transcript (ML051590198)
81	Linda Lewison		Transcript (ML051590198)
82	Richard Douglas	Resident	Transcript (ML051590198)
83	Stirling Crow	Student	Transcript (ML051590198)
84	Harry Bradley	American Nuclear Society	Transcript (ML051590198)
85	Geoff Ower	Student	Transcript (ML051590198)
86	Brian Kiedrowski	Student	Transcript (ML051590198)
87	Hannah Yount	Student	Transcript (ML051590198)
88	Steve Cohn	Teacher	Transcript (ML051590198)
89	Scott Summers	Illinois Green Party	Transcript (ML051590198)
90	John Gilpin		Transcript (ML051590198)
91	Salmaan Akhtar	University of Illinois	Transcript (ML051300569)
92	Kathleen Garibaldi		Transcript (ML051300569)
93	Ben Holtzmen		Transcript (ML051300569)
94	Carolyn Treadway		Transcript (ML051300569)
95	Terry Lane		Transcript (ML051300569)
96	Charlotte Green		Transcript (ML051300569)
97	Katherine Ferguson		Transcript (ML051300569)
98	Barbara Kessel		Transcript (ML051300569)
99	Eric Ferguson		Transcript (ML051300569)
100	John Gilpin		Transcript (ML051300569)
101	Harry Bradley	American Nuclear Society	Transcript (ML051300569)

Table E-1. (contd)

Commenter ID	Commenter	Affiliation (if stated)	Comment Source and ADAMS Accession #
102	Dave Kraft	NEIS	Transcript (ML051300569)
103	Thomas Hieronymus	DeWitt County Farm Bureau	Transcript (ML051300569)
104	Naomi Jakobsson		Transcript (ML051300569)
105	Terry Ferguson		Transcript (ML051300569)
106	Thomas Edmunds	Former Clinton Mayor	Transcript (ML051300569)
107	Vera Leopold		Transcript (ML051300569)
108	David Kraft Press release	NEIS	Transcript (ML051300569)
109	Vera Leopold		Transcript (ML051300569)
110	Roy Treadway		Transcript (ML051300569)
111	North American Young Generation in Nuclear Petition		Transcript (ML051300569)
112	Linda Zoblotsky		E-mail (ML051720170)
113	Beki Lischalk		E-mail (ML051720170)
114	Linda Ferris		E-mail (ML051720170)
115	Sarah Lanzman		E-mail (ML051720170)
116	John Lischalk		E-mail (ML051720170)
117	Mark Smith		E-mail (ML051720170)
118	Robin Lorentzen		E-mail (ML051720170)
119	Katy Nicholson		E-mail (ML051720170)
120	Mha Atma S. Klalsa		E-mail (ML051720170)
121	Elena Day		E-mail (ML051720170)
122	G Hande		E-mail (ML051720170)
123	Brent Barnes		E-mail (ML051720170)
124	Faith Sadley		E-mail (ML051720170)

Table E-1. (contd)

Commenter ID	Commenter	Affiliation (if stated)	Comment Source and ADAMS Accession #
125	William Kowatch		E-mail (ML051720170)
126	Eric Bourgeois		E-mail (ML051720170)
127	Patricia Aguirre		E-mail (ML051720170)
128	Jim and Virginia Wagner		E-mail (ML051720170)
129	Donald and Connie Roux		E-mail (ML051720170)
130	Tammie Haugen		E-mail (ML051720170)
131	Christine Roane		E-mail (ML051720170)
132	Barbara Fikes		E-mail (ML051720170)
133	Cheryl Hines-Dronzkowski		E-mail (ML051720170)
134	H. Elaine Engel		E-mail (ML051720170)
135	Dean Foss		E-mail (ML051720170)
136	Michael Laird		E-mail (ML051720170)
137	Richard Linsenberg		E-mail (ML051720170)
138	Susan Emge Milliner		E-mail (ML051720170)
139	Barbara Henderson		E-mail (ML051720170)
140	Sandra Blackburn		E-mail (ML051720170)
141	Marilyn Kray	Exelon	E-mail (ML051720170)
142	Brian Lutenegger		E-mail (ML051720170)
143	Angela McComb		E-mail (ML051720170)
144	Faith Vis		E-mail (ML051720170)
145	Rosalie Hewitt		E-mail (ML051720170)
146	Gwenn Carver		E-mail (ML051720170)
147	Sandra Lindberg	No New Nukes	E-mail (ML051720170)
148	Marty Greenberg		E-mail (ML051720170)
149	Catherine Miller		E-mail (ML051720170)

Table E-1. (contd)

Commenter ID	Commenter	Affiliation (if stated)	Comment Source and ADAMS Accession #
150	Joseph Malherek	Public Citizen	E-mail (ML051720170)
151	Joseph Malherek	Public Citizen	E-mail (ML051540382)
152	Tom Lutze		E-mail (ML051720170)
153	Vic and Cindy Connor		E-mail (ML051720170)
154	Joyce Long		E-mail (ML051720170)
155	Don Cramer		E-mail (ML051720170)
156	Samuel Galewsky		E-mail (ML051720170)
157	Elizabeth Burns		E-mail (ML051720170)
158	Thomas Philips		E-mail (ML051720170)
159	Joyce Blumenshine		E-mail (ML051720170)
160	Craig Pohlod		E-mail (ML051720170)
161	Dave Kraft	NEIS	E-mail (ML051720170)
162	Barbara Tompkins		E-mail (ML051720170)
163	Thomas Connor		E-mail (ML051720170)
164	Alan Carlson		E-mail (ML051720170)
165	Katherine Jenkins-Murphy		E-mail (ML051720170)
166	Sue Wedzel		E-mail (ML051720170)
167	Marie Overall		E-mail (ML051720170)
168	Scott Jost		E-mail (ML051720170)
169	George Gore		E-mail (ML051720170)
170	Shannon Fisk	Environmental Law and Policy Center	Letter (ML051540384)
171	Dennis Nelson	NEIS	Letter (ML051590322)
172	Kenneth Westlake	U.S. Environmental Protection Agency	Letter (ML051590180)
173	Pat Dressler		Letter (ML051590209)

Table E-1. (contd)

Commenter ID	Commenter	Affiliation (if stated)	Comment Source and ADAMS Accession #
174	Beverly Cohen		E-mail (ML051720170)
175	Will Yeager		E-mail (ML051720170)
176	Paul Stein		E-mail (ML051720170)
177	D.A. Wagner		E-mail (ML051720170)
178	David Turnoy		E-mail (ML051720170)
179	George Gore		E-mail (ML051720170)
180	William Brigman		E-mail (ML051720170)
181	Timothy Stebler		E-mail (ML051720170)
182	Joe Salazar		E-mail (ML051720170)
183	Wally Taylor		E-mail (ML051720170)
184	Joyce Blumenshine		E-mail (ML051720170)
185	Smoky Mountain		E-mail (ML051720170)
186	Jeanne Thatacher		E-mail (ML051720170)
187	Walter Ballin		E-mail (ML051720170)
188	Marguerite Joan Galimitakis		E-mail (ML051720170)
189	Clark Mleynek		E-mail (ML051720170)
190	Lydia Garvey		E-mail (ML051720170)
191	Connie and Donald Roux		Letter (ML051720170)
192	David and Jennifer Nolfi		E-mail (ML051720170)
193	James Scurrah		E-mail (ML051720170)
194	James Clarke		E-mail (ML051720170)

E.2 Comments and Responses

Table E-2 presents the categories in the order in which they are presented in this appendix.

Table E-3, which is an index to the comment categories, arranges the categories alphabetically and provides the commentor ID for each category.

Table E-2. Order of Comment Categories in Appendix E, by Section Numbers and Title

Section #	Section Title
E.2.1	Comments Related to the ESP Process
E.2.2	General Comments in Support of NRC and its ESP Process
E.2.3	General Comments in Opposition to NRC and its ESP Process
E.2.4	General Comments in Support of the Applicant and its ESP Application
E.2.5	General Comments in Opposition of the Applicant and its ESP Application
E.2.6	Comments Concerning NEPA Compliance
E.2.7	Comments Concerning Land Use
E.2.8	Comments Concerning Air Quality
E.2.9	Comments Concerning Surface Water Use and Quality
E.2.10	Comments Concerning Groundwater Use and Quality
E.2.11	Comments Concerning Aquatic Ecology
E.2.12	Comments Concerning Terrestrial Ecology
E.2.13	Comments Concerning Threatened or Endangered Species
E.2.14	Comments Concerning Socioeconomics
E.2.15	Comments Concerning Environmental Justice
E.2.16	Comments Concerning Cultural Resources
E.2.17	Comments Concerning Human Health and Radiological Impacts
E.2.18	Comments Concerning the Uranium Fuel Cycle and Waste Management
E.2.19	Comments Concerning Postulated Accidents
E.2.20	Comments Concerning Alternatives and Alternative Sites
E.2.21	Comments Concerning the Site Redress Plan
E.2.22	Comments Concerning Editorial Issues
E.2.23	Comments Concerning the Safety Review for the ESP
E.2.24	Comments Concerning Safeguards and Security
E.2.25	Comments Concerning Emergency Preparedness
E.2.26	Comments Concerning Decommissioning

Table E-2. (contd)

Section #	Section Title
E.2.27	Comments Concerning the Cost of Power
E.2.28	Comments Concerning the Need for Power
E.2.29	Comments Concerning Operational Safety
E.2.30	Comments Concerning Other Issues
E.2.31	Comments Concerning NRC's Administrative Process
E.2.32	General Comments in Support of Nuclear Power
E.2.33	General Comments in Opposition to Nuclear Power
E.2.34	Comments that are Outside the Scope of Early Site Permitting

Table E-3. Comments Indexed Alphabetically by Comment Category with Corresponding Section Numbers and Commenters' Identification Numbers (ID)

Comment Category	Commenter ID
Air Quality (Section E.2.8)	47, 55, 61, 66, 74, 75, 77, 84, 86, 101, 153, 172
Alternatives and Alternative Sites (Section E.2.20)	5, 6, 9, 10, 11, 12, 17, 21, 23, 24, 25, 29, 32, 33, 35, 36, 37, 39, 40, 41, 42, 44, 47, 55, 59, 61, 62, 64, 66, 68, 71, 75, 78, 79, 80, 81, 83, 87, 89, 93, 102, 104, 105, 109, 112, 113, 114, 115, 116, 117, 118, 119, 120, 121, 122, 123, 124, 125, 126, 127, 128, 129, 130, 131, 132, 133, 134, 135, 136, 137, 138, 139, 140, 142, 143, 144, 145, 146, 147, 149, 150, 151, 152, 154, 155, 158, 159, 161, 162, 163, 164, 165, 166, 167, 168, 169, 170, 171, 172, 173, 174, 175, 176, 177, 178, 179, 180, 181, 182, 183, 184, 185, 186, 187, 188, 189, 190, 191, 192, 193, 194
Aquatic Ecology (Section E.2.11)	112, 113, 114, 115, 116, 117, 118, 119, 120, 121, 122, 123, 124, 125, 126, 127, 128, 129, 130, 131, 132, 133, 134, 135, 136, 137, 138, 139, 140, 141, 142, 143, 144, 145, 146, 147, 149, 150, 151, 153, 154, 155, 157, 158, 159, 162, 163, 164, 165, 166, 167, 173, 174, 175, 176, 177, 178, 180, 181, 182, 183, 184, 185, 186, 187, 188, 189, 190, 192, 193, 194

Table E-3. (contd)

Comment Category	Commenter ID
Concerns Related to the ESP Process (Section E.2.1)	25, 31, 32, 34, 49, 67, 68, 69, 71, 80, 92, 107, 110, 111, 112, 113, 114, 115, 116, 117, 118, 119, 120, 121, 122, 123, 124, 125, 126, 127, 128, 129, 130, 131, 132, 133, 134, 135, 136, 137, 138, 139, 140, 141, 142, 143, 144, 145, 146, 147, 149, 150, 151, 152, 153, 154, 155, 156, 158, 159, 161, 162, 163, 164, 165, 166, 167, 169, 172, 173, 174, 175, 176, 177, 178, 179, 180, 181, 182, 183, 184, 185, 186, 187, 188, 189, 190, 191, 192, 193, 194
Cost of Power (Section E.2.27)	7, 9, 11, 12, 21, 22, 24, 26, 42, 48, 59, 61, 65, 75, 81, 84, 86, 87, 88, 89, 90, 100, 101, 102, 106, 161, 169, 170, 179
Cultural Resources (Section E.2.16)	8, 29, 42, 105, 141, 150, 151
Decommissioning (Section E.2.26)	141
Editorial Issues (Section E.2.22)	141, 150, 151, 153, 172
Emergency Preparedness (Section E.2.25)	85, 106, 157
Environmental Justice (Section E.2.15)	35, 87, 150, 151
Groundwater Use and Quality (Section E.2.10)	141
Human Health and Radiological Impacts (Section E.2.17)	1, 6, 7, 10, 17, 27, 32, 33, 38, 44, 46, 52, 55, 60, 65, 73, 86, 94, 96, 97, 98, 104, 109, 110, 141, 150, 151, 152, 153, 156, 157, 172, 183
Land Use (Section E.2.7)	77, 141, 150, 151, 153, 161, 169, 172, 179
Need for Power (Section E.2.28)	1, 24, 26, 28, 29, 42, 48, 50, 51, 66, 68, 78, 79, 84, 87, 93, 99, 101, 105, 112, 113, 114, 115, 116, 117, 118, 119, 120, 121, 122, 123, 124, 125, 126, 127, 128, 129, 130, 131, 132, 133, 134, 135, 136, 137, 138, 139, 140, 142, 143, 144, 145, 146, 147, 149, 150, 151, 154, 155, 158, 159, 162, 163, 164, 165, 166, 167, 169, 170, 171, 172, 173, 174, 175, 176, 177, 178, 179, 180, 181, 182, 183, 184, 185, 186, 187, 188, 189, 190, 191, 192, 193, 194
NEPA Compliance (Section E.2.6)	150, 151, 169, 170, 179
NRC's Administrative Process (Section E.2.31)	17, 35, 46, 49, 55, 56, 85, 94, 104, 108, 157
Operational Safety (Section E.2.29)	1, 9, 10, 12, 16, 22, 27, 43, 67, 76, 95, 378, 106

Table E-3. (contd)

Comment Category	Commenter ID
Opposition to NRC and its ESP Process (Section E.2.3)	25, 49, 56, 73, 108, 150, 151, 157, 161, 58
Opposition to Nuclear Power (Section E.2.33)	4, 5, 9, 10, 13, 22, 46, 47, 56, 60, 65, 71, 89, 94, 148
Opposition to the Applicant and its ESP Application (Section E.2.5)	2, 3, 4, 9, 10, 11, 12, 13, 15, 17, 18, 19, 20, 21, 23, 25, 32, 37, 71, 83, 90, 100, 102, 104, 109, 110, 148, 157, 161, 184, 191, 1, 48, 62, 82, 84, 101, 111, 172
Outside the Scope of Early Site Permitting (Section E.2.34)	25, 49, 52, 168, 191
Other Issues (Section E.2.30)	25, 88, 141, 169, 170, 172, 179
Postulated Accidents (Section E.2.19)	5, 15, 24, 46, 58, 62, 65, 81, 90, 94, 100, 141, 148, 150, 151, 169, 172, 179
Safeguards and Security (Section E.2.24)	4, 9, 10, 18, 28, 67, 68, 71, 80, 89, 99, 102, 112, 113, 114, 115, 116, 117, 118, 119, 120, 121, 122, 123, 124, 125, 126, 127, 128, 129, 130, 131, 132, 133, 134, 135, 136, 137, 138, 139, 140, 142, 143, 144, 145, 146, 147, 148, 149, 150, 151, 154, 155, 157, 158, 159, 162, 163, 164, 165, 166, 167, 169, 170, 173, 174, 175, 176, 177, 178, 179, 180, 181, 182, 183, 184, 185, 186, 187, 188, 189, 190, 192, 193, 194
Safety Review for the ESP (Section E.2.23)	55, 72, 172
Site Redress Plan (Section E.2.21)	169, 179
Socioeconomics (Section E.2.14)	27, 28, 29, 32, 35, 42, 43, 45, 47, 57, 58, 66, 72, 75, 79, 82, 85, 91, 93, 97, 99, 105, 106, 110, 112, 113, 114, 115, 116, 117, 118, 119, 120, 121, 122, 123, 124, 125, 126, 127, 128, 129, 130, 131, 132, 133, 134, 135, 136, 137, 138, 139, 140, 141, 142, 143, 144, 145, 146, 147, 149, 150, 151, 153, 154, 155, 157, 158, 159, 162, 163, 164, 165, 166, 167, 173, 174, 175, 176, 177, 178, 180, 181, 182, 183, 184, 185, 186, 187, 188, 189, 190, 191, 192, 193, 194
Support of NRC and its ESP Process (Section E.2.2)	1, 48, 62, 82, 84, 101, 111, 172
Support of Nuclear Power (Section E.2.33)	14, 26, 29, 42, 48, 51, 53, 61, 64, 66, 74, 75, 76, 77, 78, 86, 93, 103, 105, 111

Table E-3. (contd)

Comment Category	Commenter ID
Support of the Applicant and its ESP Application (Section E.2.4)	14, 16, 24, 26, 27, 28, 29, 42, 43, 45, 48, 51, 53, 54, 57, 63, 66, 70, 74, 76, 82, 87, 95, 97, 99, 103, 105, 106, 111, 160
Surface Water Use and Quality (Section E.2.9)	47, 68, 112, 113, 114, 115, 116, 117, 118, 119, 120, 121, 122, 123, 124, 125, 126, 127, 128, 129, 130, 131, 132, 133, 134, 135, 136, 137, 138, 139, 140, 141, 142, 143, 144, 145, 146, 147, 149, 150, 151, 153, 154, 155, 157, 158, 159, 162, 163, 164, 165, 166, 167, 169, 172, 173, 174, 175, 176, 177, 178, 179, 180, 181, 182, 183, 184, 185, 186, 187, 188, 189, 190, 191, 192, 193, 194
Terrestrial Ecology (Section E.2.12)	27, 97, 141, 150, 151, 172
Threatened or Endangered Species (Section E.2.13)	30, 172
Uranium Fuel Cycle and Waste Management (Section E.2.18)	2, 4, 5, 6, 9, 10, 11, 12, 13, 15, 19, 20, 25, 37, 46, 55, 58, 59, 66, 67, 68, 70, 71, 75, 77, 83, 86, 89, 93, 94, 106, 109, 110, 112, 113, 114, 115, 116, 117, 118, 119, 120, 121, 122, 123, 124, 125, 126, 127, 128, 129, 130, 131, 132, 133, 134, 135, 136, 137, 138, 139, 140, 141, 142, 143, 144, 145, 146, 147, 148, 149, 150, 151, 153, 154, 155, 158, 159, 160, 162, 163, 164, 165, 166, 167, 169, 170, 172, 173, 174, 175, 176, 177, 178, 179, 180, 181, 182, 183, 184, 185, 186, 187, 188, 189, 190, 191, 192, 193, 194

The comments that are considered in the evaluation of the environmental impact in this EIS are summarized in the following pages. Parenthetical notations after each comment refer to the commenter's ID letters and the comment number. Comments can be tracked to the commenter and the source document through the ID letter and comment number listed in Table E-1.

E.2.1 Comments Related to the ESP Process

Comment: They actually have a foundation for a second one. Isn't some of this work redundant? (31-1)

Response: The designs being considered for a future nuclear plant would be significantly different from the original design. In addition, codes used in the original may be different from future codes, so the original foundation may be unuseable. No change was made to the EIS as a result of the comment.

Comment: Approving generic designs in what they call the plant parameter envelope does not protect the people in this room. The NRC's slavish adherence to its carefully engineered regulations flies in the face of its mission statement. (49-5)

Response: *The NRC's mission is to regulate the nation's civilian use of by-product, source, and special nuclear materials to ensure adequate protection of public health and safety, to promote the common defense and security, and to protect the environment. Any Commission decision to grant an ESP to Exelon would be consistent with this mission. No change was made to the EIS as a result of the comment.*

Comment: On page 1.2, it talks about the construction that's allowed, and I'm not a lawyer, but just reading that, it sounds like you can essentially construct just about everything. And perhaps if you got a creative lawyer, you could construct just about everything because it doesn't, it says that you can't do anything that would reduce the amount of impact, if there were a major accident or something to that effect. Major security problem. But it's an incredibly vague statement, and it sounds like it could be very loosely interpreted and essentially you could build the whole thing and have it all done, and then apply for the construction and operating permit. (80-2)

Response: *The ESP does not authorize construction or operation of a nuclear power plant. An early site permit is a Commission approval of a site or sites for one or more nuclear power facilities. However, as discussed in Section 4.11 of this EIS, certain site-preparation activities and preliminary construction activities are allowed provided that a site redress plan is submitted by the applicant and the final ESP EIS concludes that the activities will not result in any significant adverse environmental impacts that cannot be addressed.*

The filing of an application for an ESP is a process that is separate from the filing of an application for a construction permit (CP) and operating license (OL) or a combined operating license (COL) for such a facility. The ESP application makes it possible to evaluate and resolve safety and environmental issues related to siting before the applicant makes large commitments of resources. If the ESP is approved, the applicant can "bank" the site for up to 20 years for future reactor siting. If an ESP holder decides to pursue construction of a nuclear power plant beyond any approved limited activities identified in Section 4.11 of this EIS, it must obtain a CP or a COL, the issuance of which would be a major Federal action requiring preparation of an EIS under 10 CFR 51.20 that, among other things, would address the benefits of the proposed action, such as the need for power and cost of power. No change was made to the EIS as a result of the comment.

Comment: I feel as if my question was successfully dodged. Kudos. But I would like to press the point and I ask that you all answer truthfully as the people of Clinton and its surrounding areas deserve to know. Why did you select Clinton as the site for this power plant? What attributes drew you to this area when you were determining where you wanted to place a nuclear power plant? What made you think of Clinton when you first generated ideas for a location? (92-1)

Response: Exelon chose the preferred site for business reasons. Exelon, and NRC in its independent review in the EIS, undertook a site-by-site comparison of alternative sites with the proposed site (Clinton Power Station) to determine if there were any alternative sites environmentally preferable to the proposed site. Not all possible alternative sites were considered, just a "reasonable" subset of possible alternatives. The review process involved the two-part sequential test outlined in NUREG-1555 ("Standard Review Plans for Environmental Reviews of Nuclear Power Plants" [NRC 2000]). At the first stage of the review the applicant used reconnaissance-level information to determine whether there were environmentally preferable sites among the alternatives. If the applicant identified environmentally preferable sites during the second stage of the review, it would have considered economics, technology, and institutional factors for the environmentally preferred sites to see if any of these sites was obviously superior to the proposed site. None of the alternative sites proved to be obviously superior to the ESP site. The staff performed an independent review and verified the acceptability of the applicant's review. Just because an alternative site is not obviously superior to the preferred site does not mean that the alternative site cannot be considered for future nuclear development. No change was made to the EIS as a result of the comment.

Comment: These issues will supposedly be dealt with at a later permitting stage, but more properly examined early in the siting process. (112-3)(113-3)(114-3)(115-3)(116-3)(117-3) (118-3)(119-3)(120-3)(121-3)(122-3)(123-3)(124-3)(125-3)(126-3)(127-3)(128-3)(129-3)(130-3) (131-3)(132-3)(133-3)(134-3)(135-3)(136-3)(137-3)(138-3)(139-3)(140-3)(142-3)(143-3)(144-3) (145-3)(146-3)(147-3)(149-3)(154-3)(155-3)(158-3)(159-3)(162-3)(163-3)(164-3)(165-3)(166-3) (167-3)(173-3)(174-3)(175-3)(176-3)(177-3)(178-3)(180-3)(181-3)(182-3)(185-3)(186-3)(187-3) (188-3)(189-3)(190-3)(192-3)(193-3)(194-3)

Comment: Finally, NRC should reconsider the validity of its EIS in the context of its decision to grant an ESP valid for twenty years. The EPA noted in recent comments that "the twenty year horizon allotted under the proposed ESP does not have any protective assurance that unforeseen population growth and/or additional stressor on the Air or Water resources will be accounted for. Typically an action that has not occurred within three years of an EIS requires at minimum a supplemental EIS." I urge NRC to take EPA's advice. (112-11)(113-11)(114-11) (115-11)(116-11)(117-11)(118-11)(119-11)(120-11)(121-11)(122-11)(123-11)(124-11)(125-11) (126-11)(127-11)(128-11)(129-11)(130-11)(131-11)(132-11)(133-11)(134-11)(135-11)(136-11)

| (137-11)(138-11)(139-11)(140-11)(142-11)(143-11)(144-11)(145-11)(146-11)(147-11)(149-11)
| (154-11)(155-11)(158-11)(159-11)(162-11)(163-11)(164-11)(165-11)(166-11)(167-11)(173-11)
| (174-11)(175-11)(176-11)(177-11)(178-11)(180-11)(181-11)(182-11)(185-11)(186-11)(187-11)
| (188-11)(189-11)(190-11)(192-11)(193-11)(194-11)

Response: For an ESP, the NRC prepares an EIS that resolves numerous issues based on existing environmental site characteristics, as well as bounding values of power plant design parameters postulated in the application. These issues are candidates for issue preclusion in a proceeding on an application referencing the ESP (i.e., such an issue would not be subject to litigation in the later license proceeding). NRC regulations allow an ESP applicant to defer an issue, e.g., the benefits assessment, as Exelon has elected here, but also require that a COL applicant referencing such an ESP address the issue in its application. An application referencing an ESP must also demonstrate that the design of the facility falls within the parameters specified in the ESP. In addition, the application should indicate whether the site is in compliance with the terms of the ESP.

For example, in this EIS, the staff set forth population growth estimates and reached certain conclusions based upon such estimates. If the Commission issues the requested ESP and it is later referenced in a CP or COL application, the staff will consider then-current (new) population information to determine if that information is significant. If that new population information is significant, the staff will revisit any conclusions in the ESP EIS that rest upon population growth estimates. If the new information is not significant, the conclusions documented in the ESP EIS that rest upon population growth remain valid with respect to such estimates, and the COL or CP EIS will tier off the conclusion reached in the ESP EIS.

To summarize, if the Commission issues the requested ESP and it is later referenced in a CP or COL application, that application should identify whether there is new and significant information on any issue resolved in the ESP proceeding. Issuance of either a CP or a COL is a major Federal action. Therefore, 10 CFR 51.20 requires the preparation of an EIS for such a proposed action. In its review of such a CP or COL application, the staff will consider any new information developed up until the time such an application is submitted. Accordingly, issues resolved in an ESP proceeding need not be reconsidered at the COL stage even though the ESP is valid for a 20-year period.

EPA stated in a letter (ADAMS Accession No. ML050630407) that "typically an action that has not occurred within three years of an EIS requires at a minimum a supplemental EIS." If an application to construct a nuclear power reactor on the Exelon ESP site is submitted, the staff will prepare an EIS on that application regardless of whether or not the application references an ESP. Therefore, the NRC review will not exceed the minimum specified by EPA. The

Executive Summary and Section 3.0 of this volume include an explanation of the ESP process and the interaction between the ESP EIS and the environmental review at the COL stage, if the requested ESP is granted and is referenced in a COL application. No change was made to the EIS as a result of these comments.

Comment: The arbitrary separation of the ESP and COL compromises the ability of the U.S. Nuclear Regulatory Commission (NRC) to perform a thorough and adequate evaluation at either stage or in total of the potential environmental impacts from new reactor development. Under this regime designed to "provide stability in the licensing process" (EIS, § 1.3) far too many environmental impact considerations have been deferred to the COL stage of the licensing process. In comments to the NRC regarding a draft EIS for a similar ESP sought by the energy company Dominion at its North Anna Power Station, the U.S. Environmental Protection Agency (EPA) registered its reservations with this licensing scheme: "EPA has concerns with this approach since it ignores the justification for the power plant addition in the early stage of project development as well as biases the subsequent energy alternative analysis toward nuclear power under the second EIS since the NRC would have approved the suitability under the ESP." The EPA underscored its concerns by pointing out the artificial twenty-year horizon allotted under the ESP, during which time circumstances and technologies may change dramatically, rendering the conclusions of the EIS moot. The EPA further noted that, typically, if an action has not taken place within three years of an EIS, a supplemental EIS is required. Public Citizen agrees with the EPA's concerns about this problematic licensing disjunction. This discordant licensing structure is also evident in the need for a "Site Redress Plan" (EIS, § 4.11), which addresses the activities that would be required to restore the ESP site to its present state in the case that Exelon is granted an ESP but fails to seek or acquire a CP or COL within twenty years to consummate the preparatory activities allowed under the ESP. The breadth of site-preparation activities allowed under the ESP (considered a "partial construction permit" under 10 C.F.R. 52.21) is remarkable, including clearing, grading, and excavating the site; building roads, service and support facilities; and even the construction of ancillary plant components such as cooling towers, intake and discharge structures, and a transmission system (EIS, pp. 4-42 to 4-43). This degree of construction activity and the financial investment it would require would appear to compel the construction of a nuclear unit, yet this reality is not appreciated at this stage of the licensing process, indicating the bizarre division between the ESP and the COL. Clearly, the specific site and the specific reactor are one in the same project, and the division into the separate ESP and COL licensing processes is completely arbitrary, compromising the NRC's ability to perform an adequate evaluation of the potential environmental impacts from the project. (150-3) (151-3)

Response: *As stated in NRC's ESP Review Standard RS-002 (online at: http://www.nrc.gov/reactors/new-licensing/esp/esp-public-comments-rs-002.html), the purpose of the ESP regulations in 10 CFR Part 52 is, in part, to make it possible to resolve safety and environmental issues related to siting before an applicant needs to make large commitments of resources. Having obtained an ESP, an applicant for a combined operating license (COL) for a* *nuclear power plant or plants can then reference it in the COL application. In accordance with 10 CFR 52.39, site-related issues resolved at the ESP stage will be treated as resolved at the COL stage unless a contention is admitted that a reactor does not fit within one or more of the site parameters in the ESP, a petition alleges that the site is not in compliance with the ESP, or a petition alleges that the terms and conditions of the ESP should be modified. The public had an opportunity to comment on the Part 52 ESP regulations prior to their adoption. For additional information, see the previous response to a similar comment concerning how the staff would review a COL application should new and significant information be identified after an ESP is issued for a site. No change was made to the EIS as a result of the comment.*

Comment: While Exelon has not firmly committed to constructing a new nuclear unit at the Clinton Power Station (CPS) of even selected a specific reactor design (EIS, pg. 1-5), it is part of an industry consortium called NuStart Energy Development that plans to apply for a COL. If granted an ESP, Exelon could be permitted to begin an extensive construction operation while numerous important issues, such as the need for power and the indefinite storage of additional waste onsite, have not been addressed. Simply declaring that NRC is not required to look at these issues does not make them go away. (150-4)(151-4)

Response: *Need for power need not be addressed as part of the NRC's review of an ESP application but would be addressed in a subsequent EIS if an ESP holder elected to apply for a CP or a COL for a new nuclear power plant (10 CFR 52.18). The environmental impacts of radioactive waste are discussed in Section 6.1.1.6 of the EIS. No change was made to the EIS as a result of the comment.*

Comment: Section 2.2 defines the region as within 50 miles without any justification, making it appear arbitrary and capricious, especially since the major cities of Springfield and Peoria are not fully within the boundary, resulting in those cities not being included in the draft EIS for impact analysis. Section 2.2.1 defines the vicinity as 6 miles without any justification, making it appear arbitrary and capricious, especially since several species are evaluated at 10 mile ranges (end of section 2.7.11 and others). If 10 miles is proper to evaluate endangered or threatened species, it should also be used for every other evaluation, especially those affecting humans. A 10 mile vicinity would certainly include all of Clinton and perhaps Farmer City, whereas a 6 mile vicinity does not. (169-13)(179-13)

Response: *The 10-km (6-mi) and 80-km (50-mi) radii used to evaluate impacts of routine operations and accidents are specified in NRC review guidance. The distances are based on evaluation of impacts of many reactors at many sites. Design basis accidents are events that are considered credible and sufficiently likely that the reactor is designed to minimize impacts of the accident through defense-in-depth. Severe accidents are extremely unlikely, worst-case events. The impacts of normal operations, design-basis, and severe accidents are described in detail in Chapter 5 of the EIS. No changes were made to the EIS as a result of the comment.*

Comment: The early site permit process is not supposed to examine radioactive waste issues or reactor design, not in detail anyway. The NRC also refuses to analyze studies that challenge existing radiation standards, instead trodding out its favorite pro-nuc studies without examining new data in a substantive way. (49-3)

Comment: And that may give you the false impression that, just because of its sheer bulk, it's got all the answers. And if, if that's the impression that you've been left with, then I, I have to inform you that you're mistaken. In fact, all of the important questions are either postponed until after Exelon is granted this early site permit, or they're left out entirely. (68-2)

Comment: There are major gaps in this environmental impact statement. And I would, I would request that not only are those gaps filled in, before the permit is granted, but there be another draft version of this statement put out that then people can, can re-evaluate. (68-8)

Response: *The Atomic Energy and Energy Reorganization Acts establish the specific mission of the NRC to protect the public health and safety in permitting the utilization of nuclear material. The National Environmental Policy Act (NEPA) applies to all Federal agencies to ensure that environmental values are considered in fulfilling the mission of each Federal agency. The NEPA process focuses on potential environmental impacts resulting from the proposed action rather than on issues related to safety. That said, certain safety issues are relevant to the environmental review when they could potentially result in environmental impacts, which is why the environmental effects of postulated accidents are considered in the EIS.*

Some issues have been resolved generically by the Commission, such as the environmental impacts of the uranium fuel cycle, waste confidence, and the impacts of transporting spent fuel and waste; consequently, they need not be analyzed further unless the bases do not apply, such as for other-than-light-water reactors.

Some issues will not be discussed in an EIS, such as terrorism, security, and emergency planning, because they are addressed elsewhere in the regulatory process. Other issues are not addressed in this EIS, such as the benefits assessment (e.g., need for power) and severe accident mitigation alternatives because they may be more appropriately considered at the time an applicant selects a design and requests a CP or COL. Except for selected activities listed under a site redress plan, if approved, construction cannot begin until a CP or COL is issued. A CP or COL cannot be issued until all identified environmental issues have been evaluated.

Safety issues and emergency preparedness are addressed in the Exelon ESP Safety Evaluation Report. This report is available on the NRC's website at www.nrc.gov. The safety issues that are raised during the environmental review are forwarded to the appropriate NRC safety project manager for consideration and appropriate action. No change was made to the EIS as a result of these comments.

Comment: Section 3.2, Page 3-7. "During the review of a CP or COL application referencing an ESP, the staff will assess the environmental impacts of the construction and operation of a specific plant design. If the environmental impacts addressed in the ESP EIS are found to be bounding by the staff, no additional analysis of these impacts will be required, even if the ESP applicant employed the PPE approach. However, environmental impacts not considered or not bounded at the ESP stage have to be assessed at the CP or COL stage. In addition, measures and controls to limit adverse impacts will need to be identified and evaluated for feasibility and adequacy in limiting adverse impacts at the CP or COL stage. The inputs and assumptions that were used or considered during the staff's evaluation of the ESP application (listed in Appendices J and K) will provide the basis for the staff's verification review in which the staff must determine whether or not a specific design in a CP or COL application falls within the PPE, and the environmental impacts of the construction and operation of that specific design fall within the bounds of environmental impacts estimated by the staff at the ESP stage."

This paragraph is confusing and imprecise and should be reworded. At the CP/COL stage, Exelon and the NRC will determine if the plant-specific design falls within the PPE in the ESP EIS. If the design is bounded by the PPE, the findings in the ESP EIS remain valid. If the design is not bounded by the PPE, it will then be necessary to determine if the new information significantly effects the environmental impacts as described in the ESP EIS and to identify mitigation measures for any significant increases in environmental impacts. (141-44)

Response: The comment is noted; however, the staff does not agree that the text is imprecise. No change was made to the EIS as a result of the comment.

Comment: I guess what's standing out to me here tonight, is this process. I'm a Quaker, and so I'm not used to solving problems using such competitive, I guess, forms of debate, etcetera. I've heard some wonderful minds here tonight. And I think to myself, in the process that is often used in Quaker meeting is, is more collaborative. It's where we take everybody's sources, great information of everybody and collaborate. We come together and see where, what truth we find. Because I believe not just one person has the truth. Each one of us carries a part of the truth. And the more people we bring together, the closer we come to a greater truth. And I see people just going at each other, having their minds made up, and not listening to each other, to see where we come together, and then work. And work to solve what we need. (69-1)

Comment: The U.S. Nuclear Regulatory Commission's Draft Report for the Environmental Impact Statement (EIS) for the proposed second Clinton nuclear plant is excruciatingly detailed. (110-1)

Comment: We support the ESP process as the means to guarantee an open and thorough evaluation of future nuclear projects, while ensuring the timeliness and predictability of the process. (111-3)

Comment: I think his remarks [an NRC staff member's] are an egregious example of pro-nuclear bias trumping a thorough study of alternatives--and an especially obvious bias at a time when the largest wind-farm in the world is planned for construction just 25 miles away from Clinton! One of the speakers at the meeting, Sandra Lindberg, voiced her concern--mine, too-- that the ESP process is a sham. This committee member's action was a perfect example of what Sandra was talking about. It was very troubling. (152-3)

Comment: The entire DEIS review is developed, analyzed and written in such a manner as to divorce it from the real world. It may represent a "necessary" slavish adherence to regulatory details. However, it is totally insufficient to protect the public health and safety. Frequently, the assumptions put forth by Exelon are simply those that serve its narrow interests. Whether they are accurate or not is rarely if ever challenged. The NRC staff seem to accept most of these uncritically; if there were any criticisms, these are not well documented or provide rationale. What we end up with is a largely self-serving set of GIGO inputs from Exelon, to which NRC staff seem to nod positively as if they were dash-board dollies. Significant matters are often left out of the discussion, because they are "regulatorily" outside the scope of these proceedings. We would submit that, for example, while the crash of a 500+ton Airbus A-380 Jumbo jet loaded with 300,000 liters of aviation fuel coming out of the world's busiest airport only 27 minutes away into the poorly protected spent fuel pool of the Clinton-1 reactor might have some environmental significance worth analyzing. But is seems, regulations preclude this possibility.

So, what we end up with here is a process that largely satisfies the regulatory "necessities" of conducting these proceedings, without sufficiently contributing to providing environmental protection. We do hope that NRC abandons this practice in the future. (161-1)

Response: *The comments provide no new information and were not assessed further. No change was made to the EIS as a result of these comments.*

Comment: The scope of reactor types considered within the PPE - including five light water reactor (LWR) and two gas-cooled reactor types, not all of which have been approved by the NRC (EIS, § 3.2) - is far too broad, making it impossible to provide a reasonably precise judgment of the environmental impact of a new nuclear unit at the CPS, especially considering that Exelon is not even required to employ any one of these designs if it ultimately decides to build a new nuclear unit at the CPS (EIS, pg. 3-3). The EPA, in commenting on the draft EIS for a similar new nuclear development, criticized the NRC for this imprecision, noting that "There is inadequate design information available for some of the proposed units from which to make accurate environmental assessments of the impacts." Exelon did not provide any specific design information on a heat dissipation system or radioactive waste-management system for a new nuclear unit at the CPS (EIS, pg. 3-10). Furthermore, the inaccuracy of this review system is belied by the NRC staff's admission that they neglected to review Exelon's PPE values for correctness (EIS, pg. 3-5). (150-5) (151-5)

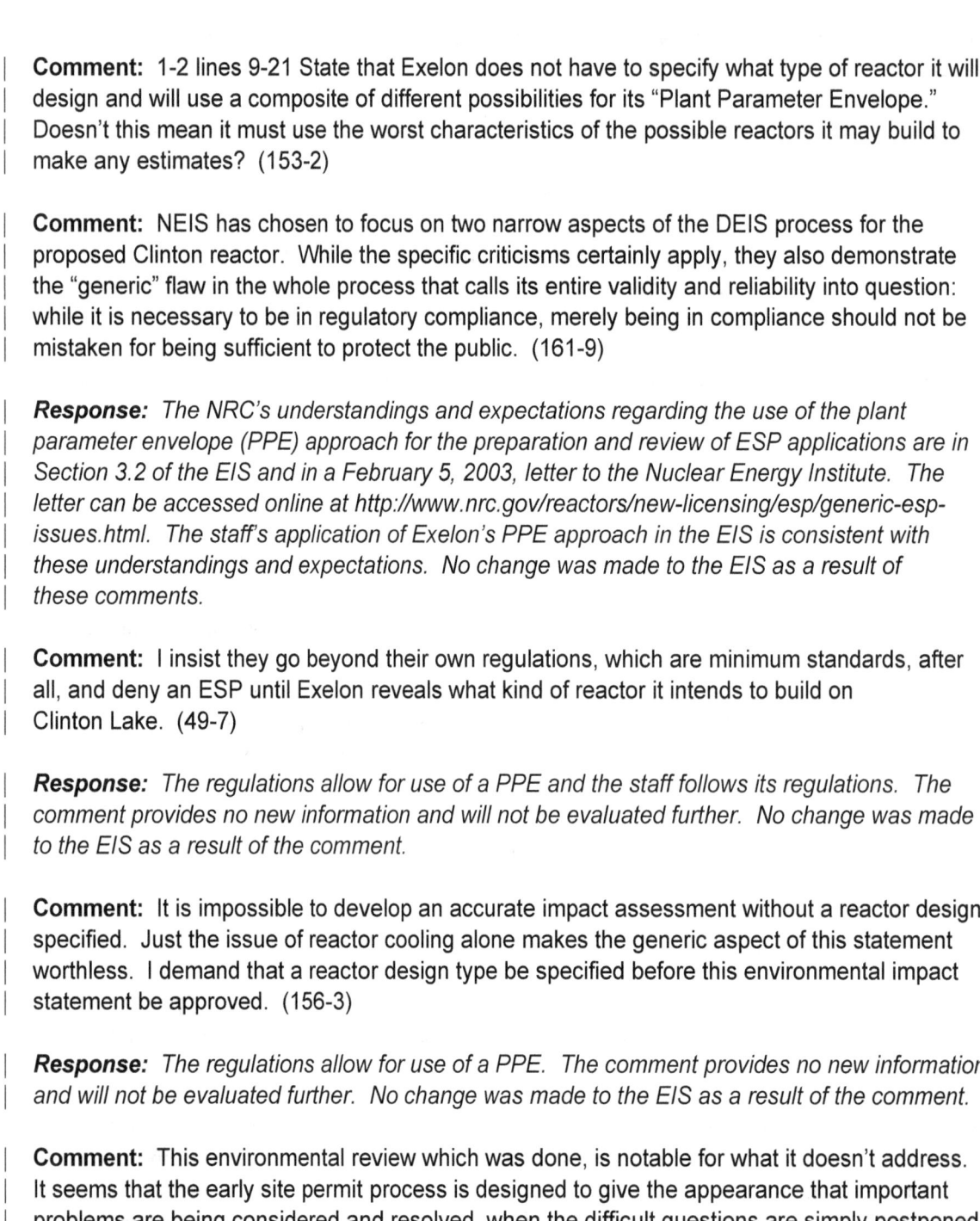

| **Comment:** 1-2 lines 9-21 State that Exelon does not have to specify what type of reactor it will design and will use a composite of different possibilities for its "Plant Parameter Envelope." Doesn't this mean it must use the worst characteristics of the possible reactors it may build to make any estimates? (153-2)

| **Comment:** NEIS has chosen to focus on two narrow aspects of the DEIS process for the proposed Clinton reactor. While the specific criticisms certainly apply, they also demonstrate the "generic" flaw in the whole process that calls its entire validity and reliability into question: while it is necessary to be in regulatory compliance, merely being in compliance should not be mistaken for being sufficient to protect the public. (161-9)

| *Response: The NRC's understandings and expectations regarding the use of the plant parameter envelope (PPE) approach for the preparation and review of ESP applications are in Section 3.2 of the EIS and in a February 5, 2003, letter to the Nuclear Energy Institute. The letter can be accessed online at http://www.nrc.gov/reactors/new-licensing/esp/generic-esp-issues.html. The staff's application of Exelon's PPE approach in the EIS is consistent with these understandings and expectations. No change was made to the EIS as a result of these comments.*

| **Comment:** I insist they go beyond their own regulations, which are minimum standards, after all, and deny an ESP until Exelon reveals what kind of reactor it intends to build on Clinton Lake. (49-7)

| *Response: The regulations allow for use of a PPE and the staff follows its regulations. The comment provides no new information and will not be evaluated further. No change was made to the EIS as a result of the comment.*

| **Comment:** It is impossible to develop an accurate impact assessment without a reactor design specified. Just the issue of reactor cooling alone makes the generic aspect of this statement worthless. I demand that a reactor design type be specified before this environmental impact statement be approved. (156-3)

| *Response: The regulations allow for use of a PPE. The comment provides no new information and will not be evaluated further. No change was made to the EIS as a result of the comment.*

| **Comment:** This environmental review which was done, is notable for what it doesn't address. It seems that the early site permit process is designed to give the appearance that important problems are being considered and resolved, when the difficult questions are simply postponed or ignored altogether. (191-3)

Response: The staff assumes that the commenter is referring to the need for power and waste disposal concerns. The need for power need not be addressed as part of the NRC's review of an ESP application but would be addressed in a subsequent EIS if an ESP holder elected to apply for a CP or a COL for a new nuclear power plant (10 CFR 52.18). The environmental impacts of radioactive waste are discussed in Section 6.1.1.6 of the EIS. No change was made to the EIS as a result of the comment.

Comment: I felt that questions asked during the presentation were not answered to anyone's satisfaction but rather were avoided or redirected. In order to believe that the NRC has developed a careful and informed EIS, I would need their reasoning, their value judgments and quantifying process explained much more clearly. (107-1)

Response: The staff attempts to answer all questions during its presentations. The staff's basis for its environmental evaluation is found throughout the EIS. Specific questions on the draft document have led to changes to the EIS where appropriate. No change was made to the EIS as a result of the comment.

Comment: My question is specifically about the NRC's solicitation of public comments. I'm curious as to by what process the NRC decided to not hold hearings in other potentially impacted communities beyond Clinton, especially considering that the reactor will be located practically as close to Farmer's City as Clinton. (32-1)

Comment: Now, again, where is all this confidence coming, that would state that the public is not even allowed to raise these issues, in a licensing proceeding. (67-5)

Comment: This process is extremely important. The process of the public coming out and discussing these issues, and debating the merits and demerits of adding another plant, and what are our alternatives here...I would submit that it's important for us to do more hearings of this nature, more here in Clinton. Obviously, people have a lot to say about this, and there should be more opportunities for them to do that. We also should do it around the state, other places like Peoria, Bloomington, Decatur, Springfield, Chicago, Champaign, Urbana, all of those places. All of those people have an interest in this, and a stake in what's happening tonight. And they should all have an equal opportunity to come out and give comments and, and debate the issue, the same way that you guys are tonight. (68-1)

Response: The staff sets meeting dates and times so as to be convenient for the public as well as the staff. The staff held two public meetings to discuss the environmental review in Clinton, a public scoping meeting on December 18, 2003, and a public meeting on the draft EIS on April 19, 2005. Members of the public who cannot attend a public meeting had the opportunity to submit comments by mail or e-mail. Such written comments received the same attention from the staff as oral comments presented at a public meeting. The comment period for scoping was

60 days, while the period for comment on the draft EIS was 75 days. No change was made to the EIS as a result of these comments.

Comment: As far as the "letter of the law" (i.e., the National Environmental Policy Act) is concerned, the USNRC is operating within the technical legal parameters when it addresses "alternatives to Clinton Unit 2" in the DEIS just for the local region in downstate Illinois. But using different (broader) assumptions, I have been considering the impacts of Clinton Unit 2 on The STATE LEVEL. (25-9)

Response: *The comment expresses opposition to the NRC ESP process. Because it did not provide new information, no change was made to the EIS as a result of the comment.*

Comment: Station Operation Impacts, Section 5.12, Summary of Operational Impacts, pages 5-79 through 5-83. The actual impact designation may vary based on the type of reactor(s) chosen for the proposed ESP at the Exelon Site. While a good attempt was made to provide adequate bounding of the issues, further evaluation under the CP or COL process will provide a more adequate assessment of these impacts. (172-42)

Response: *This comment provides no new information and will not be evaluated further. For more information on the ESP process, see Section 3.2. No change was made to the EIS as a result of this comment.*

Comment: The environmental impact is performed utilizing a surrogate, as I understand. My question is if ten, 15, 20 years down the road, an actual design for a plant is approved, then is the environmental impact statement refined? (34-1)

Response: *With respect to environmental matters, the NRC's ESP process is as follows: The NRC regulations governing an ESP application require that an applicant for an ESP must provide the NRC with an ER that meets the requirements of 10 CFR 51.45 and 51.50. As described in 10 CFR 52.17, the contents of an application must focus on the environmental effects of construction and operation of a reactor or reactors that might be built at the proposed site, even though an ESP does not authorize such construction and operation. Additionally, Section 52.18 requires that the staff prepare an EIS on the application that focuses on the same matters. Both the ER and the EIS must include an evaluation of alternative sites to determine whether there is any obviously superior alternative to the site proposed. Certain issues, however, such as the benefits of the action and alternative energy sources, may be deferred to a later licensing stage.*

For the ESP, the NRC prepares an EIS that resolves numerous issues based on existing environmental site characteristics, as well as values of power plant design parameters set forth in the application. These issues are candidates for issue preclusion in a proceeding on an application referencing the ESP (i.e., such an issue would not be subject to litigation in a later licensing proceeding). If an applicant chooses the PPE approach, as Exelon has done here, the application postulates bounding values for these plant design parameters.

NRC regulations allow an ESP applicant to defer an issue (e.g., the benefits assessment), as Exelon elected here, but also requires that a COL applicant referencing such an ESP address the issue in its application. An application for a CP or COL referencing an ESP must also demonstrate that the design of the proposed facility falls within the parameters specified in the ESP. In addition, an application referencing an ESP should indicate whether the site is in compliance with the terms of the ESP. Such an application should also identify whether there is new and significant information on any issue resolved in the ESP proceeding.

The EIS prepared for the COL will tier off the ESP EIS, should one be issued. If there is no new and significant information on an issue, the COL EIS will bring forward the conclusion reached in the ESP EIS. If there is new and significant information, then a conclusion will be reached based on the analysis of the new and significant information. No changes were made to the document as a result of the comment.

E.2.2 General Comments in Support of NRC and its ESP Process

Comment: We concur with the NRC's conclusion that environmental impacts would not prevent issuing an ESP for the Clinton site. (111-1)

Response: *The comment expresses support for the Exelon ESP. Because it did not provide new information, no change was made to the EIS as a result of the comment.*

Comment: Please consider continuing the permit process with out undo delay. (01-4)

Comment: The U.S. Nuclear Regulatory Commission's new licensing process, which we are taking part in now, demonstrates how predictable and timely this process can be, while assuring that it is thorough. The Nuclear Regulatory Commission's mandate is to protect our health and safety. The American Nuclear Society believes that the new process provides us with confidence that the NRC meets its mandate. (101-4)

Comment: Following the basic structure for an environmental impact statement for re-licensing is a good idea, since many of the activities that are being evaluated are either the same or similar. (172-1)

Comment: NAYGN supports the ESP process and a means to guarantee an open and thorough evaluation of future nuclear projects while ensuring the timeliness and the predictability of the process. (48-4)

Comment: The second thing I'd like to point out is several people have asserted that the NRC is a lap dog of the industry. Now, I'm sure the NRC is not perfect but that's when you saw the flow chart up here in this review process, it had loops or repetitions or extra steps for corrections and environmental impact statements and so forth. So they understand that they're not perfect and that's why they ask for input and comment and so forth.

Furthermore, if the NRC were the lap dog of industry, I'm very puzzled by the fact that the Clinton power station was shut down for three years. I think the NRC had something to do with that. So I don't think we need to worry too much with the NRC doing the beck and call of the industry. (62-2)

Comment: The U.S. Nuclear Regulatory Commission's new licensing process, which we are taking part in now, demonstrates how predictable and timely this process can be, while assuring that it is thorough. The Nuclear Regulatory Commission's mandate is to protect our health and safety. The American Nuclear Society believes that the new process provides us with confidence that the NRC meets its mandate. (84-4)

Response: *These comments express support for the NRC ESP process. Because they did not provide new information, no change was made to the EIS as a result of these comments.*

Comment: When the power plant was shut down, back in '96 or '97, I do have one comment in support of the NRC. I had one of those representatives out at my motel during that time, which I own the, the hotel next door here....and I approached him, I said, you know, do you know how much longer you're going to be here? And he told me, he says, until they, until it's perfect out there. That's the only way. And after, over two years, he stayed with us. And then I knew it was perfect. ...I'm in support of the NRC, looking out after us. (82-2)

Response: *The comment expresses support for the NRC. Because it did not provide new information, no change was made to the EIS as a result of the comment.*

E.2.3 General Comments in Opposition to NRC and its ESP Process

Comment: You come to our home state of Illinois tonight to preside over a process that will ultimately have real consequences for real people. We do not view it as another dry statistical run. We are not data. We are not interested in satisfying irrelevant or inadequate regulatory requirements. We're here to address the bottom line as it will affect us. (108-3)

Comment: Public Citizen views the draft EIS for the Exelon ESP as deficient, and we disagree with the NRC staff's recommendation that the ESP should be granted. (150-1)(151-1)

Comment: The purpose of this Early Site Permit (ESP) process is ostensibly to "assess whether a proposed site is suitable should Exelon decide to pursue a [construction permit (CP)] or [combined construction and operating license (COL)]" (EIS, page xxv). Yet, this draft Environmental Impact Statement (EIS) fails to consider or to fully acknowledge numerous environmental issues that could demonstrate that the Clinton site is not suitable for an additional nuclear unit. (150-2)(151-2)

Comment: Time and again the NRC has stated that we can't consider the waste, we can't consider the type of reactor to be built, we can't consider the actual need for this type of energy production. We can't. But what it will boil down to, one way or another, is that we SHOULD. (157-10)

Comment: We therefore urge NRC to reject the Exelon request for an Early Site Permit at the Clinton site. (161-12)

Comment: The Staff's preliminary recommendation that the ESP should be issued (Draft EIS at 10-8) is undermined by a number of serious shortcomings in the Draft EIS. (170-1)

Comment: As demonstrated above, the Draft EIS simply fails to satisfy the basic requirements of NEPA or provide the information necessary for the NRC to ensure that its licensing decision is not "inimical to the common defense and security or to the health and safety of the public," 42 U.S.C. § 2133(d). (170-15)

Comment: The initials "N...R...C" obviously mean "Nuclear Regulatory Commission." This time, let's make sure that the NR also means "Responsible," and NOT "Reassurance." THE NRC SHOULD "GET RESPONSIBLE," AND NOT MERELY "RUBBER-STAMP" EXELON'S "EARLY SITE PERMIT" APPLICATION FOR AN UNNEEDED SECOND CLINTON REACTOR. (25-11)

Comment: I'm afraid to say that the experts in this room do not appeal to me much. Nor does this sham of a process. (49-1)

Comment: You come to our home state of Illinois tonight to preside over a process that will ultimately have real consequences for real people. We do not view it as another dry statistical run. We are not data. We are not interested in satisfying irrelevant or inadequate regulatory requirements. We're here to address the bottom line as it will affect us. (56-5)

Comment: This document does contain a lot of good information. But at the same time, the way it emphasizes and de-emphasizes information is really curious. And some of the statements they make are quite questionable. In fact, there's so many questionable statements in this document, that it would probably take me on the order of 10 hours to talk with your employees. (73-1)

Response: *These comments express opposition to the Exelon ESP. Because they did not provide new information, no change was made to the EIS as a result of these comments.*

Comment: With the USNRC granting ("rubber-stamping?") an "Early Site Permit" for Clinton Unit 2, this action will merely perpetuate this "political stranglehold." Yet another seemingly insurmountable "political barrier" will be erected to the more widespread use of truly sustainable non-nuclear energy choices throughout our "Land of Lincoln" (where there are no insurmountable technical barriers). (25-8)

Response: *The comment expresses opposition to the NRC ESP process. Because it did not provide new information, no change was made to the EIS as a result of the comment.*

E.2.4 General Comments in Support of the Applicant and its ESP Application

Comment: I am writing to support the ESP application at the Excelon ESP Site (Tac #MC1125). (105-1)(27-1)(28-1)(29-1)(97-1)(99-1)

Comment: Constructing a second reactor (or more) at the Clinton Power Station would be a good idea for several reasons. The area would benefit economically is many different ways. In addition, the reactor would provide a source in a safe and efficient manner for needed electric power. (106-1)

Comment: The benefits of building a second reactor at Clinton Power Station are many. In addition to the economic benefits the local area receives, the other benefactors will be all of those electric power users throughout the power grid. We all seem to want to use more and more electric power each day. Nuclear power is a clean and economical source of energy. Amergen and Exelon need an opportunity to start building the new generation of nuclear reactors at the Power Station. (106-9)

Comment: We commend Exelon for being proactive and farsighted when looking for reliable methods of addressing expected increases in energy demand over the coming years, while minimizing the environmental footprint of the selected energy sources, as well as the economic burden to Exelon's customers. (111-4)

Comment: The DeWitt County Farm Bureau Board of Directors voted unanimously to support the granting of a permit to construct the second unit at Amergen's Clinton, IL Nuclear Power Station. (14-2)

Comment: I'd have no objection to another unit being built. (16-2)

Comment: Now, I'm very happy to support the adoption of the, or the issuance of the early site permit, as well as the ultimate environmental impact statement. (160-1)

Comment: One of the things that nobody has mentioned here, and there have got to be people that have lived here for a long time, and that's what it looks like around Clinton Lake, and what you can do. And what the environment is there. Has anybody seen any environmental impact at the, at the Clinton Power Station? I've been over there dozens of times in the, ensuing 30 years since they first started working on building it. I have not seen that. (160-2)

Comment: I'm in favor of a second nuclear plant near Clinton. (24-1)

Comment: Illinois Farm Bureau supports the construction of a second reactor at Exelon's nuclear power station near Clinton, Illinois. (26-1)

Comment: And I think we can come to the conclusion that Clinton is a fine site for the next nuclear power plant. (42-10)

Comment: And that is I think a proposed second reactor out here would have the same impact as the first, and that has been nothing, in my opinion, but positive things for our community. (43-1)

Comment: So I think it [a second reactor] would have the same impact [as the first reactor] and that would be positive. (43-4)

Comment: And so on behalf of the city and the chamber, we fully support and encourage the selection of Clinton for the second reactor. (45-2)

Comment: The environmental report of Exelon's ESP application and the NRC's draft environmental impact statement demonstrate in great deal what has become obvious in the area of increasing concerns about global warming, air pollution, environments of protection and industrial safety. (48-5)

Comment: So as nuclear professionals and concerned local citizens, we in NAYGN concur with the NRC's conclusion that the environmental impacts would not prevent an early site, will not prevent issuing an early site permit in the Clinton site. To that end, we have with us today a petition with over 360 signatures collected in the last two days supporting Exelon's application. (48-8)

Comment: For me, the community has been well served by the power plant. And I personally would support moving on with this application. (51-3)

Comment: On behalf of the Economic Development Committee, I just want to say that we strongly support the Unit 2. (53-2)

Comment: I support Exelon. I salute nuclear power. And I think the permit should be permitted. (54-1)

Comment: What is the alternative if we don't get this? Well, I mean, our community, things, businesses just keep leaving and leaving. There is nothing. There isn't. We need this. We have to stand together in this community and take this risk. (57-2)

Comment: We support the construction of a second nuclear generating unit at the Clinton power station. We appreciate the support and concern of Exelon and the Clinton power station and look forward to a long relationship. (63-1)

Comment: And that is why I applaud Exelon for being a pioneer and taking this step toward a proven, safe, clean and reliable and important part of the future energy mix of this country. (66-8)

Comment: Now, I'm very happy to support the adoption of the, or the issuance of the early site permit, as well as the ultimate environmental impact statement. (70-1)

Comment: One of the things that nobody has mentioned here, and there have got to be people that have lived here for a long time, and that's what it looks like around Clinton Lake, and what you can do. And what the environment is there. Has anybody seen any environmental impact at the, at the Clinton Power Station? I've been over there dozens of times in the, ensuing 30 years since they first started working on building it. I have not seen that. (70-2)

Comment: Clinton stands at the crux of our energy future. This situation here is incredibly important. (74-4)

Comment: I'm excited and happy to see Exelon applying for this early site permit. (76-1)

Comment: And no matter which reactor Exelon ultimately chooses to construct, I'm confident that these new reactors will adhere to these principles, and deliver this area with more safe, clean, affordable and reliable nuclear generated electricity. (76-4)

Comment: And I'm for this second reactor. (82-1)

Comment: And, in turn, I also support the early site. (82-3)

Comment: I would like to say that I fully support this early site permit and I hope it happens. (87-5)

Comment: I think it would be great for all...And I think it would be a very good thing for DeWitt Co. (95-1)

Comment: The DeWitt County Farm Bureau Board of Directors voted unanimously to support the granting of a permit to construct the second unit at Amergen's Clinton, IL Nuclear Power Station. (103-2)

Response: *These comments express support for the Exelon ESP. Because they did not provide new information, no change was made to the EIS as a result of these comments.*

E.2.5 General Comments in Opposition to the Applicant and its ESP Application

Comment: As a citizen of Central Illinois, I strongly oppose the proposed Clinton 2 nuclear reactor. (02-1)

Comment: NO, NO, NO and NO! (03-1)

Comment: I'm writing to express my disapproval of a new nuclear reactor at Clinton, IL. (04-1)

Comment: I am writing to ask that you reject the plan to build a new reactor at Clinton. (09-1)

Comment: As a mother and as a citizen of Illinois, I do not want this reactor to be built. Please deny Exelon's request. (10-8)

Comment: But things can go wrong, as the recent years long shut down of Clinton I confirms. And lying in the background is the New Madrid earthquake fault. The pool is full of radioactive waste. And the ingenuity and dedication of terrorists. (100-2)

Comment: Nuclear Energy Information Service calls Exelon's plans for additional nuclear plants unnecessary, unsafe and unwise at a public hearing convened by federal regulators in downstate Clinton. (102-1)

Comment: Due to the environmental and health risks to the citizens in my district, I must oppose efforts to build an additional nuclear reactor in Clinton. (104-3)

Comment: I am deeply concerned about the proposed new nuclear reactor in Clinton. (109-1)

Comment: As a citizen of Illinois, I wish to express my feelings on granting a permit for a new reactor at Clinton, Illinois. Because of environmental, health, and safety issues I am against it. (11-1)

Comment: This proposed site will have disastrous long-term environmental impacts. This report should be rejected. Besides conservation, safe and clean alternatives exist to generating needed electricity - such as wind - that should be considered for the Clinton site and this entire area. (110-5)

Comment: I strongly oppose the expansion of the Clinton, IL., plant. (12-1)

Comment: I live in Champaign and I wish to go on record as being OPPOSED to a 2nd nuclear reactor in Clinton, IL. (13-1)

Comment: Another reactor in Clinton would be unsafe, and therefore, you should make the right decision to deny the permit. (148-5)

Comment: This is written to voice my opposition to the second building of another nuclear power plant at the Clinton Power Plant site. I was opposed to the first power plant and definitely am against another power plant. (15-1)

Comment: I do not agree with the NRC's evaluation and recommendation that the ESP be approved and permitted. (157-1)

Comment: According to the current analysis of this project, DeWitt County is an expendable county, as are the people within that geographic region. That is what "Low Risk" means. Approval of this ESP is simply a license for exploitation of human, economic, and natural resources. Exelon may still make money, but we will be paying the bill. (157-11)

Comment: We believe that the aggregate of the criticisms, if thoroughly, genuinely and objectively examined would lead a reasonable person to conclude "no need" for the proposed Clinton reactor. (161-10)

Comment: Due to the environmental and health risks to the citizens in my district, I must oppose efforts to build an additional nuclear reactor in Clinton. (17-3)

Comment: I want to add my voice to those opposed to the construction of a new nuclear power plant near Clinton, IL. (18-1)

Comment: Heart of Illinois Sierra Club, representing its 900 members within central Illinois, is opposed to the Early Site Permit for the proposed second reactor at the Exelon Clinton site. Concerns for groundwater safety, costs to the public both ratepayers and taxpayers, inadequate storage for radioactive waste at the site, and the fact that wind energy and other sources of sustainable and safe energy are being developed should take precedence for Illinois. (184-1)

Comment: I must strongly protest the building of another nuclear power plant in Dewitt County. (19-1)

Comment: It is our view that building more nuclear reactors at the existing Clinton site poses far more risks than benefits to Illinois residents. An early site permit for the reactors there should be denied. (191-1)

Comment: An Early Site Permit for the Clinton, Illinois reactor should be emphatically denied. (191-10)

Comment: I am totally opposed to new nuclear plants-in Clinton, Illinois or elsewhere. (20-1)

Comment: I don't trust the utility to make better decisions about a second reactor. (21-2)

Comment: I oppose the building of a second reactor at Clinton. (21-4)

Comment: Don't build a second nuclear reactor at Clinton! (23-1)

Comment: THESE MORE DETAILED COMMENTS ARE IN FAVOR OF DENYING EXELON'S "EARLY SITE PERMIT" (ESP) APPLICATION FOR A SECOND CLINTON REACTOR (Submitted by Exelon to the USNRC on Sept. 25, 2003). (25-1)

Comment: These comments DISAGREE with the staff's preliminary recommendation to the USNRC that Exelon's "Early Site Permit" (ESP) should be issued for a new nuclear reactor to be sited adjacent to the existing Clinton Power Station (CPS), Unit 1. (25-2)

Comment: The "pronuclear cheerleaders" (especially those at Exelon) are hyping up a so-called "nuclear renaissance" (what they consider to be a "nuclear rebirth" of what I consider to be a FAILED TECHNOLOGY). I say "so-called" because this nonsense is more accurately described as a "NUCLEAR RELAPSE" (like a reoccurring "nuclear nightmare" from a B-science fiction movie where current unresolved difficulties are perpetuated and new problems are created). EXELON SEES THE PROPOSED CLINTON UNIT 2 REACTOR AS A CRUCIAL "TEST CASE" IN THE NUCLEAR INDUSTRY'S CAMPAIGN TO MAKE THIS VERY THING HAPPEN. THE POSITION OF NEIS IS THAT THIS SHOULD NOT BE ALLOWED TO HAPPEN!!! (25-4)

Comment: And I think that that's a serious enough thing to give you pause not only about the construction of a new reactor but to seriously think that maybe it's time to shut that one down. (32-5)

Comment: I do not want to see another reactor, not for me, not for Clinton, but for the future. (37-4)

Comment: I am in favor of denying Exelon's early site permit application for the second Clinton reactor. (71-2)

Comment: Exelon sees Clinton II as a crucial test case, in the nuclear industry's campaign to make this very thing happen. This should not be allowed to happen. In the matter of Clinton II, Exelon's total and blatant arrogance is twofold. (71-5)

Comment: Please say no to Clinton II. (71-9)

Comment: I believe we should not take a step in the wrong direction for our Nation's energy needs. Therefore, I oppose any permit, proclaiming that a site is suitable for nuclear power. (83-4)

Comment: But things can go wrong, as the recent years long shut down of Clinton I confirms. And lying in the background is the New Madrid earthquake fault. The pool is full of radioactive waste. And the ingenuity and dedication of terrorists. (90-2)

Response: *These comments express opposition to the Exelon ESP. Because they did not provide new information, no change was made to the EIS as a result of these comments.*

E.2.6 Comments Concerning NEPA Compliance

Comment: The draft EIS fails to adequately execute the requirements of the National Environmental Policy Act (NEPA) by not adequately providing a "detailed statement" of (1) alternatives to the proposed action, (2) unavoidable environmental impacts, (3) irretrievable commitments of resources, and (3) the relationship between short-term uses of the environment and long-term productivity [42 U.S.C. § 4332(C)]. (150-7)(151-7)

Comment: Dividing the project into multiple parts (Early Site Permit, Construction and Operating License, and Site Safety Analysis Report) to limit the scope of each part and telling the public that comments on environmental impacts of safety or operation are not being considered for this EIS is arbitrary and capricious. NEPA requires a comprehensive assessment of the impact of the entire project on the local environment and this Draft EIS was only intended for a nuclear reactor (based on the application and NRC's lead), so the comprehensive impacts of an operating nuclear reactor must be considered. (169-7)(179-7)

Comment: The Draft EIS is also insufficient under NEPA because it fails to adequately consider the environmental impacts of new nuclear power. As part of the NEPA process, the NRC is required to take a "hard look" at the environmental consequences of a proposed action. Robertson v. Methow Valley Citizens Council, 490 U.S. 332, 350 (1989). The discussion of environmental impacts is designed to provide a "scientific and analytical basis" for comparing the various alternatives for achieving the project's goals. 40 C.F.R. 1502.16; DuBois v. U.S. Dep't of Agriculture, 102 F.3d 1273, 1286 (1s' Cir. 1996). A proper analysis of the alternatives, therefore, can be carried out only if the NRC provides a complete and accurate compilation of the environmental consequences of all reasonable alternatives. Unfortunately, the Draft EIS does not do so in a number of key areas. (170-12)

Response: Section 102 of the National Environmental Policy Act (NEPA) directs that an EIS be prepared for major Federal actions that significantly affect the quality of the human environment. The NRC has implemented Section 102 of NEPA in 10 CFR Part 51. Subpart A of 10 CFR Part 52 contains the NRC regulations related to early site permits (ESPs). It is the NRC EIS rather than the applicant's environmental report (ER) that is used as the basis for the decision on the ESP application.

As set forth in 10 CFR 52.17, the ESP applicant must submit a complete ER focusing on the environmental effects of construction and operation of a reactor or reactors. However, the applicant need not include an assessment of the benefits (for example, need for power). In addition, in its denial of a petition for rulemaking, the Commission stated that the consideration of alternative energy sources may be deferred until the COL stage (68 FR 55911). The ER is intended to assist the Commission in complying with Section 102 of NEPA. The ER may be used extensively by the NRC staff as a starting point in its review. However, the Commission staff independently evaluates information contained in the ER and develops its own bases and analyses. Ultimately, the NRC staff is responsible for the reliability of any information used. As set forth in 10 CFR 52.18, the Commission has determined that an EIS will be prepared during the review of an application for an ESP. An applicant for a CP or COL for a nuclear power plant or plants to be located at the site for which an ESP was issued can reference the ESP. A CP or COL to construct and operate a nuclear power plant is a major Federal action that requires its own environmental review in accordance with 10 CFR Part 51.

To guide its assessment of environmental impacts for a proposed action or alternative actions, the NRC has established a standard of significance for impacts using Council on Environmental Quality (CEQ) guidance (40 CFR 1508.27). Using this approach, NRC has established three significance levels – SMALL, MODERATE, or LARGE – which are defined below:

- *SMALL – Environmental effects are not detectable or are so minor that they will neither destabilize nor noticeably alter any important attribute of the resource.*

- *MODERATE – Environmental effects are sufficient to alter noticeably, but not to destabilize, important attributes of the resource.*

- *LARGE – Environmental effects are clearly noticeable and are sufficient to destabilize important attributes of the resource.*

Among the areas included in the EIS, the NRC staff considered the No Action Alternative or denial of the ESP, mitigation measures to further reduce environmental impacts, alternative sites, unavoidable adverse environmental impacts, irreversible and irretrievable commitments of resources, the relationship between short-term uses and long-term productivity, cumulative impacts, construction impacts, and the impacts of operation.

In summary, the staff has complied with the requirements of NEPA by following the NRC's implementing regulations (10 CFR Parts 51 and 52) and related review guidance. No change was made to the EIS as a result of these comments.

Comment: The Draft EIS also fails to comply with NEPA because it blindly accepts Exelon's goal of creating baseload power as the purpose for the project, and then uses that purpose to reject various reasonable alternatives to new nuclear power. This approach violates NEPA because, regardless of an applicant's goal for a project, the agency carrying out the NEPA review must still ensure that the purpose of the project is defined broadly enough so as to allow for the consideration of reasonable alternatives. The Draft EIS states that "any feasible alternative" to the proposed Clinton 2 plant "would need to generate baseload power," and then proceeds to reject energy efficiency and other reasonable alternatives as inconsistent with this purpose. (Draft EIS at 8-3, 8-15). Yet the siting of a new nuclear power plant in Illinois could only be justified if it is necessary for meeting the future energy needs of Illinois customers. Energy efficiency (both individually and in combination with clean energy sources) is plainly a reasonable alternative to new base load energy generation for meeting those needs...energy efficiency is a technologically and economically feasible alternative - alone - and in combination with other energy resources - to the siting of a new nuclear power plant at Clinton. Therefore, the Draft EIS must be revised to rigorously explore and objectively evaluate the reasonable energy efficiency alternative. (170-7)

Response: The proposed plant at the ESP site is what is called a "merchant" generating facility, which means it can sell generating power anywhere, not just in Illinois. It is not within the purview of an ESP EIS to justify the proposed plant on demand for electricity.

Section 102 of the National Environmental Policy Act directs that an EIS be prepared for major Federal actions that significantly affect the quality of the human environment. The NRC has implemented Section 102 of NEPA in 10 CFR Part 51. Subpart A of 10 CFR Part 52 contains the NRC regulations related to ESPs. It is the NRC EIS rather than the applicant's ER that is used as the basis for the decision on the ESP application.

As set forth in 10 CFR 52.17, the ESP applicant must submit a complete environmental report focusing on the environmental effects of construction and operation of a reactor or reactors, however, the applicant need not include an assessment of the benefits (for example, need for power). In addition, in its denial of a petition for rulemaking, the Commission stated that the consideration of alternative energy sources may be deferred until the COL stage (68 FR 55911). The ER is intended to assist the Commission in complying with Section 102 of NEPA. The ER may be used extensively by the NRC staff as a starting point in its review. However, the Commission staff independently evaluates information contained in the ER and develops its own bases and analyses. Ultimately, the NRC staff is responsible for the reliability of any information used. As set forth in 10 CFR 52.18, the Commission has determined that an EIS will be prepared during the review of an application for an ESP. An applicant for a CP or COL for a nuclear power plant or plants to be located at the site for which an ESP was issued can reference the ESP. A CP or COL to construct and operate a nuclear power plant is major Federal action that requires its own environmental review in accordance with 10 CFR Part 51.

To guide its assessment of environmental impacts for a proposed action or alternative actions, the NRC has established a standard of significance for impacts using Council on Environmental Quality (CEQ) guidance (40 CFR 15088.27). Using this approach, NRC has established three significance levels – SMALL, MODERATE, or LARGE – which are defined below:

- *SMALL – Environmental effects are not detectable or are so minor that they will neither destabilize nor noticeably alter any important attribute of the resource.*

- *MODERATE – Environmental effects are sufficient to alter noticeably, but not to destabilize, important attributes of the resource.*

- *LARGE – Environmental effects are clearly noticeable and are sufficient to destabilize important attributes of the resource.*

Among the areas included in the EIS, the NRC staff considered the No Action Alternative or denial of the ESP, mitigation measures to further reduce environmental impacts, alternative sites, unavoidable adverse environmental impacts, irreversible and irretrievable commitments of resources, the relationship between short-term uses and long-term productivity, cumulative impacts, construction impacts, and the impacts of operation.

No change was made to the EIS as a result of the comment.

E.2.7 Comments Concerning Land Use

Comment: Page 2-4 has a map of the Clinton area and it is clearly outdated, by at least ten years. For example, Highway 51 no longer goes thru Clinton. Why use such outdated information? (153-3)

Response: *In general, the staff agrees with the comment. The print quality of the EIS document impairs the legibility of the maps in Figures 2-2 and 2-3, which provide the most recent information available. Based on this comment, those figures have been revised.*

Comment: At the end of this section geological information references Exelon 2003a to claim there are no known significant mineral resources. This should reference a Geological Survey report and define "significant" as economically viable for extraction. (169-14)(179-14)

Response: *The citation in the EIS refers to the ER where the citation of the Illinois State Geological Survey (regarding known mineral resources) can be found. No change was made to the EIS as a result of these comments.*

Comment: Nuclear power also uses less land than a lot of other energy sources. And often, the land that it does use can double as nature preserves, protecting the local wildlife. (77-3)

Response: *The comment, which expresses support for nuclear power, is a matter of opinion and is general in nature. The comment provides no new information relevant to the EIS and will not be evaluated further. No change as made to the EIS as a result of the comment.*

Comment: Exelon speaks of the need for 4 new transmission lines, resulting in broadening - nearly doubling the size of - the rights of way through the surrounding land. While this may be within some abstract regulatory guidelines and limits, it is of significant consequences to the immediate land use in the area. Further, once these alterations are made, they are more or less permanent depending on terrain, whether the reactor gets finished or not; or whether the reactor lives out its expected useful life. The environmental degradation is not easily reversed or mitigated. (161-8)

Comment: The DEIS does not provide a comprehensive description of impacts associated with the anticipated widening of the existing transmission lines rights-of-way (from 130 feet to 250 feet). Such impacts would be evaluated more closely after an Early Site Permit is issued. Therefore, it is not possible to conduct a NEPA evaluation for these impacts before the project proponents decide to decide on the proposed project. The U.S. Nuclear Regulatory Commission (USNRC) should provide a more comprehensive description of right-of-way impacts in future environmental documentation. (172-2)

Comment: Site Layout. Section 3.3 Power Transmission System, page 3-13, paragraph 2. With a need to expand the width of the transmission line right-of-way, the potential for litigation related to right-of-way acquisition may increase and should be explained in the final EIS. (172-17)

Response: *The comments are a matter of opinion and are general in nature. The comments reflect the staff's position in the EIS–that the exact configuration of planned transmission expansion cannot be known until the applicant engages the Federal Energy Regulatory Commission (FERC) process for connecting new large generation to the grid. No change was made to the EIS as a result of these comments.*

Comment: Section 2.2.2, Page 2-8, Lines 21-22. The southern section is approximately 30 km (20 mi) long with a width of 76 m (250 ft) (an area of 246 ha [610 ac]). ER Sections 2.2.1,Site and Vicinity, and 2.2.2 Transmission Corridors and Off-Site Areas, 1st paragraph: "The southern section is approximately 8-mi long with a width of 250 ft (an area of 238 ac)." (141-10)

Comment: Section 2.2.2, Page 2-8, Lines 23-24. The southern section runs southwest of the ESP site past Clinton Lake, and then turns south and terminates at the Oreana substation, just north of Decatur. ER Section 2.2.2 Transmission Corridors and Off-Site Areas, 1st paragraph: "The southern section runs southeast of the EGC ESP Site past Clinton Lake and then turns south and runs toward the southern boundary of DeWitt County." (141-11)

Response: ER Figure 2.2-4 has been revised to show the pathway of the assumed southern transmission corridor. The corridor proceeds west of Clinton Dam, then south to the junction point of the Latham-Rising line. This information was clarified by Exelon (Exelon 2006a) and the numbers in the EIS have been adjusted to reflect a total run of about 19 km (12 mi) from the CPS switchyard to the junction point.

Comment: Section 2.2.2, Page 2-9, Line 26. McLean County published a regional comprehensive plan in August 1999 (McLean County 1999). ER-Section 2.2.2 Transmission Corridors and Off-Site Areas, 11th para. should state that McLean County published a regional comprehensive plan in August 2000. (141-14)

Response: The plan was published in 1999, not 2000. The citation in the EIS is correct. No change was made to the EIS as a result of the comment.

Comment: Section 2.1, Page 2-1, Line 35. "The ESP site is approximately 5 km (3 mi) northeast of the dam," Section 2.1.1.2, Site Area Map, final paragraph states that the CPS cooling water intake is about 3 mi northeast of this location. It does not say the ESP site is there. (141-6)

Response: The staff made its own assessment of the distance from the dam to the ESP site, based on reviewing Figure 2.1 of the environmental report (ER) and using the scale provided. The staff did not cite the ER for this sentence in the EIS. The proposed structures appear to be roughly 5 km (3 mi) from the face of the dam. No change was made to the EIS as a result of the comment.

Comment: Section 2.2.2, Page 2-8, Line 36. "approximately 270 m (900 ft) above MSL in the north-central portion of the transmission." ER Section 2.2.1 Site and Vicinity, 6th paragraph: "Elevations range from approximately 800-ft above MSL in the north-central portion of the vicinity." (141-12)

Response: The text of Section 2.2.2 has been revised to state that the elevation of that portion of the transmission line is 240 m (800 ft).

Comment: Section 2.2.2, Page 2-9, Lines 8-9. The private airports include the Martin Airport, and the Thorp Airport, discussed previously in Section 2.1. ER-Section 2.2.2 Transmission

Corridors and Off-Site Areas, 7th para. should state that "The private airports are the Martin RLA Airport, Throp Airport, and Baker's Strip Airport discussed above in Section 2.2.1 (Bureau of Transportation Statistics, 2000)." (141-13)

Response: The text has been revised to include Baker's Strip Airport in the list of private airports.

Comment: Section 2.4, Page 2-17, Lines 4, 5 & 6. Statement regarding best management practices. Items were left out in the sentence beginning "Assuming." The idea is that if best management practices are used, excavation and disposal of site soils and the placement of imported fill, such as erosion and transport of sediments, should result in minimal impacts. The last part of the sentence, "the low relief terrain and geotechnical properties make landslides in the region of the site unlikely" is correct. The sentence should be changed to "Assuming best management construction practices would be employed, excavation and disposal of site soils and the placement of imported fills, should result in minimal impacts from erosion and transport of sediments. The low-relief terrain and geotechnical properties of the surficial materials make significant landslides in the region of the site unlikely." (141-17)

Response: The text in Section 2.4 has been revised as recommended.

E.2.8 Comments Concerning Air Quality

Comment: I noticed in the draft environmental impact statement that looking at temperature data, they took a period from 1972 to 1977 and used that as a basis. This fails to account for global warming. (55-2)

Response: Long-term temperature data for the region are presented for Springfield and Peoria in Section 2.3.1. These data, which include temperatures data through 2003, are representative for the region. Comparison of the Clinton temperature data described in Section 2.3.1.3 with data from Springfield and Peoria confirms that the data from those stations are representative of the Clinton site. The global warming phenomenon is sufficiently large- scale that it will not affect the Clinton site differently than Springfield and Peoria. The comment provides no new information and was not evaluated further. No change was made to the EIS as a result of the comment.

Comment: Comments have been made about, in a statement about nuclear being a good answer to global warming and being a cleaner source. It turns out that much worse than the carbon dioxide that comes from fossil fuels, for example, are chloro fluoro carbons. Most of the CFC114 released in the world comes in the nuclear fuel cycle down in Paducka and Metropolis. So nuclear is not clean. CFC's have a global warming potential on the order of 10,000 times more than carbon dioxide. So five orders of magnitude. And there's tons of this stuff coming out every year just for the nuclear fuel cycle. This is a huge problem and is not addressed in this document. (55-4)

Response: Section 6.0 of the EIS evaluates the impact of the uranium fuel cycle (including mining, milling, conversion, enrichment, and fuel fabrication) and transportation impacts. Impacts from carbon emissions were determined to be small. The Paducah Gaseous Diffusion Enrichment Facility, the only enrichment facility currently operating in the United States, uses Freon (a CFC) as a coolant. Freon does leak from pipe joints, valves, coolers, and condenser in the facility, but the leak rates are within the level allowed under U.S. Environmental Protection Agency regulations. If the proposed new enrichment facilities, using an alternate technology, are licensed by the NRC, Freon emissions would be reduced. (References: http://usec.com/v2001_02/Content/Investors/2004pdf/USEC2004AnnualReport-Financial.pdf and Environmental Impact Statement for the Proposed National Enrichment Facility in Lea County, New Mexico, NUREG-1790.) No change was made to the EIS as a result of this comment.

Comment: 2-11 lines 31-38 state the prevailing wind at the Clinton Power Plant in ALL months is from the South? But isn't it really from the West or Southwest? Did they make a mistake? Also, wind speeds are 8 mph in summer and 11 mph in winter. Are these averages? (153-4)

Comment: Affected Environment, Section 2.3.1.1, Wind, pages 2-11, 2-12. Providing a windrose of the last years wind data would assist in evaluating the relative direction of air plumes for the site. (172-8)

Response: The discussion of wind in Section 2.3.1.1 of the EIS has been revised in response to these comments.

Comment: Affected Environment, Section 2.3.1.4, Atmospheric Moisture, pages 2-12-2-13. The moisture date cited was from the 1972-1977 period. More recent data needs to be evaluated and included in assessments. The last five year period should be used for this purpose. (172-9)

Response: The moisture information presented for the site is for the 1972-1977 time period. However, as stated in the EIS, these data are consistent with data from Peoria and Springfield. The Peoria and Springfield records contain information through 2002. The data presented are adequate as a description of the climate. More detailed information on precipitation is used in evaluation of the impact of the cooling system for the postulated nuclear plant (see Section 5.3.2). No change was made to the EIS as a result of the comment.

Comment: To control the increase of emission of greenhouse gasses or harmful particulates in our atmosphere, we must increase the share of renewables, such as nuclear, hydro-power, solar, wind in our electrical mix. (101-2)

Comment: Nuclear life cycle emission factors of greenhouse gases ranks below solar cells, hydro power, biomass and wind power. This includes the releases from the mining and from the reprocessing and the enrichment processes. Furthermore, the technology is available now to use different enrichment processes that have even lower greenhouse gas releases using centrifuge technology instead of gas diffusion technology. (61-3)

Comment: Measurable climate change has occurred as a result of our desire for energy. Each year brings more people, more cars, more pollution and even worse effects on our environment. (66-3)

Comment: Will there be an environmental impact from the use of nuclear power in this country? The answer is most definitely yes. There will be a profound environmental impact. In Illinois alone, in the year 2003, 50 percent of the energy that was generated was provided by nuclear power. This means that nuclear power avoided the emission of over 150,000 metric tons of nitroxide, 400,000 tons of sulfur dioxide and nearly 100 million tons of carbon dioxide. That's in Illinois in one year alone. Imagine the pollution savings that nuclear power has provided in the last 40 years. (66-6)

Comment: I was intubated for four days in 1991, intubated again in 1996, for four days, almost died. So I take nuclear power, I mean I take clean air very seriously. (74-3)

Comment: I not only love Clinton, I also love Illinois, because you get 50 percent of your electricity from nuclear power. So, you're not sending smog, you're not sending nitrogen oxide, sulfur dioxide, mercury. You're not sending any of these things over to us in the east. (74-5)

Comment: I like that nuclear energy is not susceptible to changes in weather and climate. (75-4)

Comment: I'd like to say that there was some discussion earlier of the emissions that come from nuclear Z. Nuclear Z is truly a near zero emissions energy in comparison to other energy forms, including renewables. (75-8)

Comment: I'm going to talk to you about why I feel that nuclear power is the best choice for our environment. Nuclear power composes over 70 percent of our non-greenhouse gas emitting power. This is very important because our energy sources, such as coal and gas produce enormous amounts of carbon dioxide, sulfur oxides, nitrous oxides and mercury. All of these are being put up into the atmosphere, into the air that we breathe every day. They are also contributing to global warming, which is becoming a major concern throughout the world. (77-2)

Comment: To control the increase of emission of greenhouse gasses or harmful particulates in our atmosphere, we must increase the share of renewables, such as nuclear, hydro-power, solar, wind in our electrical mix. (84-2)

Comment: The only benefit that I see is a clean energy future. (86-2)

Comment: And remember about radioactive emissions, well the effluence they're radioactive so they'd decay away. Unlike gas, like CO2 from your car, and OX which never go away. These are here forever. Whereas radioactive byproducts do. (86-6)

Response: These comments generally support nuclear power as a clean energy alternative. They contain no new information, and no change was made to the EIS as a result of these comments.

Comment: We take our weather for granted. But the weather comes and goes over the decades. I happen to believe that global warming seems to be a very likely thing that's happening. I mean, it's not, a hundred percent of the scientists don't agree, but theres a large and emerging consensus that do. I don't think that was addressed in the NRC. And if global warming is true, then we are going to have more droughts. (47-5)

Response: This comment was made in the context of power plant cooling. The impacts of lower than average precipitation on Clinton Lake, which would be the source of cooling water for the postulated unit, are discussed in Section 5.3.2 of the EIS. The comment provides no new information. No change was made to the EIS as a result of the comment.

E.2.9 Comments Concerning Surface Water Use and Quality

Comment: Section 3.2. Statement regarding cooling tower blowdown. Section 3.2.1.1, third paragraph should be corrected as follows: A new nuclear unit would normally withdraw 2829 L/s (44,853 gpm) through the intake structure. Blowdown from the cooling tower(s) would return approximately 769 L/s (12,144 gpm) as blowdown to Clinton Lake via the discharge flume. ER Table 3.3-3 needs to be corrected as noted below to show the correct blowdown total. The blowdown flow in the ER text is based on the total from ER Table 3.3-3, which is incorrect since the total row is a repeat of the first row not the total. 12,000 + 144 = 12,144 and not 12,000 gpm. The table needs to be corrected for temperature since the revised wet bulb provided in the response to RAI 8-8 increases the discharge by 1 degree to 101 degrees F. (141-42)

Response: Table J-1 in this EIS states that the maximum blowdown flow rate from the normal plant heat sink would be 760 L/s (12,000 gpm) (PPE, Sections 2.4.4 or 2.5.4). Also, the normal ultimate heat sink blowdown flow rate would be 9.1 L/s (144 gpm), and the maximum blowdown flow rate would be 44.3 L/s (700 gpm) (PPE, Section 3.3.4). Section 3.2.1.1 of the EIS has been revised to clarify these values.

Comment: Section 3.2. Discussion of PPE. Section 3.2.2.1, second paragraph on Normal Cooling, should be revised as noted below: During normal operation at full power, based on the PPE, the cooling tower system is required to reject a heat load of 4420 MW (15.1 x 109 Btu/hr) to the environment. The new unit will reject this heat load using cooling towers. Based on the maximum wet bulb temperature of 86°F, the maximum blowdown temperature is 38.3°C (101°F). (141-43)

Comment: Sections 3.2.1 & 3.2.2. Discussion of PPE. ER Section 3.4.2.3, fifth paragraph should be revised as noted below; the CPS discharge flume will be modified to accommodate the EGC ESP Facility outflow. Engineering evaluations have not been performed to estimate the extent of the modifications but will be performed at the COL phase. The discharge from cooling tower blowdown will normally be 12,000 gpm with a maximum flow of 49,000 gpm (see Table 1.4-1 of the SSAR). The temperature of the blowdown discharge to the CPS discharge flume is estimated to be a maximum of 101°F. The blowdown temperature is dependent on the wet bulb temperature and will decrease with wet bulb temperatures less than 85°F. (141-45)

Response: Table J-1 in this EIS states that the normal plant heat sink would reject up to 15.08 x 10^9 BTU/hr to the environment (PPE, Section 2.3.2). As discussed in Section 3.2.1.1 of the EIS, the staff assumes that this heat would be dissipated using either mechanical or natural draft cooling towers. In Exelon's response to RAI 8-8 (Exelon 2004), Exelon revised the maximum blowdown temperature from 100° to 101°F. However, the PPE does not state a maximum wet bulb temperature associated with the maximum blowdown temperature. Therefore, the staff expects the 101°F temperature to be limiting regardless of the atmospheric wet bulb temperature. The text has been revised to reflect the revised maximum blowdown temperature.

Comment: Section 5.3.2, Page 5-6, Line 34. Statement regarding water-use impacts. The staff selected an adjacent stream for its analysis. Use of an adjacent stream would be proper when there are no meteorological or stream flow data available in the studied watershed. However, in this case, EGC had both records for a period before the lake is in place, after the lake is in place without the plant operating, and after the lake is in place with the plant operating. Therefore, the adjacent stream should not have been used for this analysis. (141-70)

Response: Exelon's approach in its ER for determination of inflows into Clinton Lake is based on monthly runoff coefficients that ignore snow accumulation and melt. Since snow is likely to carry over from month to month during the winter, mean runoff estimates on a monthly basis in the ER are not accurate, especially during warm, dry years. Exelon's monthly runoff coefficients were estimated based on post-dam data at the Rowell streamflow gauge, which is affected by (1) regulation of the dam and (2) an additional catchment area below the dam that contributes flow to the Rowell gauge location. Staff used an adjacent, minimally regulated, gauged watershed as a basis for estimation of inflows into Clinton Lake according to standard engineering practice. No change was made to the EIS as a result of the comment.

Comment: Since Clinton Lake may be used for cooling, both the temperature effects and the drawdown amounts should be considered more seriously than has been done. One needs only look back to 1988 to witness the effects drought can have on the Lake. In the North Fork, Salt Creek dried up completely for nearly half a mile, cutting off water supply near the transmission pole crossing. I do have pictures to attest to this. The dam is not allowing for the normal "flushing out" of silt from the creek that used to occur, thus silt deposition occurs in the lake. Dredging may be an answer, but it also carries with it some negative environmental consequences. (157-6)

Response: As proposed, Clinton Lake will be used for supplying makeup water for normal and emergency operations of the ESP facility. The staff's analysis considered the bounding (i.e., most severe or maximum) impact to the lake-water level and temperature during a sustained drought period. The staff disagrees with the comment that temperature and drawdown impacts of the proposed ESP facility are not seriously considered. The staff performed an independent assessment to ensure that the impacts were taken into account in the EIS. Pursuant to Section 404 of the Clean Water Act, any dredging of Clinton Lake would be regulated by the U.S. Army Corps. of Engineers to protect the environment. No change was made to the EIS as a result of the comment.

Comment: The plant parameter envelope (Section 1.1.1) is another example of trying to limit the scope of the entire project to a vague set of parameters for the reactor and claiming the environmental impacts of higher water use to remove wasted heat energy can't be considered. Yet, higher water use may require raising the height of the dam, flooding a bigger area, which is definitely a significant environmental impact of the project. (169-8)(179-8)

Response: As stated in EIS Section 1.1.1, the applicant for an ESP need not provide a detailed design, but should provide a sufficient set of bounding parameters and characteristics so that an assessment of site suitability can be made (i.e., a PPE). Sections 4.3 and 5.3 discuss the impacts associated with the increase in water use for the Exelon ESP. As required by 10 CFR Part 52, a separate safety evaluation report (SER) was also prepared by the staff (NRC 2005). Section 2.4 of the SER covers site hydrology, elevation of the site, and operation of Clinton Dam. If the dam were to be raised, the calculations performed in SER Section 2.4 would need to be updated to reflect these values. However, in no part of the SER, EIS, or ESP application, has raising the height of Clinton Dam or impounding a larger volume behind the dam by changing the existing operating rules been discussed. No change was made to the EIS as a result of these comments.

Comment: Section 5.3.2, Page 5-7, Line 4. Statement regarding cooling tower discharge. We agree with the staff that a cooling tower will discharge approximately 80% of its heat load in the form of evaporation. However, EGC analysis indicates that Clinton Lake discharges 71% (average of the monthly values used in our period of record model) of its total heat load (heat from solar radiation as well as condenser heat load) by way of evaporation. This is 9% less than the staff's estimate. Thus, it is suggested that the staff use 71% rather than 80% for its value. (141-72)

Response: Average monthly observed values are not appropriate for the evaporation analysis in the EIS, since the stated purpose of the staff's analysis was to compute a realistic, although conservative, value. Although not stated, the 71 percent monthly-average heat load value may include periods of plant outages and plant operations at power levels less than full output. As such, the staff's analysis will continue to assume that the CPS unit discharges 80 percent of its heat load in the form of evaporation. No change was made to the EIS as a result of the comment.

Comment: Site Layout, Section 3.2.2.2, Component Descriptions, Heat Dissipation Systems, page 3-10. A clarification between the ultimate heat sink (UHS) reservoir and Clinton Lake Reservoir needs to be provided. (172-16)

Response: Based on the comment, the staff added a brief description in Section 3.2.2.2 of the EIS to distinguish the UHS pond from the Clinton Lake Reservoir.

Comment: Section 3.2.4.1, Page 3-11. "In the PPE approach, specific quantities and concentrations of chemicals or biocides used for proper water chemistry in the reactors are not identified and will need to be revisited in the CP or COL stage." In this same page, the DEIS states that Exelon did provide bounding values for the blowdown. Therefore, at the CP/COL stage, Exelon will only need to demonstrate that those values in the PPE remain bounded by the plant-specific design. (141-51)

Response: Based on the comment, the EIS has been revised to clarify the water quality requirement of the blowdown.

Comment: Section 2.6.1.3, Page 2-20, Lines 10-12. Exelon collects flow measurements directly associated with current site operation that are required under the terms of the Exelon's existing NPDES permit. This statement is incorrect. AmerGen Energy Company, LLC, holds the NPDES permit for the CPS. Exelon does not collect any flow measurements associated with the operation of CPS. The monitoring currently conducted by Exelon is limited to collecting quarterly water level measurements from three peizometers installed at the EGC ESP Site in July and August, 2002. (141-21)

Comment: Section 2.6.2.1, Page 2-21, Line 9. "When the CPS unit is operating, pumps draw water from Clinton Lake at a rate of 35,700 L/s (566,000 gpm)." The 35,700 L/s (566,000 gpm) reported in the second sentence is the summer intake. During the winter, the intake is less (about 28,075 L/s or 445,000 gpm). The sentence should be revised to read "at a rate of 35,700 L/s (566,000 gpm) in the summer and 28,075 L/s (445,000 gpm) in the winter." (141-25)

Comment: Section 2.6.3.1, Page 2-22, Line 9. Discussion of operational impacts of a new nuclear unit on Clinton Lake water quality. Operational impacts of a new nuclear unit on Clinton Lake water quality are discussed in Section 5.3.3 of this EIS and not 5.2.2. (141-26)

Response: Sections 2.5.1.3, 2.6.2.1, and 2.6.3.1 of the EIS have been revised to reflect the three preceding comments.

Comment: Section 2.6.3.3, Page 2-22, Lines 39-41. Discussion of thermal monitoring. The last two sentences of this paragraph read, "Clinton Lake is also part of the IEPA Bureau of Water's Ambient Lake Program. Additionally, thermal lake data is collected as part of the environmental monitoring program for the CPS (BOW 2004)." The BOW document (i.e., the "Draft Illinois 2004 Section 303(d) List") does not discuss the thermal data collection for the CPS. The reference citation should be moved to the end of the previous sentence. The sentence should read "IEPA Bureau of Water's Ambient Lake Monitoring Program (BOW 2004)." The reference should actually be (IEPA) and not (BOW). The second sentence should also be revised to "thermal lake data are collected as part of the monitoring program for Clinton Lake." (141-29)

Comment: Section 2.6.3.3, Page 2-22, Lines 39-41. "Clinton Lake is also part of the IEPA Bureau of Water's ambient lake program. Additionally, thermal lake data is collected as part of the environmental monitoring program for the CPS (BOW 2004)." The BOW document (i.e., the "Draft Illinois 2004 Section 303(d) List") does not discuss the thermal data collection for the CPS. The reference citation should be moved to the end of the previous sentence. The sentence should read "IEPA Bureau of Water's ambient lake program (BOW 2004)." The second sentence should also be revised to "thermal lake data are collected as part of the monitoring program for Clinton Lake." (141-30)

Response: The staff revised Section 2.6.3.3 of the EIS to reflect this comment.

Comment: Section 5.3.2, Page 5-5, Line 39. Statement regarding water-use impacts. Outflows also include water over and through the dam. (141-65)

Comment: Section 5.3.2, Page 5-6, Line 10. Statement regarding water-use impacts. Outflows also include direct evaporation from the ESP unit. (141-66)

Comment: Section 5.3.2, Page 5-6, Line 23. "Evaporation estimates were based on calculations with Exelon's lake temperature model, discussed in Section 5.3.2 of the ER (Exelon 2003b)." The temperature model is discussed in ER Section 5.2.1. (141-68)

Response: The staff revised Section 5.3.2 of the EIS to reflect the three preceding comments.

Comment: Section 5.11, Page 5-78, Line 34. It is stated that the discharge rate of 5 cfs is a NPDES permit condition of the existing CPS. The 5 cfs discharge rate is specified in the CPS dam permit (No. DS2001236). It is actually part of the approved O&M Plan and EAP attached to the permit rather than a 'condition' directly specified in the permit. The reference to the 5 cfs can be found (on page 7) in the Operation Plan (Section 1 General, Subsection 3) Outlet Works, that reads, "The lake outlet works is provided primarily to maintain a minimum flow of 5 cfs to the creek downstream of the dam". The minimum reservoir release of 5 cfs is necessary to satisfy commitments made in the CPS Final Environmental Statement. (141-91)

Response: *Based on the comment, the staff revised Section 5.11 of the EIS.*

Comment: Section 10.1, Page 10-5, Line 38. The NRC states, "Hydrological, water use, and water quality impacts during operation would primarily be the result of the operation of the proposed wet cooling power system during periods of reduced water supply in Clinton Lake and downstream." It should be noted that the wet cooling power system has been used as the bounding condition. (141-122)

Response: *Based on the preceding comment, the staff has revised Section 10.1 of the EIS.*

Comment: Affected Environment, Section 2.6.3.3 Thermal Monitoring, pages 2-22, 2-23. The requirements of the current permit should be stated and not just cited. (172-12)

Response: *The purpose of Section 2.6.3.3 of the EIS is to describe any pre-application and pre-operation thermal monitoring activities, not to list the limits defined in the current NPDES permit. Exelon provides the current NPDES limits in Section 6.1 of the ER. No changes were made to the EIS as a result of the comment.*

Comment: Section 3.2.4.2, Page 3-12, Lines 7-12. "Sanitary systems during pre-construction and construction activities will include the use of portable toilets. During operation, sanitary system wastes will likely be handled through the existing CPS sanitary sewage treatment plant. Discharges from this plant will be controlled in accordance with an approved NPDES permit issued by the IEPA. Exelon (2003b) provided a bounding sanitary discharge rate to Clinton Lake of 3.8 L/s (60 gpm) normal and 6.2 L/s (98 gpm) maximum. As stated in Section 3.6 text of the ER, "The normal and maximum amount of sanitary discharges to Clinton Lake for the selected composite reactor are presented in Table 3.6-2 and were obtained from Table 1.4-1 of the SSAR". Upon review of Table 3.6-2 of the ER the maximum discharge rate from the sanitary sewer system is stated as 198 gpm so there is a disparity between the numerical values reported in the DEIS and the ER. The numbers in the DEIS should be revised to reflect those in the PPE table and the ER. (141-52)

Response: *Based upon the PPE and the preceding comment, Section 3.2.4.2 of the EIS has been revised to state a maximum sanitary discharge rate of 198 gpm.*

Comment: Section 2.6.3.1, Page 2-22, Line 18-19. Before a new nuclear unit could begin to operate, Exelon would be required to obtain a NPDES permit for the discharge. As stated in the ER, the Exelon ESP facility would maintain the current limits specified in the CPS NPDES permit. A new NPDES permit would not be required but a modification to the existing permit would be required to add the Exelon ESP facility to the permit. (141-27)

Comment: Section 2.6.3.4, Page 2-23, Lines 22-23. "Chemical monitoring of a variety of constituents is required, including pH, chloride, mercury, nitrate, suspended solids, and dissolved oxygen." This sentence should be revised to identify if the constituents listed are monitored under the current CPS NPDES permit, or those that will be required as part of the chemical monitoring programs for the ESP Facility. (141-32)

Comment: Section 5.3.3, Page 5-8, Line 24. It is stated that the water quality impacts are SMALL, with the exception of water temperature. As stated in the ER, the Exelon ESP facility would maintain the current limits specified in the CPS NPDES permit. A new NPDES permit would not be required but a modification to the existing permit is required. Based on this information, the staff should have enough information to perform its assessment of impacts of water temperature. (141-76)

Response: *The Illinois Environmental Protection Agency advised the NRC staff, that since the existing CPS and proposed EGC ESP facilities could have separate legal owners, separate NPDES permits and monitoring programs would be required to ensure that if a compliance issue arises, the appropriate legal entity would be identified. Consequently, no change was made to the EIS as a result of these comments.*

Comment: Section 5.3.2, Page 5-6, Line 29. Discussion of snowfall in period of record analysis. The applicant did not exclude snowfall in the period of record analysis. The values for precipitation in the analysis include both rainfall depth (in inches) and the liquid equivalent depth of snow fall (in inches). This is the value that is reported directly in the source meteorological document MRCC (2002a). The EIS uses the perceived exclusion of snowfall as justification for using data from an adjacent watershed. With that issue now set aside there should be no reason for dismissing the well documented records from the Salt Creek watershed rather than the records from a considerably different adjacent watershed. If this adjacent watershed is used, any differences in the model results must first be considered differences in the watersheds and then as deficiencies in either one of the modeling approaches. Precipitation data that were used in the period of record analysis were obtained from the reference MRCC 2002a. Precipitation values included in this reference are the sum of rainfall depth and the water equivalent depth of snow fall. The inclusion of snow in the hydrologic analysis is stated in the Technical Memorandum, Clinton Lake Period of Record Analysis – Spreadsheet Column by Column Explanation, July 7, 2004 in the section "Model Limitations". (141-69)

Response: Section 5.3.2 of the EIS refers to the lack of snow accumulation and melt processes in Exelon's water budget and drawdown calculations for Clinton Lake. Snow is likely to carry over from month to month during the winter, thus affecting the mean runoff estimated by Exelon on a monthly basis (see Table 2.3-2 of the ER), especially during warm dry years. Since streamflow at Rowell is affected by regulation due to the presence of Clinton Dam, the NRC staff determined that Exelon's method is not appropriate for accurate determination of inflows into Clinton Lake above the dam for a drawdown analysis during a drought period. Staff used an adjacent, minimally regulated, gauged watershed as a basis for estimation of inflows into Clinton Lake according to standard engineering practice. No change was made to the EIS as a result of the comment.

Comment: Section 2.6.1.3, Page 2-20, Line 25. "The lack of these measurements (water velocity) limits detailed process modeling of lake temperature and elevation levels." The sentence stating that, "The lack of these measurements (water velocity) limits detailed process modeling of lake temperature and elevation levels" is not entirely accurate. There are other ways to model the potential thermal impacts of the station operation on Clinton Lake such as the hydrothermal model of the lake developed in 1989 by J.E. Edinger Associates Inc. The Edinger model examined lake temperature changes in Clinton Lake with changing lake levels and was calibrated with lake temperatures measured during the summer of 1988. (141-23)

Response: Physically-based computational fluid dynamics models of the lake must be calibrated and validated against observed water velocity and lake elevation data before the models can be credibly used to predict impacts of the ESP unit on Clinton Lake. Suitable water velocity measurement techniques could be used to collect time-series profiles of water velocity at several stations around the lake. These data (time-series of water velocity and lake elevation) are important for verifying computed travel times and cooling of plant thermal effluents to the atmosphere. Lack of these observed measurements limits the validation of numerical models of the lake. In addition, as pointed out by the Illinois Environmental Protection Agency (June 22, 1989 ruling with the Illinois Pollution Control Board) in its review of the Edinger model, the lack of inflow data to the lake limits the accuracy of any heat budget model. No change to the EIS was made as a result of the comment.

Comment: We are concerned about the proposed project's impacts on Clinton Lake. According to the DEIS, Clinton Lake (and several connected reaches) are on Illinois EPA's Draft 2004 list of impaired waterbodies under Section 303(d) of the Clean Water Act. Low dissolved oxygen is one of the attributes of one or more of these impairments. The DEIS also states that the proposed project would increase the water temperature of Clinton Lake, which could exacerbate the low oxygen levels of the already impaired waterbodies. The USNRC should provide future environmental documentation that evaluates the impact of the proposed project on the impaired status of Clinton Lake and its connected reaches. Such environmental documentation should include commitments to mitigate these impacts. (172-5)

Response: Pursuant to the Clean Water Act, the U.S. EPA is responsible for protecting the nation's water quality. In Illinois, the U.S. EPA has delegated this responsibility to the Illinois Environmental Protection Agency (IEPA). Prior to operation, the ESP facility would be required to have an NPDES permit from IEPA, which will include water quality parameter limits. Water quality limits set by IEPA are presumed to be protective of the environment. No change to the EIS was made as a result of the comment.

Comment: Section 2.6.3.4, Page 2-23, Line 16. "Many of these same monitoring activities would be continued if the ESP unit was completed and would likely become part of the operational monitoring." As the operation monitoring for the CPS was discontinued after 1991, the statement is not accurate. The sentence should be revised to read, "Many of these same monitoring activities will be considered in the development of the operational monitoring program to be implemented if the ESP unit were completed." (141-31)

Response: The staff revised Section 2.6.3.4 of the EIS in response to the comment.

Comment: General Comments on Water Impacts in Sections 5, 7, 9, and 10. "The results of the staff analysis were that the frequency and magnitude of low water conditions are more frequent and deeper than those predicted by the applicant. However, the lack of pool elevation data made it impossible for the staff to perform an adequate calibration and verification of the approach. The analysis must be revisited at the construction permit (CP) or combined license (COL) application. The applicant has, however, committed to collect the pool elevation data that would be required to calibrate and verify the model results. Therefore, based on the Exelon ER and the staff's independent review, the staff concludes that during normal water years the water-use impacts would be SMALL, and mitigation would not be warranted. During low water years, however, the impact to the water level could be MODERATE until normal water conditions return." As page 5-7 of the DEIS indicates, some of the assumptions in the staff's analysis are "very conservative." Additionally, page 5-37 of the DEIS states that the occurrence of a drought severe enough to impact the lake level is a "rare event." NEPA mandates that the EIS use realistic assumptions, not "very conservative" assumptions. Furthermore, in determining the environmental impacts, the EIS should account for the low probability of severe drought conditions in determining the overall environmental impacts. Additionally, the EIS should give greater weight to the fact that the impacts are temporary (see DEIS, p. 10-6). When all of these factors are taken into account, the impact should be designated as SMALL. (141-71)

Response: The staff agrees that NEPA does not require a "worst case" analysis. However, the staff does not believe that the droughts considered, which are part of the historical record at the site, represent a "worst case." Additionally, the staff's conclusion refers to "low water years" as opposed to "drought years." "Low water years" refers to conditions far more common than a "severe drought." No change was made to the EIS as a result of the comment.

| Comment: Section 5.3.2, Page 5-7, Line 25. Statement regarding water-use impacts. In this summary paragraph, it is important to note that the model results presented represent the most consumptive cooling process being considered in the ESP application and that other less consumptive processes are also being considered. (141-75)

Response: NRC staff carried out bounding calculations based on the PPE values to verify Exelon's assertions with respect to water-use impacts in the ER. "Less consumptive" processes were not considered in the staff's analysis (see Section 3.2.1.1 of the EIS). No changes were made to the EIS as a result of the comment.

Comment: Section 2.6.1.1, Page 2-19, Lines 13 – 21. Discussion of surface-water hydrology. The context of these two paragraphs should be clarified to indicate it relates to the lake surface area and not the total lake watershed. (141-19)

Response: A change was made in Section 2.6.1.1 of the EIS to clarify the area under consideration.

Comment: Section 5.4.1.4, Page 5-12, Line 6-7. It is stated that it is unknown where and how much lakebed would be exposed, potential impacts could range from minimal to substantial. It is also stated that the issue would be evaluated in greater detail at CP/COL. The staff has not asked for additional information that it felt would be needed to obtain to assess an impact level. In addition, it is unclear how the staff would evaluate this issue at CP/COL any differently than can be evaluated at ESP. With the known minimum lake level assumed in the ER, there should be sufficient information to conclude that the impacts would be considered SMALL. (141-77)

Response: The staff evaluated the total acreage exposed at various lake levels based on the lake storage/lake elevation relationship provided by Exelon. While this information is adequate to estimate the total acreage exposed, it does not define the impact at a specific location within the lake. In its ER, Exelon committed to collecting lake bathymetry information prior to an application for a construction permit (CP) or combined operating license (COL). The NRC staff will ensure that the impacts of lake drawdown at specific locations would be disclosed in any EIS prepared in conjunction with a CP or COL application. No change was made to the EIS as a result of the comment.

Comment: Section 5.3.2, Page 5-7, Line 13. Discussion of modeled results. To put these modeled results in perspective, it would be beneficial to include actual low flow percentages (flow less than 5 cfs) measured at Rowell for the period without CPS (1978 –1987) and the period with CPS (1988-present). These values show the "very conservative" assumptions the staff has used in the NRC model when compared to results with measured values at Rowell. Looking at the percentages at or below low flow at the Rowell gauge, EGC values are considerably lower in the range of pre-dam (4%), pre-CPS (<1%) and CPS (1%). The NRC model results are pre-CPS (23%) and CPS (43%). The watershed adjustment factor stated in the DEIS would not account for that much difference. (141-73)

Response: The staff's independent assessment of impacts to Clinton Dam outflows used a relationship that was directly proportional to Clinton Lake's water surface elevation. The assessment assumed that outflows from the dam were 5 cfs when the lake elevation was between elevation 650 and 690 ft. At lake elevations greater than 690 ft, discharges would follow the rating curve shown in the Updated Safety Analysis Report (USAR) Figure 2.4-8 (CPS 2002). Although the Rowell gage is significantly downstream of Clinton Dam, outflows at the dam were compared to observed Rowell gage data. NRC staff observed that the overall trends (rate of discharge decrease, period of elevation discharge, etc.) for both the existing CPS facility (only) and the CPS plus the ESP facility are quite similar to the Rowell gage data.

The NRC staff also considered the low daily discharge range, observing that there are differences between the staff's independent assessment for releases from Clinton Dam and the observed discharges several miles downstream at the Rowell gage. These differences are slight and may originate from (1) accretions and/or depletions from Salt Creek downstream of Clinton Dam, (2) operational differences between the staff's assessment and actual operation, and (3) differences in computed and actual lake level elevation.

The staff's independent assessment computes outflow from the dam based upon an assumed Clinton Dam operating rule curve. Also, the assessment does not address accretions and/or depletions in Salt Creek discharge downstream of Clinton Dam. Therefore, the staff's computed outflows should not be translated downstream to the Rowell gage. Although the staff found a significant correlation between computed and observed values, the staff's assessment is not a stream flow model that is intended for calculating discharge at the Rowell gage. The text in Section 5.3.2 of the EIS has been modified to clarify the staff's approach as well as reflect the inclusion of the recent meteorological record.

Comment: One pressing issue is how the additional nuclear capacity would affect the health and vitality of Clinton Lake. The Clinton nuclear reactor relies on water from the lake to cool it, but additional generation capacity would require more water and may overtax and -deplete the lake, especially-in drought years when water levels are low. Such overuse may force the plant to shut down, since the loss of coolant is a serious safety problem that could lead to meltdown, and could make the lake less desirable as a source of recreation due to high water temperatures. The precise impact is unclear, since neither Exelon nor NRC has done a full analysis of how a new reactor would affect the lake temperature. (191-2)

Comment: I don't think I've heard anyone talk tonight about what the, what the specific impact is going to be on, on Clinton Lake. And while there are certainly major problems with that draft environmental impact statement, there's a few valuable nuggets in there. One being that "the consumptive water loss of the atmosphere, from the cooling tower of a new nuclear unit, could lower the water level of the lake significantly, during times of drought." Which, as we heard, are likely to become only more prevalent with, with future climate change. This could impact both boating and fishing at the lake, because of increases in temperature, and lower lake levels for more evaporation. And I would also point out, while the NRC has, has tentatively approved this permit, the impact of, on temperature, is still unclear. No one knows exactly, just because that data doesn't exist yet. (68-7)

Response: In Section 5.3.2 of the EIS, the staff discloses that during times of low water surface elevations in Clinton Lake, the increased water withdrawals for the ESP unit will cause a further decline in water surface elevation. During times of relative water excess, withdrawals for the ESP unit will not affect lake water surface elevations. The effects of climate change on Clinton Lake are uncertain. The staff believes that the public's interest will be adequately protected by the water use permit (issued by IEPA) in the event that significant changes in drought frequency cause adverse effects. Except during extreme drought conditions, it is unlikely that boating would be adversely affected. Low water levels could impede access to boats but are unlikely to adversely affect recreational fishing. The staff finds that because the ESP unit discharge is relatively small compared to the CPS unit (approximately 1 to 3 percent), the relative temperature impacts of the ESP unit discharge would not be significant.

The EIS limits the cooling system discussion to the wet tower option only (Section 3.2.1.1). The estimated discharge to Clinton Lake from the ESP unit during normal operations is 760 L/s (12,000 gpm). By comparison, discharge from the CPS unit at 100-percent load is approximately 38,950 L/s (615,000 gpm or 1373 cfs) (Edinger 1989). In other words, if both the CPS and ESP units are operating, the percent of discharge originating from the ESP unit is expected to be less than 2 percent of the total discharge passing through the discharge canal. Exelon's RAI ID R3-26 to NRC RAI No. E5.2-3 (Exelon 2004) confirms that the "blowdown discharge rates are relatively small (1 to 3 percent of existing CPS discharge)." The staff, therefore, finds that the incremental increase in lake temperature caused by the ESP unit would be almost undetectable. No change was made to the EIS as a result of the comment.

Comment: 5-19 lines 24-28 show that the average temperature of Lake Clinton has gone up 14 degrees Fahrenheit since the CPS became operational. Won't a second reactor heat up the lake to dangerous bacteria producing levels? (153-7)

Comment: 5-44 lines 3-22 state that "lake temperatures from the plant intake to the discharge appear to be about 5 degrees Fahrenheit warmer on average," however, they stated in 5-19 that "the average temperature of Lake Clinton has gone up 14 degrees." This means that the water by the discharge must be more than 20 degrees above normal. In addition, they state that increased temperatures can greatly increase the number of thermophilic microorganisms, which can be "causative agents of potentially serious human infections." Won't the second nuclear reactor's heat cause for the growth of more of these microorganisms? (153-11)

Response: The EIS limits the cooling system discussion to the wet tower option only *(Section 3.2.1.1). The estimated cooling tower blowdown during normal plant operational would be 760 L/s (12,000 gpm). By comparison, discharge from the CPS unit at 100-percent load is approximately 38,950 L/s (615,000 gpm or 1373 cfs) (Edinger 1989). In other words, if both the CPS and ESP units are operating, the percent of discharge originating from the ESP unit is expected to be less than 2 percent of the total discharge passing through the discharge canal. In its response to an RAI, Exelon states (Exelon 2004) that "because the blowdown discharge rates are relatively small (1 to 3 percent of existing CPS discharge) and the blowdown water temperatures are low, lake temperature increases due to boiler cooling tower blowdown are expected to be negligible." The staff finds that the incremental increase in temperature caused by the ESP unit would be small and would not significantly increase the abundance of thermophilic microorganisms in Clinton Lake (see Section 5.8.1 of the EIS). No change was made to the EIS as a result of these comments.*

Comment: Sections 3.2.1 & 3.2.2. Discussion of PPE ER Section 3.4.2.4, fourth paragraph should be revised as follows: The maximum discharge flow from the UHS cooling system to the UHS cooling towers is 26,125 gpm during normal operation and 52,250 gpm during shutdown (see Table 1.4-1 of the SSAR). The maximum heat load on the UHS cooling system is 2.25E+08 Btu/hr during normal operation and 4.11E+08 Btu/hr during shutdown. The discharge from UHS cooling tower blowdown is normally 144 gpm with a maximum blowdown of 700 gpm. The maximum temperature of the UHS blowdown discharge is 95°F. (141-46)

Comment: Section 3.2.2.1, Page 3-9, Lines 19-23. "Based on the PPE, during shutdown, the UHS system for each unit would reject 123 MW (420 x 106 Btu/hr) to the environment. Makeup water for the mechanical draft UHS cooling towers is withdrawn from the UHS reservoir. The reservoir is required to maintain an adequate supply of water for 30 days of emergency operation. Based on the PPE, the maximum blowdown discharged to the discharge canal is 54 L/s (850 gpm)". In Section 3.4.2.4 of the ER – Ultimate Heat Sink – it is stated that, "The maximum discharge flow from the UHS cooling system to the UHS cooling towers is 26,125 gpm during normal operation and 52,250 gpm during shutdown (see Table 1.4-1 of the SSAR). The maximum heat load on the UHS cooling system is 2.25E+08 Btu/hr during normal operation and 4.11E+08 Btu/hr during shutdown. The discharge from UHS cooling tower blowdown is normally 100 gpm with a maximum blowdown of 700 gpm. The maximum temperature of the UHS blowdown discharge is 95°F". There is a slight disparity between the numbers reported in the ER versus those reported in the DEIS. It should be noted that the numbers reported in the ER are consistent with those reported in the PPE table. The numbers in the DEIS should be revised to reflect those in the PPE table and the ER. (141-50)

Comment: Site Layout, Section 3.2.2.1 Description and Operational Modes, Ultimate Heat Sink, page 3-9. There appears to be a conflict in the blowdown discharge values for the cooling towers of 760 liters per second versus 54 liters per second. A clarification of this apparent discrepancy is needed. (172-15)

Response: *Section 3.2.2.1 of the EIS has been revised to reflect the current PPE values for the maximum blowdown discharge.*

Comment: Section 2.6.1.1, Page 2-18, Line 18. Two small gates near the service spillway are able to provide small releases to maintain minimum downstream flows. CPS documents (e.g., USAR Section 2.4.8.1.4 Outlet Works and ER-OLS Section 2.4.1.4.1) indicate that there are three sluice gates that regulate the downstream releases of water from the lake. (141-18)

Response: *Section 2.6.1.1 of the EIS has been revised to state that there are three sluice gates that provide the minimum releases from Clinton Lake.*

Comment: Section 2.6.1.3, Page 2-20, Line 28-30. These measurements would become part of Exelon's pre-application monitoring program. This should be clarified to mean that the measurements taken would become part of the pre-construction monitoring program. The rationale for this clarification is that there could be a significant time period between CP/COL application and the commencement of construction activities. (141-24)

Response: *Section 2.6.1.3 of the draft EIS states (lines 28-30): "These measurements would become part of Exelon's pre-application (referring to the construction permit [CP] or combined operating license [COL] application) monitoring program." The staff finds that this statement is clear because these data should be collected and analyzed before construction activities are initiated so that a determination of plant thermal load impacts on the lake can be computed and potentially mitigated, if necessary. No change was made to the EIS as a result of the comment.*

Comment: Site Layout, Section 3.2.1.2, Plant Water Treatment, Page 3-8. With bounding of potential situations and emissions being an integral portion of this document, the water quality of effluents should be bounded so that any of the choices of systems would be covered in the basic analysis and, in later documents, could be system specific. (172-14)

Response: *The NRC does not have authority to set discharge requirements. Pursuant to the Clean Water Act, this responsibility is assigned to the U.S. EPA. In Illinois, the U.S. EPA has delegated this responsibility to the Illinois Environmental Protection Agency (IEPA). Prior to operation, the COL applicant would be required to obtain an NPDES permit from IEPA. No changes were made to the EIS as a result of the comment.*

Comment: If there is a wet cooling system for this, my concern is that is it going to be sufficient for like what we might say is a worse case scenario, to cool both, what would be both units here at Clinton. (47-4)

Comment: They had some tons of rain in L.A. but before that, I mean, the water table had just dropped and dropped and dropped. The reservoirs had dropped and dropped and dropped. And that can very easily happen at this lake. And I don't think that's given due consideration because we tend to think of, well, lately it's been hunky-dory and it probably has. But, you know, we have to plan. (47-6)

Response: The water budget for Clinton Lake is based on the historical record of precipitation and is believed to be representative of future conditions. The staff has determined that there will be adequate water to safely operate the unit. The plant parameter envelope (PPE) contains specifications for both normal plant cooling and cooling under emergency conditions, e.g., the ultimate heat sink (UHS). Appendix J of this EIS lists the entire PPE values. According to the PPE Section 3.3.9, under emergency conditions the makeup flow rate is 555 gpm (1400 gpm maximum). Given the size of Clinton Lake, these makeup flow rate needs can easily be satisfied by the lake even under drought conditions (CPS 2002, Section 2.4.11). No change was made to the EIS as a result of these comments.

Comment: Section 5.3.2, Page 5-7, Line 18. Statement regarding minimum flow values. EGC agrees with the comparison to minimum flow values with one plant operating. A comparison to minimum flows without a power plant does not appear to be relevant as there is a permitted and operating power plant on the site. If a comparison to natural conditions is desired it would seem appropriate to show minimum flow values for the time period before the plant and dam were in place. (141-74)

Response: The staff agrees with this comment and all references to the staff modeling a "no units" scenario have been deleted from the EIS.

Comment: Section 2.1, Page 2-5, Line 10 "around the lake up to the expected 212-m (697-ft) high-water mark." Unable to find these elevation data in the ER or SSAR. A reference for this information or how this number was calculated should be provided. (141-7)

Response: The staff concurs with the comment. The ER states that the elevation of Clinton Lake at normal pool is 210 m (690 ft). The EIS text has been revised to reflect the comment.

Comment: The DEIS is also incomplete in its analysis of the effects a new reactor will have on Clinton Lake, the only source of cooling water for the existing and proposed reactors. The DEIS does note that "the consumptive water loss to the atmosphere from the cooling tower of a new nuclear unit could lower the water levels of the lake significantly during times of drought...However, it fails to note that drought conditions in the Midwest are predicted to become more prevalent in coming decades due to climate change. This must be factored into the lake impact analysis. (112-8)(113-8)(114-8)(115-8)(116-8)(117-8)(118-8)(119-8)(120-8) (121-8)(122-8)(123-8)(124-8)(125-8)(126-8)(127-8)(128-8)(129-8)(130-8)(131-8)(132-8) (133-8)(134-8)(135-8)(136-8)(137-8)(138-8)(139-8)(140-8)(142-8)(143-8)(144-8)(145-8)(146-8) (147-8)(149-8)(154-8)(155-8)(158-8)(159-8)(162-8)(163-8)(164-8)(165-8)(166-8)(167-8)(173-8) (174-8)(175-8)(176-8)(177-8)(178-8)(180-8)(181-8)(182-8)(185-8)(186-8)(187-8)(188-8)(189-8) (190-8)(192-8)(193-8)(194-8)

Response: Although there is wide spread acknowledgment that long-term global climate change is occurring, it is very difficult to predict the magnitude of change and climate factors that will be affected. The NRC staff reviewed the most recent national assessment "Climate Change Impacts on the United States: The Potential Consequences of Climate Variability and Change," published by the National Assessment Synthesis Team of the U.S. Global Change Research Program in 2000 (USGCRP 2000). For the Midwestern U.S. region, including the region of the proposed Exelon ESP, both models used in the National Assessment predict both increases in temperature and precipitation. The primary difference between the two models is the amount of summer precipitation. Despite the increase in precipitation, the increase in temperature could offset the increased precipitation with increased evaporation, thereby increasing the frequency of drought conditions. Although both of the models produce reasonably similar predictions of climatic changes, considerable uncertainty remains in the predictive skill of such climate models. Any subsequent change in drought frequency due to climate change is further confounded by changes in land use and water use patterns that would occur through the 21st century. Given the authority of the State of Illinois to regulate water use on an ongoing basis, the NRC staff believes that the public's interest will be adequately protected if significant changes in drought frequency were to cause adverse effects. The NRC's authority to regulate the safety of the plant would consider changes in water availability as part of its continuous, ongoing evaluation of plant safety. No change was made to the EIS as a result of these comments.

Comment: Section 2.2.2, Page 2-7, Lines 20-21 "to 210 m (690 ft) above MSL and 212 m (697 ft) above MSL along Clinton Lake (Exelon 2003a)." The ER used the numbers 700-ft and 696-ft above MSL, respectively. ER data are referenced as USGS, 1990. ER Section 2.2.2 uses the 700 ft. (141-9)

Response: The text in Section 2.2.2 has been modified to state, "Elevations range from approximately 244 m (800 ft) above mean sea level (MSL) in the north-central portion of the vicinity to 210 m (690 ft) above MSL along Clinton Lake (USGS 2001)." To avoid confusion over the various elevations presented in the ER, the staff chose to report the USGS maximum reported elevation in the area and the spillway elevation of Clinton Dam.

Comment: Will Clinton Lake be able to support this significant additional withdrawal, even in years of severe drought? How would the safe operation of the plant be affected, in such a situation? Could lower lake levels cause or contribute to the severity of a loss-of-coolant accident? The final EIS should demonstrate a trenchant investigation into these questions, considering the desirability of preserving Clinton Lake and the critical importance of a healthy water supply to the safe functioning of the plant. (150-15)(151-15)

Response: This comment raises safety issues related to availability of cooling water. Safety issues are addressed in NRC's Safety Evaluation Report for the Clinton ESP site (NRC 2006). Adequate water storage would be maintained in the UHS for the continued safe shutdown of the *facility. Should water levels in Clinton Lake drop too low, the facility would be derated or shutdown long before it would be a safety concern. No change was made to the EIS as a result of the comment.*

Comment: According to the EIS, Exelon has yet to provide site-specific data for the chemistry of groundwater under the ESP site (§ 2.6.3.2), nor has it reported velocity measurements within Clinton Lake, which are essential to understand the hydrodynamics of the lake (§ 2.6.1.3). How can the NRC adequately consider the impact of the operation of CPS's existing nuclear unit-much less an additional one-without this important information? (150-18)(151-18)

Response: *While lack of velocity data does limit the ability to predict changes in the hydrodynamics in Clinton Lake, given the relative simple geometry of Clinton Lake, the general pattern of flow is well understood. Since the ESP facility is proposed to utilize wet cooling with only minor discharges of heated blowdown water to Clinton Lake (relative to the existing CPS discharge), thermal conditions in Clinton Lake would only be indirectly impacted by operation of the ESP facility. The primary direct impact of the operation of the proposed ESP facility would be a reduction in the lake level elevation and downstream releases at certain times due to consumptive water loss. Using numerical models to predict the extent and location of the thermal plume in Clinton Lake would require collecting velocity data. In Section 6.3.1.2 of the ER, Exelon has committed to collecting monthly velocity data prior to COL application. No changes were made to the EIS regarding velocity measurements.*

Regarding the lack of site-specific groundwater chemistry data, the DEIS was in error and Section 2.6.3.2 of the EIS has been revised.

E.2.10 Comments Concerning Groundwater Use and Quality

Comment: Section 4.3.1, Page 4-6, Line 10. The second sentence indicating that "the dewatering system would possibly change the available capacity of local wells." This sentence is not entirely accurate. ER Section 4.2.2.3, indicates that based on the existing information, the closest shallow residential well (30-foot deep) is located approximately 0.73 miles southwest of the CPS. Potential construction-related impacts to this well, if any, will be dependent on the final embedment depth and the continuity of the more permeable zones within the shallow glacial till. The distance and generally low permeability of the shallow glacial materials will help to minimize impacts to the shallow wells. (141-55)

Response: *In adopting the bounding philosophy of the plant parameters envelope (PPE) approach, the staff considered the impact to the local groundwater surface elevation based on dewatering of the maximum footprint to the maximum embedment depth specified in the PPE. Additionally, the staff assumed in this EIS that in the future, new wells could be placed outside the existing plant property boundaries closer to the area that would be impacted by dewatering. No change was made to the EIS as a result of the comment.*

Comment: Section 5.3.2, Page 5-6, Line 12. "Based on groundwater elevation measurements, the only time Clinton Lake would be expected to recharge the adjacent aquifer would be after the lake was refilled following an extended period of very low lake elevations." Based on the measured water levels and gradients and the occurrence of the springs, the North Fork of Salt Creek and Salt Creek have been and, as part of Clinton Lake, continue to be, the discharge zone for shallow groundwater. Therefore, it is unclear why the Clinton Lake would need to recharge the aquifer if there was an extended period of very low lake elevations. (141-67)

Response: The staff agrees that groundwater would generally discharge to Clinton Lake regardless of the prior elevation of the lake. The staff's use of the term "aquifer" in Section 5.3.2 of the DEIS likely overstates the regional extent of the subsurface that would respond to increasing lake levels after an extended period of low water levels. The text in Section 5.3.2 of the EIS was revised to clarify that the staff only expects recharge limited to the soils of the bank adjacent to the lake.

Comment: Section 2.4, Page 2-16, Line 31. Groundwater aquifers are described in Section 2.3.1.2 of the ER. Groundwater aquifers are described in Section 2.3.1.3 of the ER and not 2.3.1.2. (141-15)

Comment: Section 2.6.1.2, Page 2-19, Line 25. Groundwater aquifers are described in Section 2.3.1.2 of the ER. Groundwater aquifers are described in Section 2.3.1.3 of the ER and not 2.3.1.2. (141-20)

Comment: Section 2.6.1.3, Page 2-20, Lines 12-13. "Exelon proposes to augment its groundwater and aquifer characterization program...related to the CPS Operating License,..." Exelon did not conduct the investigation programs prior to the construction of the CPS unit or related to the CPS Operating License. Item 1 should be revised to replace "its" with "the" so the sentence reads "augment the groundwater and aquifer characterization program". Similarly, Item 2 should be revised from, "continue its ongoing groundwater monitoring program related to the CPS Operating License" to read, "design and implement a groundwater monitoring program that will be conducted prior to construction activities." (141-22)

Comment: Section 2.6.3.2, Page 2-22, Line 24. "...there are no site-specific data available for the chemistry of groundwater underlying the ESP site." This sentence is not accurate. Glacial drift groundwater chemistry data from selected site piezometers collected as part of the CPS investigations are presented in Table 2.3-20 of the ER. (141-28)

Response: The EIS was revised to reflect the four preceding comments.

E.2.11 Comments Concerning Aquatic Ecology

Comment: Section 5.4.2.2, Page 5-19, Lines 23-26. The average lake temperature, determined by monitoring during the CPS pre-operational period (1985 and 1986), was 13.3C (55.9F) (IPC 1992). The average lake temperature monitored over 5 years after CPS operation (1987 through 1991) was 21.1C (70.0F) (IPC 1992). Thus, the CPS has increased lake temperatures approximately 7.8C (14F) over pre-operational conditions (IPC 1992). Although the average temperatures presented are correct, the information presented may be overstated. Section 8 of "Environmental Monitoring Program Water Quality Report 1978-1991" also states, "the greater average temperature was partially due to a change in the sampling schedule. During the operational period, temperatures were not determined during some of the winter months" (see page 20). (141-78)

Response: Although the average lake temperature increase due to the CPS unit reported by IPC (1992) may be conservative, values serve to illustrate the range of temperature increase due to operation of the unit. These values are also useful for understanding how the new ESP unit will influence the site. Therefore, the paragraph will remain in the EIS; however the EIS has been modified to clarify these points for the reader.

Comment: It is also unacceptable that the new reactor's effect on lake temperature remains undetermined; temperature has a direct impact on water levels, enjoyment of the lake for recreational purposes, and its acceptability as habitat for various animal species. This should be rectified before granting the ESP. (112-10)(113-10)(114-10)(115-10)(116-10)(117-10) (118-10)(119-10)(120-10)(121-10)(122-10)(123-10)(124-10)(125-10)(126-10)(127-10)(128-10) (129-10)(130-10)(131-10)(132-10)(133-10)(134-10)(135-10)(136-10)(137-10)(138-10)(139-10) (140-10)(142-10)(143-10)(144-10)(145-10)(146-10)(147-10)(149-10)(154-10)(155-10)(158-10) (159-10)(162-10)(163-10)(164-10)(165-10)(166-10)(167-10)(173-10)(174-10)(175-10)(176-10) (177-10)(178-10)(180-10)(181-10)(182-10)(185-10)(186-10)(187-10)(188-10)(189-10)(190-10) (192-10)(193-10)(194-10)

Response: Because a specific design for a new nuclear unit has not been selected, a reliable estimate of increased water temperature in Clinton Lake is not available at this time. However, the current EIS discusses wet tower operation in Section 3.2.1.1. The estimated cooling tower blowdown during normal plant operation is 760 L/s (12,000 gpm). By comparison, discharge from the CPS unit at 100-percent load is approximately 38,950 L/s (615,000 gpm or 1373 cfs) (Edinger 1989). In other words, if both the CPS and ESP units are operating, the percent of discharge originating from the ESP unit is expected to be less than 2 percent of the total discharge passing through the discharge canal. Exelon's RAI ID R3-26 to NRC RAI No. E5.2-3 (Exelon 2004) confirms that the "blowdown discharge rates are relatively small (1 to 3 percent of existing CPS discharge)." NRC staff, therefore, feels that the incremental increase in temperature caused by the ESP unit would be small compared to those impacts caused by the CPS unit.

Exelon has made a commitment to collect sufficient data to calibrate a multidimensional numerical thermal plume model before a construction permit (CP) or combined operating license (COL) application would be submitted. Further analysis of potential impacts from thermal discharge would be conducted at that time. If a new nuclear unit were constructed, water discharge from the new nuclear unit would be required to meet thermal discharge limits as set by the Illinois Environmental Protection Agency in an NPDES permit. These limits would be specific to Clinton Lake and Salt Creek and would take into account potential impacts to water levels, recreational use of the lake, and the ability of the lake and creek to maintain a balanced aquatic ecosystem.

These comments did not provide new information relevant to this EIS and will not be evaluated further. No change was made to the EIS as a result of these comments.

Comment: 5-21 lines 1-5 show that because of this increase in temperature, the dissolved oxygen (DO) in the lake has gone down from 10.2 mg/L to 7.8 mg/L. Further, they state that 5.0 mg/L of DO is necessary for a healthy aquatic community. In other words, the oxygen content of the lake has gone down 23% and if it goes down 27% more of its pre-CPS level, then the aquatic life will be seriously impacted. Is this going to happen? (153-8)

Response: *Dissolved oxygen (DO) levels are affected by many variables, including water temperature. The decrease in average DO in Clinton Lake should not be entirely attributed to operation of the CPS, though this is the most conservative method for evaluating impacts from plant operation. Other factors that affect DO levels include air temperature, water volume and water flow through the system, the types and number of plants present in and around the lake, the amount of suspended solids in the water column, the amount and type of nutrients present, and the influx of groundwater into the system.*

Average DO levels in Clinton Lake should not drop to below 5 mg/L as a result of construction and operation of a new nuclear unit. The proposed new nuclear unit is expected to have a cooling-tower-based heat dissipation system, which discharges significantly less water than the existing CPS once-through cooling system. Nationwide, experience with similar systems has indicated that low DO in the discharge has not been a concern at operating nuclear power plants with cooling towers or cooling ponds (NRC 1996).

Even during periods when some regions of the lake experience low DO, other regions of the lake will have DO levels sufficient to support aquatic life. Most fish and other aquatic organisms can recover from short periods of low DO availability, and many can move from areas of low DO to areas of suitable DO.

The comment did not provide new information relevant to this EIS and will not be evaluated further. No change was made to the EIS as a result of the comment.

Comment: Would the phenomena of impingement and entrainment-described in § 5.4.2.1 of the EIS-be amplified by the addition of a new nuclear unit at the CPS? How would the EPA regulations referenced (but not described) as mitigation measures effectively reduce aquatic life mortality? How can this very significant environmental impact be judged in the absence of a specific cooling water intake design selected by Exelon (EIS, pg. 5-17)? Clearly this is an important, environmental effect, as evidenced by the study conducted in 1987-1988 at the CPS, during which it is estimated that over 43 million gizzard shad fish where killed from impingement (EIS, pg. 5-18). (150-17)(151-17)

Response: *Rates of impingement and entrainment are expected to increase slightly with addition of a new nuclear unit. Because the proposed new unit would have a cooling-tower-based heat dissipation system, it would withdraw significantly less water than the existing CPS once-through cooling system. Nationwide, experience with similar operating cooling-tower-based systems has indicated that "the relatively small volumes of makeup and blowdown water needed for closed-cycle cooling systems result in concomitantly low entrainment, impingement, and discharge effects" (NRC 1996). Studies of intake effects of closed-cycle cooling systems have generally judged the impacts to be insignificant (NRC 1996).*

However, a complete review of impingement and entrainment impacts to important aquatic species cannot be performed without a specific cooling water intake design. EPA's Phase I regulations on intake design and operation implemented by IEPA will assure adequate protection of fish and shellfish in the reservoir. No change was made to the EIS as a result of this comment.

Comment: Section 2.7.2.3. "Exelon proposes to reinstate a fisheries monitoring program based on the one established in support of the 1973 CPS ER for the CP stage." Fisheries monitoring, to the extent required pursuant to the Clean Water Act 316 regulations will be followed when developing the program. (141-33)

Response: *The environmental report (ER) seems to indicate in Section 6.5.2.1 that a monitoring program similar to that established in support of the CPS ER will be continued, with the addition of new locations within Clinton Lake, "associated with the proposed intake structure and discharge from the EGC ESP Facility to evaluate effects on fishery resources during operation."*

No change was made to the EIS as a result of the comment.

Comment: The lake itself has been placed on the IEPA's list of impaired waters and even received a violation due to temperature increases. Fish kills have happened repeatedly, one of the most recent during a routine shutdown of the current plant. (157-7)

Response: The IEPA, in 2004, listed Lake Clinton as fully supporting aquatic life and fish consumption (IEPA 2004). The listed impairments were related to primary and secondary (recreational) contact and were attributed to state-wide impairments related to metals and algal growth.

A discussion of fish kills in Clinton Lake is included in Section 5.4.2.2 of the EIS. The regulatory agencies responsible for maintaining the health of the aquatic ecosystem must consider the maintenance of a balanced aquatic ecosystem at the local scale, but must also consider the impacts within the context of a regional scale.

The EIS does not evaluate the potential impacts associated with noncompliance with regulations. The comment did not provide new information relevant to the EIS and will not be evaluated further. No change was made to the EIS as a result of the comment.

Comment: Section 5.4.2.2, Page 5-21, Lines 33-35. Statement regarding aquatic impacts. The third sentence states, "They currently range between 1.1 and 4.4C (2 and 8F) higher than those at the Rowell gauging station located 19.3 km (12 mi) downstream of the Clinton Dam (Exelon 2003b)." It should be noted that the difference is only based on measurements in the months of June, July and August (see ER Section 5.2.1.1.3). (141-79)

Response: The text of Section 5.4.2.2 has been revised to state, "Summer stream temperatures currently range between 1.1 and 4.4°C (2 and 8°F) higher than those at the Rowell gauging station located 19.3 km (12 mi) downstream of the Clinton Dam (Exelon 2003b)."

Comment: How will the addition of a new nuclear unit to the CPS, with great consumptive water use and potential thermal impacts (EIS, pg. 3-7), affect the health of the various species of fish that populate Clinton Lake, such as the striped bass, as well as threatened species such as the slippershell mussel and spike that may be present in the vicinity of the CPS (EIS, pg. 2-32, 2- 35)? How would an investigation of the hydrodynamics of the lake-something currently lacking from Exelon's environmental report for the Clinton ESP (§ 2.6.1.3) aid in knowledge of such effects? Is it possible that the effects of "cold shock" recorded instances of which occurred in 2001 and 2004, when a wintertime plant shutdown and loss of heated liquid discharge kills fish that have congregated in the warmer water (EIS, pg. 5-22) could be exacerbated by the addition of a new reactor unit at the CPS if all reactor units must shut down simultaneously? (150-16)(151-16)

Response: While cooling towers have been suggested as mitigative measures to reduce known or predicted entrainment and impingement losses, they do evaporate cooling water, making some of the water drawn from the water body unavailable downstream resulting in "consumptive loss." Aquatic species found in Clinton Lake or in the vicinity are not likely to be impacted by the relatively small amount of water consumption from a new nuclear unit. Exelon *has committed to contact the Illinois Department Natural Resources before commencement of any construction and/or operations activities to make sure that the assumptions made about important aquatic species status and locations that led to this conclusion remain valid. Heated effluent discharge of a new closed-cycle nuclear unit combined with that from the CPS would slightly increase the localized area of warm water surrounding the discharge and, therefore, would slightly increase the potential for fish to be exposed to rapidly dropping water temperatures should the CPS and new nuclear unit cease operation suddenly and simultaneously. However, the number of fish lost in such an event would likely remain small in relation to the total abundance of the species within Clinton Lake and throughout the surrounding region (see also Section 5.4.2.2). Exelon has expressed a goal of maintaining the combined CPS and new unit discharge flows and temperatures within the conditions of the current NPDES permit for the CPS (Exelon 2006b, IEPA 2000). These conditions are considered adequate to protect a balanced, indigenous population of fish, shellfish, and other aquatic organisms in the lake.*

No changes were made to the EIS as a result of the comment.

E.2.12 Comments Concerning Terrestrial Ecology

Comment: From an environmental point of view, I can say that if the best fishing in central Illinois and a deer population of over 500 in a 2-mile radius of the power plant is an indication of good environmental health, than bring on unit 2. We have a beautiful area to live and the power plant has been a good neighbor. (27-5)(97-5)

Response: The comments are noted. No change was made to the text as a result of these comments.

Comment: Section 4.4.1.1. "However, the locations of associated equipment laydown and fill disposal areas and the conduit for the new intake are currently unknown and could, thus, impact wetland and forest habitat, depending on their ultimate locations. Nevertheless, Exelon would site these so as to preclude impacts to these wetlands". The proposed power plant will not directly affect any forested areas or wetlands. The proposed new intake structure will affect an area of "Waters of the United States". The proposed transmission line has potential to affect small areas of forest and wetlands. These impacts will be avoided and/or minimized to the greatest extent practicable. (141-56)

Response: The text of Section 4.4.1.1 has been revised to reflect that although the locations of associated equipment laydown and fill disposal areas and the conduit for the new intake are currently unknown, they would not be anticipated to adversely affect wetlands and associated forest habitat onsite.

Comment: Section 4.4.1.1, Page 4-9, Line 38. It is stated that transmission system construction techniques would be determined during the CP/COL phase. It would be more accurate to state that the transmission system construction techniques would be determined before or during the CP/COL phase. (141-57)

Response: *The staff agrees with the comment. The text of Section 4.4.1.1 has been revised to reflect that the transmission system construction techniques would be determined before or during the submittal of an application for a CP or COL.*

Comment: Section 4.4.1.1, Page 4-13, Line 24-25. It is stated that the staff will conduct its own review of transmission line construction impacts at CP/COL. If routing of the transmission system for the ESP is different than evaluated at the ESP, then the staff would review the construction impacts of the different routing. (141-58)

Response: *The nature of any transmission system upgrades and associated impacts to terrestrial ecosystems is currently considered unresolved at the ESP stage for reasons presented in Section 4.4.1.1. The definitive nature of transmission system upgrades and the magnitude of associated impacts to terrestrial ecosystems would be evaluated by the transmission and distribution system owner and operator under the regulatory process described in Section 4.4.1.1 prior to or during the CP or COL phase. The NRC would disclose the results of this evaluation in future environmental documentation in response to submittal of an application for a CP or COL. No change was made to the EIS as a result of this comment.*

Comment: Section 4.4.3, Page 4-26, Line 35-37. It is stated that Exelon would determine suitability of habitat for Indiana bat. The transmission system operator, through the course of obtaining permits for any construction activities, would determine suitability of habitat for Indiana bat, not Exelon. (141-59)

Response: *The staff agrees with the comment. The text in Section 4.4.3 was changed to reflect that the transmission distribution owner and operator will determine the suitability of the Indiana bat habitat within areas that will be disturbed for transmission line improvements, corridor widening, or new corridor routing (if needed).*

Comment: About three-and-a-half acres of forest habitat would be cleared for the construction of a new nuclear unit at the CPS, but their loss is considered "negligible" (ETIS, pg. 4-7). Also, construction of electric transmission lines to serve the new generating capacity at the CPS may require the clearing of up to 74 acres of forest and may destroy habitat for the endangered Indiana Bat (EIS, § 4-16), but this impact is considered "minor" (EIS, pg. 4-10). Such impacts deserve more evaluation in the final EIS. (150-26)(151-26)

Response: *The text of Section 4.4.3 was revised to reflect that loss of the 1.4 ha (3.5 ac) of forest habitat onsite would be considered minor, contingent upon the applicant taking the* *recommended actions described in that section. The nature of any transmission system upgrades and associated impacts to terrestrial ecosystems is currently considered unresolved at the ESP stage for reasons presented in Section 4.4.1.1. The definitive nature of transmission system upgrades and the magnitude of associated impacts to terrestrial ecosystems, including the Indiana bat, would be evaluated by the transmission and distribution system owner and operator under the regulatory process described in Section 4.4.1.1 prior to or during the CP or COL phase. The NRC would disclose the results of this evaluation in future environmental documentation in response to submittal of an application for a CP or COL.*

Comment: Environmental Consequences of Proposed Action, Section 4.1.1.1, Habitat, page 4-10, paragraph 3. Clarification needs to be provided on the rationale regarding the methodology that will be used to minimize the potential wetlands degradation in the transmission line corridors. (172-19)

Response: *The following has been added after the last paragraph in Section 4.1.1.1:*

Before issuing a construction permit, the NRC would ensure that an applicant referencing the Clinton ESP in an application for a CP or COL would obtain an ACE Section 404 permit that would address such areas as wetland filling, vegetation clearing, and hydrological alterations, etc. The ACE's permitting process ensures that impacts of construction are limited by requiring that the appropriate construction best management and mitigation practices be followed. Future environmental documentation would provide sufficient information about the wetlands to support a detailed description of potential construction impacts and best management practices and mitigation that would limit impacts.

Comment: Station Operation Impacts, Section 5.4.1.6, Impacts of Electromagnetic Fields on Flora and Fauna (plants, agricultural crops, honeybees, wildlife, livestock), page 5-13. Clarification on whether or not more recent studies were included prior to evaluation of the GEIS results need to be made. (172-26)

Response: *The following has been added to Section 5.4.1.6 to show that studies that followed publication of the GEIS (NRC 1996) were utilized in the evaluation:*

Since 1997, over a dozen studies have been published that looked at cancer in animals that were exposed to power-frequency for all of, or most of, their lives. These studies have found no evidence that power-frequency fields cause any specific types of cancer in rats or mice.

Comment: The level of wetland information provided in the DEIS is insufficient. There is no wetland delineation or functions and values information provided, nor a detailed description of the wetland impacts caused by the proposed project. The EIS should include temporary and permanent impacts, such as wetland filling, vegetation clearing and hydrological alterations.

Future environmental documentation should include this information, as well as a comprehensive mitigation strategy. The USNRC should consult with the U.S. Army Corps of Engineers to ensure compliance with Section 404 of the Clean Water Act. (172-4)

Response: *The text in Section 4.4.1.1 has been revised to state that the current level of wetland information is insufficient to support a detailed description of construction impacts. This text has also been revised to state that an applicant referencing the Clinton ESP in an application for a CP or COL would obtain an ACE Section 404 permit that would address such areas as wetland filling, vegetation clearing, and hydrological alterations, etc. The NRC would disclose in future environmental documentation related to an application for a CP or COL the provisions of this permit that would include such temporary and permanent wetland impacts. The NRC would not consult with the ACE to ensure an applicant's compliance with the provisions of the Section 404 permit, rather the ACE would ensure the applicant's compliance with its permit.*

E.2.13 Comments Concerning Threatened or Endangered Species

Comment: The DEIS adequately discusses potential impacts of the project alternatives on fish and wildlife resources, as well as species protected by the Endangered Species Act. The greatest potential for impacts is associated with the possible need for modifications to transmission line rights-of-way, with a maximum loss of no more than 74 acres of forested habitat expected. These potential impacts will be addressed further in the construction permit application stage. Exelon has also agreed to contact the FWS before beginning any construction activities to ascertain whether previous determinations regarding threatened and endangered species remain valid or whether further evaluation would be needed. The Department appreciates this commitment. (30-1)

Response: *The comment is noted. The comment does not provide new information and will not be evaluated further. No change was made to the EIS as a result of the comment.*

Comment: We are concerned about project impacts to the Indiana Bat, a federally-listed endangered species. Construction in the expanded transmission lines rights-of-way could impact these bats and their habitat. The DEIS does acknowledge that forest stands in the study area should be evaluated for suitable Indiana Bat habitat, and that the project should undergo a Section 7 consultation if suitable habitat is found. However, USNRC places the responsibility for these activities on Exelon Generation Company, LLC (Exelon). As the lead federal agency for this project, USNRC must take a proactive role in mitigating impacts to the Indiana Bat. (172-6)

Response: *The staff agrees that potential impacts to the Indiana bat onsite and along the existing transmission line corridor are described in Section 4.4.3 of this current EIS. The last sentence of the third paragraph in Section 4.4.3 states that if forest habitat is found by Exelon to be suitable for and occupied by Indiana bats, that the NRC expects Exelon to undertake the FWS consultation (implying a consultation under Section 7 of the Endangered Species Act). This was corrected to state that NRC would undertake the Section 7 FWS consultation.*

Comment: Environmental Consequences of Proposed Action, Section 4.1.2, Transmission Line Rights-of-Way and Offsite Areas, page 4-3. See Comment 11 above. Potential takings issues could lead to litigation that would make this a moderate impact instead of small impact. (172-18)

Response: Section 4.1.2 concerns land use. The staff has addressed issues relating to Federally threatened and endangered species in Section 4.4.3. The only species that could potentially be "taken" would be the Indiana bat. The staff agrees that take of the species could potentially result in a MODERATE or LARGE impact. Thus, the species-specific summary in Section 4.4.3 was revised to state that because there are no known occurrences of the Indiana bat within 16 km (10 mi) of the ESP site, potential impacts to the species would be considered negligible. The summary statement of Section 4.4.3 was revised to state that the conclusion of SMALL impacts by the NRC staff is predicated on certain assumptions made by the staff. These include the current occurrence of Federally-listed threatened and endangered species and critical habitat in the project area, the current listing status of such species, and the current designation of critical habitat.

Comment: Environmental Consequences of Proposed Action, Section 4.4.1.3, State-Listed Species, pages 4-12, 4-13. Demonstrations of small impact are not provided to address this issue. Assertions are made, but facts or demonstrations are not provided to support the assertions. (172-20)

Response: Section 4.4.1.1 has been revised to include discussion of State-listed threatened or endangered species under the wildlife evaluation. It is reasonable to assume that the State-listed birds that have been sighted but are not known to rest in the area would be minimally impacted, if at all, by construction. Impacts to State-listed species are not called out as a separate issue, but are considered only as a part of overall impacts to wildlife.

E.2.14 Comments Concerning Socioeconomics

Comment: Section 2.8.2.2. Exelon is listed as the entity paying taxes from 1996 through 2002. Prior to 2000, Illinois Power owned and operated CPS. Therefore, Illinois Power paid taxes to the taxing entities. After the sale of CPS in 2000 to AmerGen Energy Company, LLC, AmerGen paid taxes to the taxing entities. (141-37)

Response: Section 2.8.2.2 was changed to reflect the different corporate entities and the fact AmerGen is a subsidiary of the utility holding company Exelon, the ESP applicant.

Comment: Section 2.8.2.7, Page 2-61, Line 1. Exelon is listed as the entity paying taxes. Prior to 2000, Illinois Power owned and operated CPS. Therefore, Illinois Power paid taxes to the taxing entities. After the sale of CPS in 2000 to AmerGen Energy Company, LLC, AmerGen paid taxes to the taxing entities. (141-39)

Response: *Section 2.8.2.7 was changed to reflect the different corporate entities and the fact AmerGen is a subsidiary of the utility holding company Exelon, the ESP applicant.*

Comment: Section 4.5.3.1, Page 4-24, Line 12. Table 4-1, Page 4-46, Line 5. Section 5.5.3.2, Page 5-33, Line 3. "the [positive] impacts of construction on the economy of the region would be beneficial and SMALL everywhere in the region except DeWitt County, where the impacts could be MODERATE, and that mitigation would not be warranted." It is more accurate to describe the impacts in that they would be "beneficial" and MODERATE. (141-60)

Response: *Section 4.5.3 states that the economic impacts are beneficial and SMALL. Additional text was added to the effect that the impacts would be "beneficially" MODERATE in DeWitt county.*

Comment: Section 4.5.3.5, Pages 4-30 & 4-31, Lines 29-35 & 1-7. Two sections discuss the potential shortage of housing in the region and the associated upward pressure on rent costs. ER-Section 4.4.2.4 Housing Information, 2nd & 3rd para. – This section of the ER discusses that no families or households will be displaced as a result of rising rent costs due to an abundance of existing vacancies in the area. This is a contradiction to the statements in the DEIS. A reference should be provided as substantiation of the staff's position of this potential for housing shortage. (141-62)

Response: *Section 4.5.3.5 of the DEIS (pages 4-28 to 4-30) discusses in great detail the potential housing impacts and generally support the conclusions of the ER based on current, available information and supports the conclusion of a SMALL impact. However, it is the purpose of the DEIS to bound the potential impacts. Looking into the future 20-plus years and trying to predict what might happen is difficult. Thus, the DEIS also analyzed potential impacts if the assumptions made in the ER do not hold and a large number of construction workers decided to live in DeWitt, Logan and Piatt counties, where there is a current shortage of rental housing. Should this occur, one could expect MODERATE impacts in these three counties. No change was made to the EIS as a result of the comment.*

Comment: Amergen has been a good corporate citizen to Clinton and DeWitt County. Tourism provided by the lake and the marina has become a large industry locally. Donations made by the company and its employees to local charities and organizations have been substantial. Work done by the power plant employees with local churches and organizations has been invaluable. The power plant has provided a good place for local people to work. It has also brought in employees that have now settled in Clinton and call Clinton home. Some of the employees of the plant are the finest you would ever have the privilege to meet. In addition to jobs created, the additional tax base the second reactor would bring to local governments would be a huge shot in the arm. The power plant property used to pay about 86% of the tax dollars received by the Clinton School District. With the change in the assessment of the plant, the last year of the agreement, the taxes paid will be a small fraction of that percentage. Richland Community College in Decatur, The Warner Library District, the County of DeWitt and other taxing bodies will receive substantial benefits. The end result of this is that the individual taxpayer will have to pay a smaller share of the pie. More importantly, our children will receive more educational opportunities from the resulting income to the schools. (106-8)

Response: *Sections 4.5.3 and 5.5.3 discuss the social, economic, and tax impacts to Clinton and the surrounding region from construction and operation, respectively. The proposed project would have SMALL to MODERATE (for construction) to SMALL to LARGE (for operation) beneficial economic impacts, depending on where in the region the impacted sites are located (e.g., Clinton and DeWitt County would have more beneficial impacts). The construction and operation of a new nuclear facility at the Clinton Power Station (CPS) site will add to the tax base of DeWitt County and other government jurisdictions receiving property tax revenues from the proposed facility. These impacts would be beneficially SMALL to LARGE, depending on the jurisdiction, with Clinton, DeWitt County and the Warner Library District being among the most beneficially impacted. No change was made to the EIS as a result of the comment.*

Comment: The construction of another reactor would result in many jobs for the construction unions. The area communities would benefit from the travel and relocation of the construction workers. Clinton, which has had its share of bad luck economically the past few years with plant closings, would be a major benefactor. After construction, additional employment would be needed by Amergen. This project would also continue the life of the plant for another significant span of time. (106-2)

Response: *Sections 4.5.3 and 5.5.3 discuss the social, economic, and tax impacts to Clinton and the surrounding region from construction and operation, respectively. The proposed project would have SMALL to MODERATE (for construction) to SMALL to LARGE (for operation) beneficial economic impacts, depending on where in the region the impacted sites are located (e.g., Clinton and DeWitt County). No change was made to the EIS as a result of the comment.*

Comment: Section 5.5.3.2, Page 5-35, Lines 5-15. Discussion of the potential shortage of housing in the region and the associated upward pressure on rent/house prices if new housing were to have to be constructed to house the construction workers. ER-Section 4.4.2.4 Housing Information, 2nd & 3rd para. – This section of the ER discusses that no families or households will be displaced as a result of rising rent costs due to an abundance of existing vacancies in the area. This is a contradiction to the statements in the DEIS. A reference should be provided to substantiate the staff's claim as to this potential for housing shortage. (141-80)

Response: See discussion under Sections 2.8.2.1 and 2.8.2.5 and Tables 2-15 and 2-16 for discussion and references supporting the fact there could be a housing shortage in DeWitt County (and possibly Piatt and Logan Counties) if certain assumptions do not hold. Text was inserted in Section 5.5.3.5 to make clear that housing impacts in the region in general would be SMALL, but could be SMALL to MODERATE in DeWitt County and possibly Logan and Piatt Counties, depending on where the operations workforce is located or might relocate.

Comment: Section 5.5.3.4, Page 5-37, Line 40. A statement is made here pertaining to reduce the units power or shutdown of CPS and the Exelon ESP units. It should be stated that there is a potential to reduce the power or shutdown of the CPS and/or Exelon ESP facility. (141-82)

Response: The change was made to the text in Section 5.5.3.4 of the revised EIS to reflect the point of the comment.

Comment: Page 4-24 section 4.5.3.2 Taxes, I would like to challenge the statement that no new property taxes would be paid during construction. During the construction of Unit 1 assessed value was increased as construction progressed. I would expect the same to happen for new construction unless waived by the local taxing bodies. (29-4)(42-3)(105-4)

Response: The commenters are correct. During construction of the Clinton Power Station there were property taxes collected during the construction phase. Section 4.5.3.2 of the EIS has been changed to reflect the correction.

Comment: Section 2.1, Page 2-1 Sentence 17: "DeWitt County, which had a population of approximately 17,000 in 2000." Unable to locate this data in ER or SSAR. (141-4)

Response: The official population for DeWitt County was 16,798 in 2000, based on the 2000 Census. The sentence is intended to indicate a general size in terms of population. There is no need to tie this number to something reported in the ER. No changes were made to the EIS as a result of the comment.

Comment: Section 7.6, Page 7-7, Lines 27-30. Conclusion that cumulative impacts on housing will be SMALL to MODERATE. This cumulative impact is based on earlier sections in the DEIS that discuss the potential for an upward trend in house/rent prices as a result of the influx of construction workers to the area. The ER contradicts these sections and therefore this statement in the DEIS is dependent on the validity of those earlier presumptions. There is insufficient information in the DEIS to support this conclusion. A reference should be provided as to this potential for housing shortage. (141-121)

Response: The purpose of the discussion of housing in the DEIS is to bound potential impacts. The DEIS presents detailed analysis (including references) to support the potential impact of MODERATE if the assumptions in the ER about the workforce (construction and operation) do *not hold. See earlier responses to this comment, particularly Sections 4.5.3.5 and 5.5.3.5 and, as supporting background, Sections 2.8.2.1 and 2.8.2.5. No change was made to the EIS as a result of the comment.*

Comment: Section 2.8.1, Page 2-40, Line 5. Total population in 2000 is listed as 764,366. ER-Section 2.5.1.2, population between 16 km and 80 km (10 mi and 50 mi), 1st para. Lists the population as 752,008. (141-34)

Response: *The commenter appears to be referencing Table 2-4 (p. 2-39) in the DEIS. The total population of 764,365 is correct as it contains the population living within the 0- to 16-km (0- to 10-mi) radius, in addition to the 16- to 80-km (10- to 50-mi) radius as found in the ESP. No change was made to the EIS as a result of the comment.*

Comment: Section 2.8.2.1, Page 2-47, Lines 11-21. Reference to information in Table 2-10 of the DEIS. Numbers in this table do not match those in Table 2.5-10 of the ER. The DEIS referenced BEA 2001; County and City Data Books, 1994a, 2000. The ER referenced USDOL 2002. (141-35)

Comment: Section 2.8.2.1, Page 2-47, Line 32. Table 2-10 Regional Employment Trends, 1990 and 2000. Numbers in this table do not match those in Table 2.5-10 of the ER. The DEIS referenced BEA 2001; County and City Data Books, 1994a, 2000. The ER referenced USDOL 2002. (141-36)

Response: *The purpose of writing the EIS is to independently verify what is in the ER from independent sources. This often means using more recent (in time) information than that referenced in the ER, which was written before the DEIS was prepared. Table 2.5-10 of the EIS references workers employed, while Table 2-10 includes workers employed full and part-time, which results in a higher number. As their source for unemployment statistics, the "County and City Data Books" use data from the U.S. Bureau of Labor Statistics (BLS). A link to the web site containing the data books is provided from the U.S. Census Web site. Since unemployment data was adjusted as a result of the 2000 Census, more recent data (July 2005) from the BLS website was used to update Tables 2-11 and 2-12.*

Comment: As the economic value of the plant declines in the region, what guarantee is there that a new nuclear unit-built to export electricity for profit-would be an economic benefit to the region? And is it not likely that the Clinton School District could be overstressed by the children of the 3150 construction workers-of whom may move to the area-required to build the CPS? A more thorough consideration of the place of Exelon and the CPS in DeWitt County, addressing these questions and investigating how the plant serves the community and how it may hurt it, should be included in the final EIS. (150-21)(151-21)

Response: The economic benefits to the local community of the proposed ESP facility would be SMALL to MODERATE (for the construction phase). Sections 4.5 and 5.5 of this EIS discuss such beneficial impacts in detail. In summary, the economic benefits to DeWitt County and other government jurisdictions receiving benefits from CPS (and by inference the ESP facility if it is built) include increased tax revenues (sales and use taxes and property taxes) from the plant itself and from the plant's workforce. Increased income taxes to the State government are based on taxable income from the plant and on the salaries/wages of the plant's employees. An increase in area employment resulting from those employed directly by the ESP facility and other jobs created by the economic multiplier effect (see Section 4.5.2 for a definition) results from expenditures associated with the facility. No change was made to the EIS as a result of these comments.

Comment: Page 4-20 section 4.5.1.3 Roads: I would like to comment that as the Harp Township Highway Commissioner, I would say that the local roads serving the power plant site are adequate and are able to handle the expected traffic. We had 14-foot wide gravel roads that served the area before construction of Unit 1. Many of these roads only had 20 cars per day prior to construction of Unit 1. During the construction the roads were upgraded to 20 fl. asphalt roadways that handled up to 700 cars per day. Currently over-weight loads are brought into the Clinton Power Plant on Harp Township roads because of weight-restricted bridges on Route 54. (105-3)

Comment: Most of the bridges leading to and from the plant either on Route 54 or 10 are posted with weight limits. This is of major concern when considering the nuclear waste that will have to transported out of DeWitt County at sometime in the future (unless our neighborhood is going to become a designated high-level nuclear waste dump). Both highway and railroad bridges are more likely than not going to be unable to support the waste casks weights...And township roads are not up to the task of handling heavy traffic weights or flows. The road which runs directly north of the entrance to the plant (for instance) is barely a two-lane road, and has many twists and turns. It definitely would not be a suitable alternate route for either construction equipment or waste removal as some parties have stated. This is a rural area, and the transportation system was built as such. This, in turn, creates a funnel effect both into and out of the plant. It is not a suitable situation for security in today's global climate. Exelon would be depending on federal troops to help "defend" the site, and this will cost taxpayers dearly. (157-4)

Comment: Page 4-20 section 4.5.1.3 Roads: I would like to comment that as the Harp Township Highway Commissioner, I would say that the local roads serving the power plant site are adequate and are able to handle the expected traffic. We had 14-foot wide gravel roads that served the area before construction of Unit 1. Many of these roads only had 20 cars per day prior to construction of Unit 1. During the construction the roads were upgraded to 20 fl.

asphalt roadways that handled up to 700 cars per day. Currently over-weight loads are brought into the Clinton Power Plant on Harp Township roads because of weight-restricted bridges on Route 54. (29-3)

Comment: Page 4-20, Section 4.5.1.3 under Roads, I'd like to comment that as the Harp Highway Commissioner, I would say that the local roads serving the power plant site are very adequate and able to handle any expected traffic. When the first power plant was built, many of these roads were 14 foot wide gravel roads. And many of them only served a dozen or so cars a day. During construction, I updated the roads. They're currently 20 foot asphalt roads. And in the past have been able to handle over 700 cars a day safely. (42-2)

Response: This EIS considers the impact of construction and operations activity on the local road and transportations system. During their site visit the week of March 1, 2004, NRC staff observed that most of the roadways within DeWitt, Logan, and Piatt Counties are rural, lightly traveled, and well maintained. Exelon stated in its ER that it would adhere to applicable local, State and Federal requirements regarding traffic control during construction. Therefore, the staff concluded that the impacts of congestion during construction would be SMALL.

In its ER, Exelon stated that none of the roads and highways near the proposed ESP Facility site at CPS would be physically impacted by construction of the new nuclear facility. Exelon provided no justification or reference in the ER for such a conclusion.

The major State highways in the area of the CPS (Routes 10 and 54) have maximum weight limitations set by the State of Illinois at 80,000 pounds. While there may be some weight limitations on some of the bridges on these routes, there are ways, using State, County, and Harp Township roads and overweight permits, of getting loads in excess of 80,000 pounds into the ESP facility site. For example, loads weighing in excess of 120,000 pounds have been brought into CPS in the past.

As to the future (potential) transport of spent nuclear fuel (SNF), the weight limitations (governed by Federal or State restrictions) are 73,000 pounds per truck or 100 tons per rail car. So it would appear that transport of SNF out of the CPS site would not be a problem. Rail is one alternative for transporting heavier loads (e.g., heavy equipment, construction materials, and SNF). However, the rail system leading to the site may need upgrading to accommodate such loads.

Therefore, NRC staff concluded that the physical impacts of heavy loads upon the roads could be SMALL to MODERATE. Some upgrading of roads and bridges may be required if the loads routinely exceed the maximum load restriction of 80,000 pounds (something that would not occur without State, DeWitt County, or Harp Township approvals). Changes were made to Section 4.5.1.3 of the EIS as a result of these comments, although the impacts (SMALL to MEDIUM) remain unchanged.

Comment: This could impact both boating (lower water levels) and fishing (lower water levels and elevated temperatures) at the lake. (112-9)(113-9)(114-9)(115-9)(116-9)(117-9)(118-9) (119-9)(120-9)(121-9)(122-9)(123-9)(124-9)(125-9)(126-9)(127-9)(128-9)(129-9)(130-9)(131-9) (132-9)(133-9)(134-9)(135-9)(136-9)(137-9)(138-9)(139-9)(140-9)(142-9)(143-9)(144-9)(145-9) (146-9)(147-9)(149-9)(154-9)(155-9)(158-9)(159-9)(162-9)(163-9)(164-9)(165-9)(166-9)(167-9) (173-9)(174-9)(175-9)(176-9)(177-9)(178-9)(180-9)(181-9)(182-9)(185-9)(186-9)(187-9)(188-9) (189-9)(190-9)(192-9)(193-9)(194-9)

Response: *The staff recognizes that both the CPS and a new nuclear unit could impact lake pool elevations and temperature, which in turn could impact boating and fishing. A drought severe enough to impact lake levels and water quality is a rare event. Mitigative actions might include cutting back on the unit's power production or shutting down one or both units. Therefore, based on this and other potential impacts of station operation, the staff concluded that potential impacts of station operation on recreation would be SMALL to MODERATE. Mitigation would be warranted only when a drought occurs and could be undertaken by changing the way in which the units are operated. No change was made to the EIS as a result of these comments.*

Comment: Preliminary projections from the 2000 census show somewhat different trends in several counties from those used in this report. Exelon has taken the 1990 projections and extrapolated population trends by a ratio method for 2030, 2040, 2050, and 2060; this method does not capture the dynamics of population growth. Since the projected populations are presented in the report by zones from the Clinton site, they cannot be checked for reasonableness for a county or township (U.S. Nuclear Regulatory Commission, 2005, 5-39). What the purpose of making projections to 2060, something only the naive would do, is unclear. They are not used to justify demand for electricity in the area surrounding the plant, nor do they consider any impact of additional labor force (which is small) at the proposed plant on the projected future population in the area. (110-2)

Response: *The EIS relied upon the Exelon ER for population projections within zones of 0 to 16 km (0 to 10 mi), 16 to 40 km (10 to 25 mi), 40 to 60 km (25 to 37 mi), and 60 to 80 km (37 to 50 mi) zones. Exelon used 2000 Census data and projected population data by obtained from Illinois State University. The projected populations within each zone in 10-year increments (starting with 2000) were presented for three reasons:*

- *To give a sense of projected population growth within the region (within a 80-km [50-mi] radius of the ESP site at Clinton Power Station), based on the latest information available*

- *To show the projected population changes over the potential licensed life of a new facility of 40 years, assuming construction takes place and the plant comes online by 2020*

- *To estimate the population in the vicinity and region for the Exelon's "Site Safety Analysis Report."*

The population projections for each of the potentially economically impacted counties in the region (Table 2-6 of the EIS) are based upon population projections prepared by the Illinois Department of Commerce and Economic Development, Office of Policy, Development, Planning, and Research.

The proposed plant at the ESP site is a type called a "merchant" generating facility, which means that it can sell generated power anywhere, not just in Illinois. It is not within the purview of an ESP EIS to justify the proposed plant in terms of demand for electricity. No change was made to the EIS as a result of the comment.

Comment: 5-35 lines 32-34 state that everyone, state of Illinois, DeWitt County and the city of Clinton, would get taxes from the CPS for at least 60 years. First, how do they know this? Page 2-53 above shows how the assessed value of CPS has gone down dramatically in less than 7 years. (153-10)

Response: *The paragraph in question states that personal and corporate income taxes would be paid to the State of Illinois. Sale and use taxes would beneficially impact the City of Clinton and property taxes would directly benefit DeWitt County and other taxing jurisdictions deriving tax revenue from CPS and, by inference, from the proposed ESP facility.*

Deregulation of electrical power generation in Illinois has resulted in a reduction in the amount of property taxes that generating facilities pay to local taxing jurisdictions. Such is the case with CPS. The deregulation legislation provided for a transition period for property tax assessment away from depreciated book value to value of the plant based on the market value of electricity generated by the plant or some other market-based approach. As such, the amount of property taxes the CPS facility pays has declined over the 7-year period referenced in the comment. While it is not certain that 60 years from now the plant would be paying property taxes, it is probably reasonable to assume that there would be some form of taxation of the plant for purposes of supporting local government jurisdictions.

The 60-year estimate of plant life, over which it was assumed property taxes would be paid, is based on the original license period for the plant (40 years) and potential license renewal for 20-years when the initial license expires. The discussion on deregulation in the EIS was updated to reflect the status as of January 2006 of negotiations on methods of valuing of the CPS for property tax purposes (see Section 2.8.2.2).

Comment: Without the hope of good jobs and adequate supplies of energy to heat and cool our homes and businesses, what kind of quality of life do we really have? (27-7)

Response: The salaries of jobs in the nuclear industry are above the average prevailing hourly wage for most jobs in the region of interest. Clinton Power Station is capable of generating nearly 1017 net megawatts and can produce enough power to support the electricity needs of about 1-million average American homes, providing them with energy to heat and cool these homes. The proposed facility could generate between 2400 and 6800 MW(t). No changes was made to the EIS as a result of the comment.

Comment: General Comments on Housing in Section 5, 9, and 10. "Based on the information provided by Exelon and the staff's independent review, the staff concludes that potential impacts of a new nuclear unit on housing would be SMALL to MODERATE in DeWitt County and potentially in Piatt and Logan Counties. Market forces, represented by increased housing demand, would result in more housing being built, which, over time, would mitigate any housing shortages." This conclusion appears doubtful, given the fact that the availability of housing in the region "could easily accommodate the expected workforce of 580 new employees." DEIS, p. 5-38. Therefore, this impact should be categorized as SMALL. (141-81)

Response: The summary paragraph (lines 8-17, p. 5-39 of the DEIS) was changed to more clearly state that if the operating workforce comes from within the region, housing impacts in the region would be SMALL. However, if this assumption does not hold and the preponderance of the operating workforce comes from outside the region, or decides to relocate to DeWitt County (and possibly Piatt and Logan Counties) to be nearer the ESP site of employment, then the impacts could be SMALL to MODERATE for these counties for the reasons stated in Section 5.5.3.5 (Housing).

Comment: Section 2.8.2.2, Page 2-53, Line 5. Pre-deregulation taxes are stated as being paid based on depreciated assessed value. Pre-deregulation taxes were based on depreciated book value not assessed value. (141-38)

Response: The text in Section 2.8.2.2 of the revised EIS was changed to reflect the comment.

Comment: Section 2.1, Page 2-5, Line 11. There were 972,616 visitors to the lake in 2000 (Exelon 2003a). Unable to find these elevation data in the ER or SSAR. A reference for this information or how this number was calculated should be provided. (141-8)

Response: The text in Section 2.1 was modified. The correct citation for the visitation number is IOC 2001, Fiscal Focus, May/June 2001, Illinois Office of the Comptroller, available online at: http://www.apps.ioc.state.il.us/ioc-pdf/FiscalFocusMayJun01.pdf.

Comment: Following September 11, 2001, the lake was shut down to all visitors. "There were 972,616 visitors to the lake in 2000 (Exelon 2003a)". It seems that there would be a significant economic problem should this have to done again. It would appear that it might have to be done if construction is ongoing at an active plant. I would hope that lake access would be tightly controlled to limit access by water. For the months that the lake was shutdown through 2002, DeWitt County and the surrounding communities claimed major economic impacts. (157-8)

Response: Clinton Lake was shut down after September 11, 2001, for security reasons. A major part of the lake was later reopened, but access to the CPS site from the lake is still prohibited. The prohibited area is larger than it was before 9-11. Unless there was a security incident, it is not likely that all of the lake would again be closed during construction or operation of the proposed new unit(s). There would be economic consequence to the local area should the lake completely shut down again. However, there are substitute water recreation opportunities similar to what can be found at Clinton Lake within an hour or so drive of Clinton (e.g., Lake Shelbyville south of Decatur or Lake Springfield near Springfield). No change was made to the EIS as a result of the comment.

Comment: And I'd just like to first start by thanking Exelon for all the stuff that they've given our communities; our schools, our equipment for our fire departments; our educational stuff for our children. They sponsor our ball teams. They sponsor all of those items. (57-1)

Response: Clinton Power Station (CPS) currently employees approximately 550 people for the operation of the plant. Some of these employees are actively engaged in community activities. The tax base provided by CPS has been used to construct new infrastructure in DeWitt County and Clinton and to support Clinton School District 15 and other jurisdictions through the property taxes collected on the facility. No change was made to the EIS as a result of the comment.

Comment: The Clinton Power Plant has provided a good job base and has provided a lovely lake that makes Clinton a tourism magnet for central Illinois. (27-6)

Response: Clinton Power Station currently employs approximately 550 people for the operation of the plant. With the addition of the proposed facility, that workforce is expected to increase to 580 employees. The construction labor force is expected to number approximately 3150 workers. The salaries and wages of these jobs are above the average, prevailing hourly wage for most jobs in the area. Clinton Lake is a recreational resource for DeWitt County and the surrounding area. No change was made to the EIS as a result of the comment.

Comment: We also need to be concerned about the long-term future of our industry and quality of life. The construction of another plant would provide 1000's of good paying construction jobs, 100's of skilled operation jobs, and countless other spin-off jobs. (28-3)

Response: Clinton Power Station currently employs approximately 550 people for the operation of the plant. With the addition of the proposed facility, that workforce is expected to increase to 580. The construction workforce of the proposed facility would require approximately 3150 workers at the height of construction activity. The salaries and wages of these jobs are above the average, prevailing hourly wage for most jobs in the area. The multiplier effect from construction and operations expenditures would serve to increase the economic well-being of the surrounding area. No change was made to the EIS as a result of the comment.

Comment: While I'm sure you recognize the significant economical benefits of Clinton station and the good corporate citizenship, I'm not so sure that you realize the positive environmental impact that Clinton has already provided. (66-1)

Response: Clinton Power Station currently employs approximately 550 people for the operation of the plant. With the addition of the proposed facility, that workforce is expected to increase to 580. The construction workforce of the proposed facility would require approximately 3150 workers at the height of construction activity. The salaries and wages of these jobs are above the average, prevailing hourly wage for most jobs in the area. The multiplier effect from construction and operations expenditures would serve to increase the economic well-being of the surrounding area. The tax base provided by Clinton Power Station has enabled DeWitt County to construct a courthouse and Clinton to construct a city hall, among other improvements. In addition, Clinton Lake, a major recreational asset to the community, was constructed as a source of cooling water for the power plant. No change was made as a result of the comment.

Comment: From the other aspect, as far as the impact on our community, economically it has been huge. We would not have the things that we have today, especially the infrastructure without that plant having been built here in our community, even down to the building that we're in here this evening. (43-3)

Comment: I've had the privilege and the honor of working with many of the emergent employees. And they have always been good neighbors and provided many much needed jobs for our area for over a generation. (45-1)

Comment: I'd like to say to the people of Clinton here, I understand that you think this is a good idea, that the risk is acceptable and this is a big part of your tax base. And if I was living here, that would be a much, much larger part of what I would focus on and I appreciate that. (47-1)

Response: Clinton Power Station currently employs approximately 550 people for the operation of the plant. With the addition of the proposed facility, that workforce is expected to increase to 580. The construction workforce of the proposed facility would require approximately 3150 workers at the height of construction activity. The salaries and wages of these jobs are above the average, prevailing hourly wage for most jobs in the area. The multiplier effect from construction and operations expenditures would serve to increase the economic well-being of the surrounding area. The tax base provided by Clinton Power Station has enabled DeWitt County to construct a courthouse and Clinton to construct a city hall, among other improvements. No change was made to the EIS as a result of these comments.

Comment: The power plant workers have been good neighbors and bring stability to our community. Many local leaders are employees of the Clinton Power Plant and have added stability to the community as many manufacturing jobs have left. (27-4)

Response: Clinton Power Station currently employs approximately 550 people for the operation of the plant. With the addition of the proposed facility, that workforce is expected to increase to 580. The salaries of these jobs are above the average, prevailing hourly wage for most jobs in the area. No change was made to the EIS as a result of the comment.

Comment: We also need to be concerned about the long-term future of our industry and quality of life. The construction of another plant would provide 1000's of good paying construction jobs, 100's of skilled operation jobs, and countless other spin-off jobs. (99-3)

Comment: The power plant workers have been good neighbors and bring stability to our community. Many local leaders are employees of the Clinton Power Plant and have added stability to the community as many manufacturing jobs have left. (97-4)

Response: Clinton Power Station has provided economic benefits and stability to the region. Clinton Power Station currently employs approximately 550 people for the operation of the plant. With the addition of the proposed facility, that workforce is expected to increase to 580. The construction workforce of the proposed facility would require approximately 3150 workers at the height of construction activity. The salaries and wages of these jobs are above the average, prevailing hourly wage for most jobs in the area. The multiplier effect from construction and operations expenditures, the taxes paid (income and sales and use taxes, in addition to property taxes) would serve to increase the economic well-being of the surrounding area. No change was made to the EIS as a result of these comments.

Comment: The Clinton Power Plant has provided a good job base and has provided a lovely lake that makes Clinton a tourism magnet for central Illinois. (97-6)

Response: Clinton Power Station has provided economic benefits to the region. Clinton Power Station currently employs approximately 550 people for the operation of the plant. With the addition of the proposed facility, that workforce is expected to increase to 580. The construction workforce of the proposed facility would require approximately 3150 workers at the height of construction activity. The salaries and wages of these jobs are above the average, prevailing hourly wage for most jobs in the area. The multiplier effect from construction and operations expenditures, the taxes paid (income and sales and use taxes, in addition to property taxes) would serve to increase the economic well-being of the surrounding area. Clinton Lake is a recreational resource for DeWitt County and the surrounding area. No change was made to the EIS as a result of the comment.

Comment: I believe that nuclear energy benefits the local communities because it does provide affordable power. It creates jobs. It contributes to local economies, and it reduces the dependence on natural resources controlled by foreign governments. (75-7)

Response: Clinton Power Station has provided economic benefits to the region. Clinton Power Station currently employs approximately 550 people for the operation of the plant. With the addition of the proposed facility, that workforce is expected to increase to 580. The construction workforce of the proposed facility would require approximately 3150 workers at the height of construction activity. The salaries and wages of these jobs are above the average, prevailing hourly wage for most jobs in the area. The multiplier effect from construction and operations expenditures, the taxes paid (income and sales and use taxes, in addition to property taxes) would serve to increase the economic well-being of the surrounding area. In addition, Clinton Power Station does afford a base of reliable power generation. No change was made to the EIS as a result of the comment.

Comment: I think this is needed for this town. And being in the business in this town, I support it. (82-4)

Comment: While it is true that internal affairs regarding public safety and publicity are in need of change within the nuclear industry; however, there is no argument for Exelon as a nuclear utility (more environmentally sound and scrutinized than any fossil fuel plants) gobbling up tax dollars and spitting it back out in to the community in the form of public programs and enhanced job creations. Keep up the good work. (91-1)

Comment: The locals who spoke tonight were all in favor of having a new plant because of the positive impact which this plant will have on the town's economic development. (93-4)

Response: Clinton Power Station has provided economic benefits to the region. Clinton Power Station currently employs approximately 550 people for the operation of the plant. With the addition of the proposed facility, that workforce is expected to increase to 580. The construction workforce of the proposed facility would require approximately 3150 workers at the height of construction activity. The salaries and wages of these jobs are above the average, prevailing hourly wage for most jobs in the area. The multiplier effect from construction and operations expenditures, the taxes paid (income and sales and use taxes, in addition to property taxes) would serve to increase the economic well-being of the surrounding area. No change was made to the EIS as a result of these comments.

Comment: Without the hope of good jobs and adequate supplies of energy to heat and cool our homes and businesses, what kind of quality of life do we really have? (97-7)

Response: Clinton Power Station is capable of generating nearly 1017 net megawatts and can produce enough power to support the electricity needs of about 1-million average American homes, providing them with energy to heat and cool these homes. The proposed facility could generate between 2400 and 6800 MW(t). No change was made to the EIS as a result of the comment.

Comment: The educational benefit of siting the power plant here, in Clinton, as opposed to any of the other places. Earlier this year, myself and several other students, from UIUC, came and toured the nuclear power plant. And we got to talk with the engineers, got to see the equipment in action. That was very valuable. (79-3)

Response: Clinton Power Station is located within approximately 60 km (40 mi) of the University of Illinois at Urbana-Champaign. Visits to the plant, to the extent they are allowed, would be educational and informative. No change was made to the EIS as a result of the comment.

Comment: 2-53 lines 26-35 shows the assessed value for CPS (Clinton Power Station) dropping dramatically from $558 million in 1996 to $165 million in 2003. This means that the tax base for this power plant has dropped to less than 30% of its value 7 years earlier and it shows no indication of stopping there. Will the new plant also end up quickly reducing tax revenue for DeWitt County? (153-5)

Comment: Taxpayers foot half the bill for license applications, yet Exelon is not a very good corporate citizen in return. Because it has taken advantage of new electricity deregulation rules, its property tax payments have declined from 80 percent of DeWitt County's total property tax revenue in 1996 to 53 percent in 2002. This resulted in an annual revenue loss of $8.8 million to the county; local officials report that their economy has "reached bottom" and Clinton School District 15 has been forced to cut its budget by $3 million and spend reserves over the past several years. (191-9)

Comment: I understand that the community does have a need. One speaker that came up here spoke about how important it was to get a nuclear plant. And I also understand that one of the reasons for this need is because you have an existing nuclear reactor that costs $4.4 billion in construction that is now valued at $100 million because Exelon has pushed and pushed for devaluation. I don't even think 78 Ford Pintos devalue quite that poorly even if it's not running.

But the current Clinton reactor is running. And it's running quite well in terms of reactors running. And I find it pretty offensive that it would be devalued like that I know two weeks ago this community voted down a school referendum. I certainly wouldn't dispute that. I understand that you feel that that wasn't necessary. But probably this wouldn't have happened at all had Exelon actually treated you fairly and not try to take money away when they still got the same reactor turning out just like it was. (32-2)

Comment: And also I want everyone to seriously think about the connection between Exelon sponsoring your schools, sports teams, between Exelon sponsoring your education system. Why does Exelon have to do this? Why can't you have public funding? Why aren't there public funds to do it? Well, because Exelon manipulated the tax base. Exelon over the past 30 years has created a dependency of people in Clinton on the corporation. And now we can't envision anything else, right? (35-6)

Comment: I'm also disgusted by Exelon's role as a, as a citizen, as a corporate citizen. They have not been a good corporate citizen. They, they cut and run on property taxes here. They undermine your property values. They're not paying their fair share. They're not, they're not paying their fair share in Zion either. We had to pass a referendum to pick up their responsibility. So now, the property tax, the property taxes, residential property taxes in Zion, Illinois, are paying what Exelon should be paying, their fair share. (85-3)

Response: *Deregulation of electrical power generators in Illinois has resulted in a reduction in property taxes that generating facilities pay to local taxing jurisdictions. Such is the case with CPS. The deregulation legislation provided for a transition period for property tax assessment, away from depreciated book value to the value of the plant based on the market value of electricity generated by the plant or some other market-based approach. As such, the amount of property taxes the CPS facility pays has declined over the 7-year period referenced in the comment. This has impacted local jurisdictions dependent upon CPS for some of their operating revenues. However, the addition of a new ESP facility would add to the local tax base and increase property tax revenue. Hence, from a tax revenue standpoint, the new proposed facility would pay taxes in addition to the taxes paid by CPS. The discussion on deregulation in the EIS was updated to reflect the current status (January 2006) of negotiations on methods of valuing the CPS for property tax purposes (see Section 2.8.2.2).*

Comment: So I looked at the projections to see how they were used. And I find many problems with them, particularly in how they extrapolate to 2026 for no reason or purpose that I can see. And I just hope the rest of the report is done better than what I see in the demographic parts of the report. (58-1)

Response: *If the Exelon ESP is approved, Exelon has 20 years from the date of approval to apply for and receive potential approval to begin constructing the new nuclear plant. The new plant, if approved for operations, may operate for up to 60 years (original license granting*

40 years and potential license renewal granting another 20 years of operation). The purpose of extrapolating demographic data to 2020 (Table 2-6 of the EIS) or 2060 (Table 2-4 of the EIS) is as follows:

- *To give a sense of projected population growth within the region (80-km (50-mi) radius from the ESP site at CPS), based on the latest information available*

- *To show the projected populations changes over the potential licensed life of a new facility of 40 years, assuming construction takes place and the plant comes online by 2020*

- *To estimate the population in the vicinity and region for the Exelon's Site Safety Analysis Report.*

These projections are undertaken in recognition of the fact that the future cannot be precisely predicted. No change was made to the EIS as a result of the comment.

Comment: 5-34 lines 23-24 state that up to 580 employees will be permanently employed at the CPS. Is this really true? The CPS was fully operational in 1987, but by 1996 the CPS was closed for two and a half years. Did those permanent employees stay in the area for two and a half years while it was non-operational? (153-9)

Response: *In 1996, CPS shut down for refueling outage and other maintenance. During that period, all operating personnel were retained and the total number of employees at the plant actually increased. During the 30-month shutdown, AmerGen purchased CPS. After the purchase was complete, AmerGen began a reorganization at CPS, which caused a reduction (to approximately 550) in the total number of operating employees at the plant. No change was made to the EIS as a result of the comment.*

Comment: Table 5.15, Page 5-81, Line 13. It is stated that the impacts on the economy would be SMALL to MODERATE. It should be noted that the impacts to the economy would be beneficial and SMALL to MODERATE. (141-93)

Response: *In Table 5-15 under "comments," for the economy it is stated that the impacts would be beneficial. No change was made to the EIS as a result of the comment.*

Comment: Table 5.15, Page 5-81, Line 14-15. It is stated that the impacts on the economy would be SMALL to LARGE. It should be noted that the impacts to the economy would be beneficial and SMALL to LARGE. (141-94)

Response: *In Table 5-15 under "comments," for taxes it is stated that the impacts would be generally beneficial. No change was made to the EIS as a result of the comment.*

Comment: Section 4.5.3.3, Page 4-27, Line 32. "Near the Exelon ESP site, 2500 cars and trucks and 1850 cars and trucks travel daily on Illinois..." ER-Section 4.4.2.8 Transportation Facilities, 2nd para. – "Near the EGC ESP Facility, 2750 cars and trucks and 2000 cars and trucks travel daily on IL Route 54 and 10, respectively (IDOT, 2003)." (141-61)

Response: In the DEIS, the actual paragraph in question appears on page 4-26, lines 36 and 37. In the EIS more recent data was used than in the ER: 2004 (DEIS) versus 2003 (ER). No change was made to the EIS as a result of the comment.

Comment: I have a question for the community. Do they really think that Exelon is providing a lot of financial donations, to here in the community for the community's benefit itself? Or does it have a corporate interest in making sure that people here in this community get money from this corporation? Clinton does get a lot of money from this company, Exelon. But a number of communities in the area outside it, which are also affected by the situation of the plant here, do not receive the donation. So, I'd just like the community to think about that. (72-1)

Response: In the socioeconomic impact analysis conducted on the proposed Exelon facility, NRC found that the economic benefits of facility construction and operation would result in positive economic benefits not only for the City of Clinton and DeWitt County, but also for surrounding counties and communities. The positive economic benefits would result from more and higher-paying jobs, a larger tax base, and the general multiplier effects resulting from facility and employee expenditures. No change was made to the EIS as a result of the comment.

Comment: Table 5-15, Page 5-82, Line 1. Housing ranked as a SMALL to MODERATE impact. There is insufficient information provided in this section supporting the impact conclusion. (141-95)

Response: Information supporting the impact levels is found in Section 5.5.3.5 (Housing). No change was made to the EIS as a result of the comment.

E.2.15 Comments Concerning Environmental Justice

Comment: I have a question about specifically in the environmental justice and how the NRC comes up with the definition of environmental justice considering that the People of Colors Caucus demands, that's the guiding document for environmental justice in the current environmental movement, demands an end to all toxic waste production, which has historically impacted people of color and the poor. So, just knowing what the definition of environmental justice is because we can all look at that document and read it, can you just clarify for me how the NRC can analyze environmental justice without green washing it? (35-1)

Comment: And the issue, I think, really goes back to that issue of environmental justice that I questioned the NRC about earlier. And I'd like to first request again that the NRC look at the People of Caucus Deceleration of Environmental Justice, which really is the guiding document for the environmental justice movement. And I think the people of Clinton are well aware of economic injustice and are affected by it. And economic and environmental injustice are interrelated. They are not mutually exclusive. And I think that the NRC, I think it's a fact that the NRC knows what environmental justice is and what the real definition of environmental justice is otherwise they wouldn't have taken it out of their list of environmental contentions in 2004. That's just absolutely inexcusable that the Nuclear Regulatory Commission would do that. So I would really like you guys to look back at that because environmental justice is important. (35-3)

Response: *Environmental justice refers to a Federal policy under which each executive agency identifies and addresses, as appropriate, disproportionately high and adverse impacts on human health or environmental effects of its programs, policies, and activities on minority or low-income populations. Executive Order 12898 (59 FR 7629) directs Federal executive agencies to consider environmental justice under the National Environmental Policy Act of 1969. The Council on Environmental Quality (CEQ) has provided guidance for addressing environmental justice (CEQ 1997). Although it is not subject to the Executive Order, the Commission has voluntarily committed to undertake environmental justice reviews. The staff uses as guidance the NRC Office of Nuclear Reactor Regulation Office Instruction Number LIC-203 ("Procedural Guidance for Preparing Environmental Assessments and Considering Environmental Issues," NRC ADAMS Accession No. ML0117100730). No change was made to the EIS as a result of these comments.*

Comment: By forcing an immature technology that cannot carry base load such as wind energy, that may be able to diversify our energy mix but not carry base load, will in fact, ultimately hurt a lot of the minorities or lower income groups that we have been talking about. Because those groups can't afford the higher energy prices that would cost. (87-3)

Response: *Wind has a high degree of intermittence, and average annual capacity factors for wind plants are relatively low (less than 30 percent). Wind power, in conjunction with energy storage mechanisms, might serve as a means of providing small amounts of base-load power. However, current energy storage technologies are too expensive for wind power to serve as a large base-load generator. If forced to go to wind-generating alternatives, the cost of power would rise relative to the costs of generating power from a baseload nuclear plant. This would impact all socioeconomic groups, potentially having more of an impact on low-income populations due to their limited incomes. No change was made to the EIS as a result of the comment.*

Comment: Exelon did not follow NRC guidance in assessing minority and low-income populations because of the presence of a single Native American person in a particular census block (EIS, pg. 2-67), and they "underemphasized" census block groups where the percentage of minority or low-income, populations was high-notably an area in Logan county that contains two prisons (EIS, pg. 2-68). To what extent were Exelon's evidently faulty evaluations relied upon by the NRC in its own consideration of environmental justice issues? (150-27)(151-27)

Response: The NRC conducted an independent environmental justice analysis (following NRC guidance on environmental justice) and did not rely on Exelon's analysis. No change was made to the EIS as a result of the comment.

E.2.16 Comments Concerning Cultural Resources

Comment: Section 2.9.2, Page 2-69, Lines 19-20. The DEIS discusses historic / archaeological sites and suggests the following, "Prior to construction, this area will need to be further investigated using appropriate methods such as tilling, surveying, and shovel-testing." ER-Section 2.5.3 Historic Properties, final paragraph provides discussion that archaeological testing of the area to be disturbed by the new construction is not necessary. However, Exelon, will follow the IL SHPO guidelines. (141-40)

Comment: Section 4.6, Pages 4-34 & 4-35, Lines 36-40 & 5-13. Discussion of the previously disturbed nature of the construction area and states that: "Therefore, archaeological testing of this area does not appear to be warranted." ER-Section 2.5.3 Historic Properties, final para., provides discussion that archaeological testing of the area to be disturbed by the new construction is not necessary. Nonetheless, Exelon, will follow the IL SHPO guidelines. (141-63)

Response: As explained in Sections 2.9.2 and 4.6, the Illinois Historic Preservation Agency does not consider shallow disturbance of any area to exempt it from further archaeological consideration. The comments did not provide new information to the EIS and will not be evaluated further. No change was made to the EIS as a result of these comments.

Comment: The project area has not been surveyed and may contain prehistoric/historic archaeological resources. Accordingly, a Phase I archaeological reconnaissance survey to locate, identify, and record all archaeological resources within the project area will be required. This decision is based upon our understanding that there has not been any large scale disturbance of the ground surface (excluding agricultural activities) such as major construction activity within the project area which would have destroyed existing cultural resources prior to your project. If the area has been heavily disturbed prior to your project, please contact our office with the appropriate written and/or photographic evidence. The area(s) that need(s) to be surveyed include(s) all area(s) that will be developed as a result of the issuance of the federal agency permit(s) or the granting of the federal grants, funds, or loan guarantees that have prompted this review. (08-1)

Comment: Since there is a "high potential for prehistoric sites" in the general area (BIS, pg. 2-5), what mitigation measures will be required in order to protect the integrity of these sites? (150-28)(151-28)

Response: *As stated in Section 4.6, before construction, consultation with the Illinois Historic Preservation Agency will identify any protective measures that should be taken. No change was made to the EIS as a result of these comments.*

Comment: And given such a substantial footprint, and the fact that no analysis of impacts on cultural and historic resources along the transmission line easement has been performed (EIS, § 4-34), how can the NRC staff judge the impact of the construction of such lines to be "small" (EIS, § 4.1.2; pg. 4-34)? (150-25)(151-25)

Response: *Evaluations of cultural resources on transmission lines were not included in the analysis. As explained in Section 4.6, the impacts can only be fully addressed after following the FERC process for connecting new large-generation sources to the grid. If existing transmission lines are used, continued maintenance of the lines will not impact cultural resources. If new lines are developed, cultural reviews will be accomplished, and any potential effect on important cultural resources will be identified and addressed. No change was made to the EIS as a result of these comments.*

Comment: On page 2-62 the Cultural Background 2-9.1 line 31 a correction should be made the Methodist Church at Birkbeck has been torn down. (29-2)(105-2)

Comment: On Page 2-62, for the cultural background, Section 2.9.1 Line 31, it's a minor thing but it states that the Methodist Church at Birkbeck is there as a historical building in the township. It's no longer there. (42-1)

Response: *This was corrected. Based on these comments, Section 2.9.1 of this EIS was changed.*

E.2.17 Comments Concerning Human Health and Radiological Impacts

Comment: Station Operation Impacts, Section 5.9.1, Exposure Pathways, page 548, paragraph 2. It is unclear whether incidental ingestion of water during swimming or boating was evaluated as an exposure route. (172-29)

Response: *Ingestion of water during swimming or boating was not evaluated as an exposure route. The staff believes that such exposures would not contribute significantly to dose. No change was made to the EIS as a result of the comment.*

Comment: Station Operation Impacts, Section 5.9.2, Radiation Doses to Member of the Public, Page 5-50. Documentation for the calculated dose to the Maximally Exposed Individual need to be provided. (172-31)

Response: *Appendix H to the EIS provides supporting information on the radiological dose assessment performed to calculate dose to the maximally exposed individual from normal plant operations. This information includes a discussion of the computer codes used in calculating the doses as well as the inputs to the codes. No change was made to the EIS as a result of the comment.*

Comment: When I went back and looked at some of the previous comments, somebody asked the question, and as near as I could find, they did not answer it. Said what the NRC does not point out is that the background radiation includes the emissions from radioactive chemicals which occur naturally and on and on. But it says, in fact, emissions released by a nuclear reactor are still considered background radiation after one year. So, I don't know if that's true or not, the response to this series of questions didn't answer it.

But if that's true, we had an initial background radiation, we added the initial power plant, that added some level. And now, we're now saying that that increased amount is now the background, so now we can go up incrementally from that.

And then we go up from that. So, can you kill us slowly, incrementally? (33-2)

Response: *As discussed in Section 2.5 of the EIS, the licensee conducted a pre-operational environmental operating program from 1980 to 1987 to establish a baseline to observe fluctuations of radioactivity in the environment after operations began. CPS Unit 1 began operations in 1987 and since that time results of their environmental monitoring program continue to be compared to the preoperational study. The same preoperational study will be used as a basis to compare any impacts from the proposed ESP unit. No change was made to the EIS as a result of the comment.*

Comment: Station Operation Impacts, Section 5.9.3.2 Population Dose, page 5-55, paragraph 3. The information is this paragraph is misleading at best. The National Academy of Science has reviewed all studies through 1998 on low level exposures to radiation, with the results published in the Biological Effects of Ionizing Radiation report VI (BEIR VI), on Health Effects of Exposure to Radon. The conclusion drawn for the studies was that the Linear No Threshold Theory was supported by the data from studies conducted world-wide to that point in time. These results were also concurred with by the National Council on Radiation Protection and Measurement (NCRP), as well as the International Commission on Radiological Protection (ICRP). Assertions that there is no unequivocal data is misleading. USNRC rules and regulations meet this viewpoint and are not used merely for conservatism, as implied by this statement. (172-35)

Response: As stated in the EIS, the staff accepts the linear, no-threshold dose response model. In its recent report, the BEIR VII Committee of the National Research Council concluded that the current scientific evidence is consistent with the hypothesis that there is a linear, no-threshold dose-response relationship between exposure to ionizing radiation and the development of cancer in humans (National Research Council 2006). Having accepted this model, the staff does feel that this model is conservative when applied to workers and members of the public who are exposed to radiation from nuclear power plants. This is based on the fact that numerous epidemiological studies have not shown increased incidences of cancer at low doses. Some of these studies included: (1) the 1990 National Cancer Institute study of cancer mortality rates around 52 nuclear power plants, (2) the University of Pittsburgh study that found no link between radiation released during the 1979 accident at the Three Mile Island nuclear power station and cancer deaths among residents, and (3) the 2001 study performed by the Connecticut Academy of Sciences and Engineering that found no meaningful links associated from exposures to radionuclides around the Haddam Neck nuclear power plant in Connecticut to the cancers studied. In addition, a position statement entitled "Radiation Risk in Perspective" by the Health Physics Society (revised August 2004) made the following points regarding radiological health effects:

- *Radiological health effects (primarily cancer) have been demonstrated in humans through epidemiological studies only at doses exceeding 0.05-0.1 Sv (5-10 rem) delivered at high dose rates. Below this dose, estimation of adverse effects remains speculative.*

- *Epidemiological studies have not demonstrated adverse health effects in individuals exposed to small doses (less than 0.1 Sv [10 rem] delivered in a period of many years).*

No change was made to the EIS as a result of the comment.

Comment: Station Operation Impacts, Section 5.9.1, Exposure Pathways, page 5-48, paragraph 4. Documentation or published studies that demonstrate the N–16 data need to be provided. (172-30)

Response: Environmental thermoluminescent dosimeter results from site boundary locations as part of the CPS radiological environmental monitoring program did not show any contribution to dose to the public located at the site boundary from nitrogen-16. Exelon assumes that contained sources of radiation at a new nuclear unit would be shielded and would not contribute to the external dose of the maximally exposed individual or the population. Because the reactor design for the proposed unit on the Exelon ESP site is not known at this time, an evaluation of dose to the maximally exposed individual from nitrogen-16 was not performed by the staff. No change was made to the EIS as a result of the comment.*

Comment: Recent research by Dr. Given Harper of my school found that, in 5 out of 6 deer carcasses from the Clinton area, there were levels of strontium-90 radioactive, (a dangerous substance) significantly higher than acceptable "background levels." As mammals that breathe the same air and drink the same water as these deer, the levels of radioactive substances in our bodies are surely rising as a result of exposure to emissions from the currently operating Clinton plant. The increased risk of cancer and genetic malformation's, and increased infant mortality rates, as a result of this exposure is completely unacceptable. As someone who has lost a family member to cancer, I will not support any increased risk of cancer for the surrounding families and communities, no matter how small. (109-2)

Response: As part of its Radiological Environmental Monitoring Program (REMP) for the CPS, Exelon analyzes for strontium-90 in the environment. The CPS Radiological Environmental Operating Reports for 1999, 2000, 2001, and 2002 found radioactivity levels in the environment, including strontium-90, similar to the pre-operational levels found prior to CPS start-up. The strontium-90 found in the deer carcasses is likely due to weapons testing.

No change was made to the EIS as a result of the comment.

Comment: ...suffer from higher % of cancer. (07-2)

Comment: Please look at the example of the Chernobyl disaster. Skyrocketing cancer rates and radioactive food still plague that region of the world. I have seen photographs of abandoned towns and children born with horrible birth defects directly caused by nuclear power. (10-5)

Comment: The proposed nuclear reactor in Clinton poses several potential health and safety risks to the citizens of neighboring communities. Even with measures to reduce the amount of radiation allowed into the air, radiation escapes from the plants and pollutes the air in nearby towns. According to the Radiation and Public Health Project, communities within 50 miles of a nuclear reactor experience an average yearly increase in breast cancer cases of between 14 and 40 percent. The average yearly rate for communities without reactors is one percent. Communities near reactors are also at a higher risk of birth defects in their children and a higher risk of thyroid problems, including hypothyroidism and thyroid cancer. (104-1)(17-1)

Comment: An energy source that increases the risk of childhood cancer and other personal tragedies? (109-5)

Comment: The increased risk of cancer and genetic malformation's, and increased infant mortality rates, as a result of this exposure is completely unacceptable. As someone who has lost a family member to cancer, I will not support any increased risk of cancer for the surrounding families and communities, no matter how small. (109-7)

Comment: I hope the rest of the report on something which is very important - the risks to the human population from radioactivity from the plant - is much more carefully and accurately done. In great detail, the rest of the report examines some of the environmental consequences of building a second nuclear power plant at Clinton. With technical over-precision, it minimizes the risks of a second plant due to radioactive exposure to construction workers, the public, regular workers, and persons living along routes where the waste might travel. If one believes what one reads, all those risks are SMALL. Even risks to normal accidents and severe accidents are expected to be small. For instance, we are told that "the probability of a severe accident without the loss of containment ... Is estimated to be [.000000134] per reactor year..." (U.S. Nuclear Regulatory Commission, 2005, 5-67). The report also admits that "radiation [-] protection experts conservatively assume that any [italics added] amount of radiation may pose some risk of causing cancer or a severe hereditary effect and that the risk is greater for higher radiation exposures" (U.S. Nuclear Regulatory Commission, 2005, 5-55). Nothing is to be worried about, according to the report. What the NRC is doing, while admitting risks exist from nuclear power, is to claim those risks are so small that they can be ignored and rejected. Certainly in everything we do, there are trade offs between risks. This is clear in medicine where we have to chose between the risks of the medicine harming us and the risks of not benefitting from the medicine if we do not use it. In statistics, these errors are called type I and type II errors (Blalock, 1979, 110-112). If nuclear energy really harms people (as the NRC admits), in this report the NRC takes a huge risk (and makes a type II error). The consequences to humans of this error can be catastrophic, as Chernobyl showed. Those of us who oppose nuclear power could be making another error in thinking that nuclear power is not safe (a type I error). For us, however, the consequences of making a mistake to build a nuclear power plant is far greater than of not building one. The fact that a nuclear power plant can raise the risk of cancers and cause other health effects in people means that this site near Clinton is environmentally unsafe. (110-3)

Comment: Here in Illinois we are all subjected to a vast array of carcinogens, including not only agricultural chemicals but also radiation from nuclear power plants. I want to know whether my cancer--and the growing numbers of cancers in our area--have been in any way triggered by nuclear power plant "venting" before we go ahead with more exposure. Taking such chances with our health, especially when other options--like wind--appear viable, seems unreasonable at best, and--if you've suffered through cancer, you know what I mean--inhuman at worst. (152-1)

Comment: Public health records dramatically show a decrease in infant mortality rates in downwind counties when the reactor was non functional from 1996-1999. The rates jumped back to their current higher levels ill when the reactor was re-started. No such change in mortality levels was observed in the upwind counties. I demand that before an impact CD statement can be approved a complete epidemiological study of the surrounding counties be carried out that establishes rates of leukemia, autism, childhood cancers, infant mortality and compares these data with all counties within a 50 mile radius. I further insist that an independent scientific review board be established to analyze this data. (156-1)

Comment: The reality is we have numbers that demonstrate that there is a higher incidence of infant mortality in DeWitt, Hyatt and Champaign Counties when Clinton No. 1 is in operation than when Clinton No. 1 was not in operation, when it was closed down in the late '90's. We can't epidemiologically prove that the reactor causes a higher incidence of infant mortality. But I've seen the numbers and I believe it. (32-4)

Comment: What are the health risks for the surrounding towns? (44-1)

Comment: Now imagine what it would mean to you if this [well being] were lost. If that precious child or grandchild or neighbor child looked at you with hollow eyes due to leukemia or due to genetic malformation because radiation is carcinogenic. And children are especially vulnerable. Radiation is also mutagenic. It changes our genes. (46-2)(94-2)

Comment: Yet this has become a very site specific, it has become very site specific if only because this site, Clinton, brings the applicant, Exelon Corporation, before you. And you must consider whether any potential human health impact associated with living in proximity to their existing and proposed reactors are being fully investigated and addressed. (52-2)

Comment: From, what, '96 to '99 when it was shut down, which enabled studies to be made determining that there was a huge increase in infant mortality when that unit was brought back on line. We would expect to see another increase in infant mortality if the second unit was brought on line. (55-7)

Comment: From 1999 to 2003, the radiation and public health project studied environmental radiation from nuclear reactors in childhood cancer in southeastern Florida. The latest baby teeth study report issued in 2003 concluded that, I know it's hard to understand like this, just from hearing it but try to grasp this. Here's what the study concluded. Exposure to radioactive releases from nuclear reactors is a significant factor in increasing childhood cancer and other adverse effects in southeast Florida.

The report also found that radioactive levels are significantly higher in the teeth of children with cancer than in teeth of healthy children. That difference cannot be underestimated. That difference should be something we all should think very, very carefully about. (60-2)

Comment: Dr. Rosaley Purtel, an epidemiologist who's been studying effects of low level ionizing radiation for decades. She writes this. We know now that radiation exposure to one generation induces genomic instability in offspring. Induces genomic instability in offspring. What does that mean? It induces instability in our genetic coding. What is the most important thing in the world? Maybe our generic coding that allows us to be human and for humanity to be passed on from generation to generation. To induce instability in the genetic coding. It's a crime against creation to be taken very seriously. (60-4)

Comment: But it may just partially account for the fact that when the current reactor was shut down in 1996, infant mortality of the downwind counties dropped in half. And that by 1999, after the reactor was restarted, infant mortality jumped back up to its pre 1996 levels. (73-3)

Comment: I am calling on the Nuclear Regulatory Commission not to issue a permit to Exelon to build a second nuclear power plant at Clinton until the NRC can certify that the radiation from the currently existing plant does not harm the health of residents of DeWitt County, Champaign County, and other counties downwind from the Clinton plant. I'm concerned about the effect of radiation from the nuclear power plant, which is currently operating, on the health of citizens of Champaign County. A study done by Dr. Samuel Galewsky, a professor of molecular biology at Milliken University, shows a correlation between infant mortality rates and the operation of the nuclear power plant at Clinton. Dr. Galewski looked at the infant mortality rates in DeWitt County and the counties surrounding Clinton before, during, and after the Clinton plant was shut down for repairs in 1996-99. When the plant was shut down, infant mortality rates dropped in the counties downwind from Clinton. When the plant resumed operation, infant mortality rates went back up. In the counties not downwind from Clinton, infant mortality rates decreased before, during, and after the plant shut-down. This study seems to indicate that radiation from the Clinton plant may be the cause of infants dying in Champaign County. This is tragic enough, but Samuel Galewsky's study on infant mortality may be an indicator of other health problems, the canary in the coal mine, if you will. Low-level radiation may also cause pediatric cancer, breast cancer, and leukemia. The NRC has a responsibility to us in Champaign County and other downwind counties to further investigate the possible health risks to our residents. (96-1)

Comment: I call on the NRC to commission an independent study on the health impacts of radiation on counties downwind from nuclear power plants. This study should include the effect of radiation on infant mortality, cancer, leukemia, birth defects, and reproductive health. The NRC should not issue any more permits for nuclear reactors until it can produce definitive evidence that radiation emitted from currently operating plants does not harm the health of citizens living downwind. (96-2)

Comment: Five U.S. nuclear reactors, closed permanently between 1987 and 1998 were studied as to the rates of infant mortality before and after shutdown; the infant mortality went down dramatically by close to the same amount in the two years following shutdown - 15-18% at each site (while the U.S. average in that time period was 6.4% drop) in the downwind counties, 50-70 miles away. For fetuses, infants and children up to 5 years, the rate continued to drop for six years following shutdown. Why infant mortality? Because fetuses and babies are developing cells rapidly and are more intensely affected by radiation: the results are seen in miscarriages, stillbirths, malformed and low birth weight babies. Do we have to put up with this increased risk in order to have the energy we need? The answer is shocking. The purpose of this plant, according to Exelon, is to ship energy to other states for profit because Illinois has all the energy it needs. For this our children should die? (98-1)

Response: *Health effects from exposure to radiation are dose-dependent, ranging from no effect at all to death. Above certain doses, radiation can be responsible for inducing diseases such as leukemia, breast cancer, and lung cancer. Very high (hundreds of times higher than a rem), short-term doses of radiation have been known to cause prompt (or early, also called "acute") effects, such as vomiting and diarrhea, skin burns, cataracts, and even death. When radiation interacts within the cells of our bodies, several events can occur. First, the damaged cells can repair themselves and permanent damage does not result. Second, the cells may die, much like large numbers of cells do every day in our bodies, and dead cells may be replaced through normal biological processes. Third, the cells may either incorrectly repair themselves (resulting in a change in the cells' genetic structure), they can mutate and subsequently be repaired without any effect, or they can sometimes form precancerous cells that may become cancerous.*

Radiation is only one of many agents with the potential for causing cancer, and cancer caused by radiation cannot be distinguished from cancer attributable to any other cause, such as chemical carcinogens. The chances of getting cancer from a low dose of radiation is not known precisely because the few effects that may occur cannot be distinguished from normally occurring cancers. The normal chance of dying from cancer is about one in five.

The actual amount of radiation any member of the public receives from activities at nuclear power facilities is so small that scientists have been unable to make empirically based estimates of radiation risk with any precision. There are many difficulties involved in designing research studies that can accurately measure the projected small increases in cancer cases that might be caused by low exposures to radiation when compared to the rate of cancer resulting from all other causes. In the absence of a clear answer, the U.S. Nuclear Regulatory Commission conservatively assumes that any amount of radiation may pose some risk for causing cancer or having some hereditary effect and that the risk is higher for higher radiation exposures. This is called a linear, no-threshold dose-response model and is used to describe the relationship between radiation dose and the occurrence of cancer.

This model suggests that any increase in dose above background levels, no matter how small, results in an incremental increase in risk above existing levels of risk. Although the U.S. Nuclear Regulatory Commission has accepted this hypothesis as a "conservative" (i.e., cautious) model for determining radiation standards, the U.S. Nuclear Regulatory Commission, like other authoritative bodies, recognizes that this model will probably over-estimate radiation risk. The associations between radiation exposure and the development of cancer are mostly based on studies of populations exposed to relatively high levels of ionizing radiation (for instance, the Japanese atomic bomb survivors and the recipients of selected diagnostic or therapeutic medical procedures).

Although radiation can cause cancers at high doses and high dose rates, currently there are no data to establish unequivocally the occurrence of cancer following exposures to doses below about 10 rem. The average annual dose to a member of the public from a nuclear power facility is in the range of less than 1/1000th rem (1 millirem) per year. This is compared to the 10 rem (10,000 millirem) discussed previously. At doses above 10 rem, a relationship between radiation and cancer can be observed. There are no data to establish unequivocally the occurrence of cancer following exposures to doses below 10 rem. Although there is a statistical chance that radiation levels that small (i.e., less than 10 rem) could result in a cancer, it has not been possible to calculate with any certainty the probability of cancer induction from a dose this small. Because many agents cause cancer, it is often not possible to say conclusively whether the cancer was radiation-induced cancer.

Authors of various reports have stated or implied that there are cause-and-effect relationships in the statistical associations between cancer rates and reactor operations. While it is true that cancer rates vary among locations, it is very difficult to ascribe the cause of a cluster of cancers to some local environmental exposure, such as radiation from a nuclear power facility. Statistical association alone does not prove causation, and well-established scientific methods must be used to determine causation. For example, a person could say, "In the winter I wear boots, and in the winter I get colds." While there is a strong statistical association between wearing boots and getting colds, it would be inappropriate to say that wearing boots causes colds.

The scientific community adheres to several principles of good science that need to be employed before a cause-and-effect claim can be made. These principles include whether the study can be replicated, whether it has considered all the data or was selective (e.g., in the population or in the years studied), whether it evaluated all possible explanations for the observations, whether the data were valid and reliable, and whether the conclusions were subjected to independent peer review, evaluation, and confirmation.

A number of studies that conformed to these principles have been performed to examine the health effects around nuclear power facilities.

- *In 1990, at the request of Congress, the National Cancer Institute conducted a study (NCI 1990) of cancer mortality rates around 52 nuclear power plants and 10 other nuclear facilities. The study covered the period from 1950 to 1984 and evaluated the change in mortality rates before and during facility operations. The study concluded there was no evidence that nuclear facilities may be linked causally with excess deaths from leukemia or from other cancers in populations living nearby.*

- *Investigators from the University of Pittsburgh found no link between radiation released during the 1979 accident at the Three-Mile Island nuclear station and cancer deaths among nearby residents. Their study followed more than 32,000 people who lived within 8 km (5 mi) of the facility at the time of the accident.*

- *In January 2001, the Connecticut Academy of Sciences and Engineering issued a report on a study around the Haddam Neck nuclear power plant in Connecticut and concluded that exposures to radionuclides were so low as to be negligible and found no meaningful associations to the cancers studied.*

- *In 2001, the American Cancer Society concluded that, although reports about cancer clusters in some communities have raised public concern, studies show that clusters do not occur more often near nuclear plants than they do by chance elsewhere in the population. Likewise, there is no evidence linking the isotope strontium-90 with increases in breast cancer, prostate cancer, or childhood cancer rates.*

- *In 2001, the Florida Bureau of Environmental Epidemiology reviewed claims that there are striking increases in cancer rates in southeastern Florida counties caused by increased radiation exposures from nuclear power plants. However, using the same data to reconstruct the calculations on which the claims were based, Florida officials did not identify unusually high rates of cancers in these counties compared with the rest of the state of Florida and the nation.*

- *In 2000, the Illinois Public Health Department compared childhood cancer statistics for counties with nuclear power plants to similar counties without nuclear plants and found no statistically significant difference.*

In summary, there are no studies to date that are accepted by the nation's leading scientific authorities that indicate a causative relationship between radiation dose from nuclear power facilities and cancer in the general public. The amount of radioactive material released from nuclear power facilities is well measured, well monitored, and known to be very small. These comments did not result in a change to the environmental impact statement.

Comment: Environmental Consequences of Proposed Action, Section 4.8.1, Public and Occupational Health, Public Health, page 4-35. Illinois Administrative Code 35 IAC 201.146tt, is cited without a specific description of the Code's purpose. It needs to be provided as a clarification. Additionally, this needs to address whether or not this citation takes into account the particulate matter standards for respirable and fine particulates found in USEPA PM 10 and PM 2.5 regulations. (172-21)

Response: Illinois Administrative Code 35 IAC 201.146tt refers to activities associated with the construction, onsite repair, maintenance, or dismantlement of buildings, utility lines, pipelines, wells, excavations, earthworks, and other structures that do not constitute emission units. Section 201.146 of the IAC does not relieve the applicant of the obligation to obtain a permit pursuant to Section 9.1(d) and 39.5 of IAC, Sections 165, 173, and 502 of the Clean Air Act or any other applicable permit or registration requirements. No change was made to the EIS as a result of the comment.

Comment: Environmental Consequences of Proposed Action, Section 4.9.2, Radiation Exposures from Gaseous Effluents, page 440. The methodology for this evaluation is not clearly specified, nor are the necessary assumptions. (172-23)

Response: In its ER, the applicant estimated the annual dose to construction workers by taking the highest dose to a member of the public at the site boundary and scaling that value, assuming an individual was present for 2080 hr/yr (i.e., occupancy time for a typical construction worker). The applicant obtained this gaseous effluent dose from the Clinton Power Station's (CPS's) 2001 annual radioactive effluent release report. The staff considers this general approach acceptable. The site boundary location is at the nearest public access road, approximately 0.5 km (0.3 mi) southeast of the CPS. The estimated dose to a worker at this location was 3×10^{-4} mSv (0.03 mrem). In the applicant's ER, Exelon states that the ESP facilities will be located more than 300 m (1000 ft) from CPS. Although it is possible that a construction worker could be closer to the CPS than the site boundary location (0.3 to 0.5 km [0.2 mi versus 0.3 mi]), this distance would not significantly increase the dose received from gaseous effluents; therefore, direct radiation will remain the only significant contributor to dose. No change was made to the EIS as a result of the comment.

Comment: I am very opposed to the building of the nuclear power plant at Clinton Lake because nuclear plants are a highly polluting source of energy. My concern is not only for the permitted radioactive emissions, but also for the unexpected radioactive emissions. (06-1)

Comment: The admission was made that the reactor releases small, quote, unquote, small amounts of nuclear radioactive material. Well, think about that. A small release. Sounds reasonable. Maybe harmless, maybe. But another small release and another and another and another. It's the cumulative over and over and over again. That's the problem. If it was one small release and only that, it would be different. But the cumulative impact of many small releases builds up. (60-3)

Comment: I'd like to know if the nuclear power plant is going to emit radiation. And if so, which compounds will be the source of the radiation and how long will those compounds be around emitting radiation at all, if there's any radiation at all from the nuclear power plant in the surrounding area (38-1)

Comment: If this radioactivity is not controlled, long lasting hazardous radioactive materials, such as strontianite, Cesium 132, Cesium 134, Strontium 80, Strontium 90, Lanthanum 132, Barium 140, Zirconium 95, Molybdenum 90, Ruthenium 103 and 106, Neptunium 239, Plutonium 238 and 240, Cobalt and not to mention Iodine 131, which affects the thyroid glands of the children. (65-4)

Comment: And also, I never realized that nuclear reactors actually made a lot of gasses that cannot be contained, it builds up tremendous pressure, and the government allows nuclear power plants to just vent this gas off weekly, if not more often. So there are a lot of other things to consider. Now, I know the people here would like this for economic reasons. But it's just not affecting you, what's created here gets blown away 50, 100, 200 miles. (73-4)

Comment: It appears that all too often the overriding reason given for approval was that this is a "low risk" area. As a resident within the 5 mile zone of the current plant, I do not consider myself or my neighbors "not detectable or are so minor that they will neither destabilize nor noticeably alter any important attribute of the resource" if some major incident should occur at the plant site and we "disappear" due to illness or death from a radiological event. (157-2)

Response: Gaseous and liquid effluents from any new unit on the Exelon ESP site will be monitored and reported annually in an Annual Effluent Monitoring Report. Levels of radioactive material in the environment will be monitored from any new units on the Exelon ESP site as part of the Radiological Environmental Monitoring Program (REMP) similar to the program at the CPS. Results of the REMP will be reported annually in the publically available Annual Radiological Environmental Operating Report. The REMP will help identify any radioactive material that is accumulating the environment by comparing results of soil, sediment, vegetation, water, and air to those results from the preoperational monitoring conducted on the site from 1980 to 1987. The REMP for CPS has not identified any radionuclides accumulating in the environment around the plant.

The EIS evaluated the impact of radiological emissions from the proposed reactor(s) on the Exelon ESP site in Section 5.9 of the EIS and found results to be within regulatory limits.

No change was made to the EIS as a result of these comments.

Comment: 4-38 lines 14-26 talk about thermoluminescent dosimeters (TLDs) that measure gamma radiation as far as 5 kilometers (3 miles) from the power plant and their measurements are from 13 to 21 mrems and yet a dosimeter is much smaller than a person. This brings up two disturbing items. First, on page 2-17 above, they state that a member of the public will get a maximum of only 0.003 mrem per year and yet 5 kilometers away from the power plant, they are measuring over 13 mrems per year. Second, they are specifically talking about gamma radiation that the TLDs measure; however, it is the alpha radiation that is more damaging to people and animals. Why aren't they talking about alpha radiation? Gamma radiation falls off at the rate of the inverse square law, alpha particles can be breathed in or ingested and damage your body tremendously! (153-6)

Comment: 5-52 lines 24-27 talk about gamma and beta radiation. However, it is the alpha radiation that is more worrisome and incredibly damaging to biological organisms. Why don't they talk about alpha radiation? (153-12)

Response: The 3 x 10^{-5} mSv (0.003 mrem) per year referred to in Section 2.5 of the EIS is the maximum dose to a member of the public from gaseous effluents. There have been no liquid effluents released from the CPS since 1992, as documented in the CPS annual effluent release reports. The discussion in Section 4.9.1 of the EIS is referring to dose from external radiation sources measured using thermoluminescent dosimeters (TLDs). The TLDs measure background radiation and any external radiation generated by plant operations. These TLDs were placed at the site boundary to determine the impacts of external radiation sources from the plant on the public. As discussed in Section 4.9.1 of the EIS, TLDs placed near the site boundary measured from 0.13 to 0.22 mSv (13 to 22 mrem) quarterly. TLDs placed in control locations had similar readings; therefore, the dose at this location is due to natural background radiation. Thus, the CPS operation is not contributing to the external radiation dose at the site boundary. According to NCRP Report 94, (NRCP 1988) the average person in the United States will be exposed to approximately 300 mrem from naturally occurring radioactive sources, of which approximately 60 mrem is from external radiation sources, with the remaining 2.4 mSv (240 mrem) from inhaled activity and activity in the body.

Effluents released from operating nuclear plants are beta-gamma emitters, as shown in the annual radioactive effluent release reports. Alpha-emitters in reactor fuel remain bound in the spent fuel and are not released to the environment under normal operating conditions. No change was made to the EIS as a result of these comments.

Comment: Station Operation Impacts, Section 5.9.6, Radiological Monitoring, pages 5-59, 5-60. Conducting a radiological environmental monitoring program (REMP) is an excellent idea and should be pursued in as much detail as possible. Incorporation of previously collected data over the time of the current plant should be considered and used as a base from which to expand. This information would then be able to be cited and used for support of decisions made concerning these parameters. (172-39)

Comment: You can dismiss what I say but somewhere in your heart of hearts you too know the huge difference between clean air and invisible nuclear pollution. You will also know you do not want your own precious descendants to suffer the burden and the fall out of our nuclear waste. 10,000 generations should not pay the penalties so that you and I can have electricity today. (46-11)(94-11)

Comment: And here we are down here next to a nuclear power plant. Right now this place is quite contaminated. There's a lot of radiation coming out of that plant, especially the older they get. (65-2)

Response: *The comments are noted. The comments do not provide new information and will not be evaluated further. No change was made to the EIS as a result of these comments.*

Comment: Section 5.9.4, Page 5-57, Line 2-3. It is stated that the relationship between current LWR and the specific design would be verified at CP/COL. The ER provided justification as to why occupational exposure for new nuclear units are bounded by occupational exposure from currently operating LWRs. Therefore, there exists sufficient information at this time to determine the impacts from such exposures. If these parameters are not bounded at CP/COL, they would then have to be assessed. (141-83)

Response: Section 5.9.4 of the EIS was revised to state that based on the information provided by Exelon and on its own independent evaluation, the staff concludes that the health impacts from occupational radiation exposure would be SMALL based on individual worker doses being maintained within 10 CFR 20.1201 limits and collective occupational doses being typical of doses found in current operating LWR reactors.

Comment: Environmental Consequences of Proposed Action, Section 4.9.4, Total Dose to Site-Preparation Workers, page 4-40. The clarification needs to be made that the annual radiation worker occupational dose limit is 0.05 Sv (5rem); otherwise the workers would fall under the public exposure standards of 1 mSv (100 mrem) for the dose. (172-24)

Response: The following sentence was added to Section 4.9.4 of the EIS: "The annual dose estimate for the site preparation workers was approximately 0.25 mSv (25 mrem) which is less than the 1 mSv (100 mrem) annual dose limit to an individual member of the public. If the dose estimate had exceeded 1 mSv (100 mrem) annually, the site preparation workers would need to be treated as radiological workers and would be subject to the annual occupational dose limit of 0.05 Sv (5 rem) found in 10 CFR 20.1201."

Comment: Station Operation Impacts, Section 5.9.2.2, Gaseous Effluent Pathway, page 5-5 1. The models cited for calculating doses to the public were dated 1986 and 1987. More up-to date modeling programs should now be available and used for a better evaluation of dose projection to the maximally exposed individual. If more current modeling is not used in the Final EIS, provide a rationale for using outdated models for dose projections. (172-33)

Response: The GASPAR II and LADTAP II codes are based on the latest NRC guidance provided to licensees for determining compliance with the design objectives in 10 CFR Part 50, Appendix I. This guidance is provided in Regulatory Guide 1.109 (NRC 1977a), entitled "Calculation of Annual Doses to Man from Routine Releases of Reactor Effluents for the Purpose of Evaluating Compliance with 10 CFR Part 50, Appendix I." In addition, Section 5.4.2 of NUREG-1555 (NRC 2000) specifies that the GASPAR and LADTAP computer codes be used to estimate doses from gaseous and liquid radioactive releases, respectively.

No change was made to the EIS as a result of the comment.

Comment: Environmental Consequences of Proposed Action, Section 4.9.1 Direct Radiation Exposures, page 4-38, paragraph 1. This paragraph needs to be clarified as to whether or not the radiation described in this paragraph was included in the radiation evaluation. (172-22)

Response: The sources of direct radiation described in paragraph 1 of Section 4.9.1 of the DEIS were included in the applicant's evaluation of direct radiation dose. The applicant's evaluation estimated annual dose to the construction workers using readings from thermoluminescent dosimeters located at the protected area fence line. These measurements would include any contribution to dose from the cycled condensate storage tank and skyshine from nitrogen-16 present in the turbine building. Section 4.9.1 of the EIS was revised to state that the sources of radiation described in paragraph 1 were included in the dose estimate from direct radiation.

Comment: Section 5.9.5.3, first paragraph: The disclaimers about comparing 40 CFR 190 criteria to biota doses notwithstanding, these comparisons should be avoided because they are misleading. The appropriate benchmarks for biota doses are the ICRP and IAEA values identified later in this section. (172-65)

Response: The staff agrees that appropriate benchmarks for biota doses are the NCRP and IAEA values. The comparison to 40 CFR Part 190 standards was included for completeness as the applicant included this in the ER. This statement in the EIS does reference the applicant's ER. No change was made to the EIS as a result of the comment.

Comment: Station Operation Impacts, Section 5.9.5, Impacts to Biota Other than Members of the Public, Page 5-57. This short paragraph make assertions without any citation or data provided to support the assertions. Please provide this information, or state that these were assumptions without available data to use for a proper evaluation of potential impacts. (172-37)

Response: The use of surrogate species is a commonly accepted technique for evaluating dose impacts to the biota. No change was made to the EIS as a result of the comment.

Comment: In order to have adequate data to clearly assess the environmental impact of the proposed and current nuclear power station I insist that a real time radiation monitoring system showing effluent releases, amounts, radiation levels and wind direction be implemented for all perimeter detector systems. This data must be made available to the public at all times via the internet. The NRC's current effluent data collection methods and availability for public analysis is totally inadequate. (156-2)

Response: The staff believes that current regulations regarding environmental monitoring around nuclear power plants are adequate. These regulations require each commercial reactor site to have a Radiological Environmental Monitoring Program (REMP). The purpose of the REMP is to sample, measure, analyze, and monitor the radiological impact of reactor operations on the following pathways: direct radiation, atmospheric, aquatic, and terrestrial. Results of the REMP are summarized each year in the publicly available Annual Environmental Radiological Operating Report. Effluent releases are summarized annually in an Annual Radioactive Effluent Release Report. In addition, each site must monitor gaseous and liquid effluents in real time. Effluent monitors will alarm if routine release levels are exceeded.

No change was made to the EIS as a result of the comment.

Comment: Environmental Consequences of Proposed Action, Section 4.9.5, Summary of Radiological Health Impacts, page 4-41. The conclusion that the impact due to radiological exposures is small is not supported by the documentation provided. (172-25)

Response: *The staff believes that the discussion in Section 4.9 supports the conclusion that the impact of radiological exposures to site preparation workers is small. Dose from gaseous and liquid effluents would be very small (<0.01 mSv [1 mrem] annually) and dose from external radiation sources attributed to the plant operation was conservatively estimated at 0.25 mSv (25 mrem) annually. This is significantly less than the 1 mSv (100 mrem) annual limit to a member of the public from 10 CFR 20.1301. Section 4.9.1 of the EIS list several reasons why the staff considered the Exelon estimate of 0.25 mSv (25 mrem) annually to be conservative. No change was made to the EIS as a result of the comment.*

Comment: Affected Environment, Section 2.5, Radiological Environment, Page 2-17. The inclusion of 40 CFR 61, Subpart I dose requirements would be appropriate for facilities to meet the Constraint Rule Requirements under the USNRC guidances as well as incorporation of this rule by Illinois under agreement with USNRC to meet these requirements. (172-11)

Comment: Station Operation Impacts, Section 5.9.3.1, Maximally Exposed Individual, page 5-53. The USNRC constraint rule is not included, nor is 40 CFR 61, Subpart I, from which it was derived in order to minimize public exposures to radionuclide emissions from NRC facilities. While the 40 CFR 61 rule may not be strictly applicable, it is definitely relevant and appropriate to be included. The Iodine doses specified in 40 CFR 61, Part 190, are for planned emissions and do not apply to unplanned emissions. (172-34)

Comment: Station Operation Impacts, Section 5.13, References, page 5-83. Inclusion of 40 CFR 61, Subpart I, should be done to provide the appropriate reference to the USNRC constraint rule that requires facilities licensed by the USNRC to substantially meet the dose standards found at 40 CFR 61, Subpart I. (172-43)

Response: *The staff believes that the national emissions standards for radionuclide emissions in 40 CFR 61, Subpart I are not applicable to commercial nuclear power reactors. 40 CFR 61.100 states that Subpart I applies to facilities owned or operated by any Federal agency other than the Department of Energy and not licensed by the Nuclear Regulatory Commission.*

Commercial nuclear power plants are licensed by the NRC. No change was made to the EIS as a result of these comments.

Comment: Station Operation Impacts, Section 5.9.5.3, Impact of Estimated Biota Doses, pages 5-58, 5-59. Biota comparisons for radiation exposures may not be equivalent. Though ICRP and NCRP state that it would not be expected to have a major impact, the data may not exist at this time to make this type of assumption. USNRC should look at studies that have been conducted to date involving other biota, and attempting an extrapolation from them. (172-38)

Response: *The staff relies on (1) the NCRP Report No. 93 (NCRP 1987) conclusion that appreciable effects in aquatic populations would not be expected at a dose lower than 10 mGy/d (1000 mrad/d), and (2) the IAEA Technical Report 332 (IAEA 1992) conclusion that chronic dose rates of 1 mGy/d (100 mrad/d) or less do not appear to cause observable changes in terrestrial animal populations. The staff did not identify any more recent references in this area. No change was made to the EIS as a result of the comment.*

Comment: Station Operation Impacts, Section 5.9.4 Occupational Doses to Workers, pages 5-56, 5-57. The Occupational Doses to Workers are regulated on an individual basis and the person-Sv values used do not provide an appropriate or comparable value to a maximally exposed individual. The maximum dose exposures for individuals should be referenced and used for a better and more realistic exposure determination. (172-36)

Response: *The staff recognizes that the reactor licensee would need to maintain dose to workers within 0.05 Sv (5 rem) annually as specified in 10 CFR 20.1201; however, reactor design information typically provides a collective dose estimate with the understanding that individual doses would be maintained within 10 CFR Part 20 limits. Section 5.9.4 of the EIS was revised to state that the licensee of a new plant will need to maintain individual doses to workers within 0.05 Sv (5 rem) annually as specified in 10 CFR 20.1201. Facilities are also required to apply the As Low As Reasonably Achievable (ALARA) process to maintain doses below 10 CFR 20.1201 limits. The concluding statement in the EIS was revised to state that the staff's assessment of small impacts for occupational doses was based on occupational doses being maintained within 10 CFR Part 20 dose limits and annual occupational collective dose estimates being within those typical of current operating LWR plants.*

Comment: Despite a finding by the National Institute of Environmental Health Sciences (NIEHS) that "extremely low frequency-electromagnetic field (ELF-EMF) exposure cannot be recognized as entirely safe" and may pose a leukemia hazard, the staff does not consider this to be a significant environmental impact to the public (EIS, § 5.8.4). Would a stronger electromagnetic field produced by increased voltage capacity on the transmission lines from the CPS amplify this hazard? Further, Exelon is allowed to wait until the COL licensing stage to determine whether transmission lines from the site meet the requirements of the National

Electric Safety Code (NESC) regarding electrostatic effects from operation. Why is this issue not being addressed at this stage in the licensing process? (150-23)(151-23)

Response: The transmission lines serving the CPS are owned and operated by AmernIP. The lines that serve the CPS are rated at 345 kV and would remain at that voltage should the postulated unit be built. The NIEHS panel reviewing the literature on effects of ELF-EMF considered information available on the full range of transmission line currents and voltages (NIEHS 1999). They were unable to come to a conclusion that ELF-EMF is a hazard from transmission lines. Therefore, it would be improper to conclude that increasing the output of the CPS site would increase the hazard.

The process for connecting new generation sources to the transmission system is described briefly in Section 3.0 of the EIS. It involves studies by the owner of the transmission system and the owner of the generation facility to select a transmission route and transmission line design. Further, Exelon has stated in its ER that any new lines would be constructed to meet applicable standards. Therefore, the staff concludes that acute effects from transmission lines would be small.

The comments did not provide any new information. No change was made to the EIS as a result of these comments.

Comment: Overall the nuclear industry produces more power with fewer health problem than any other fossil fuel source. Even with a major accident as in Three Mile Island, very little impact is felt by the surrounding population. (01-2)

Comment: It has been discovered that people living in brick homes experience more radiation exposure than someone living next to a nuclear power plant. (27-3)(97-3)

Comment: I want to talk about radiation. And there's a lot of, I just want to put it all into perspective. I bet there's a lot more radiation from the bricks in this room, with all the uranium and thorium and that, than I'm getting, than I would get if I were to stand right on the edge of the exclusion zone from the power plants. And the amount, so we have to put this into perspective of what we get from a natural background, because there's radiation all around us in the air, the cosmic rays penetrating our bodies, doing lots of stuff to us right now. When life evolved, natural background was 10 to 20 times higher. There are places on this planet where natural background is naturally 10 to 20 times higher. And these areas observe no increase in cancer in any way. In fact, most of those areas have a decrease, people are more healthy in those areas. This is a very interesting find. So, so the amount from a nuclear power plant is less than one percent, probably closer to one tenth of one percent at the exclusion zone. (86-5)

Comment: And someone brought up that plutonium is the most deadly substance on earth. Well, chemically, caffeine is far more deadly, which is, which is shown to be true. And there have been studies of people who were at Los Alamos, back in 1943, and we know a lot more about radiation now, then we did back then. And they'd ingest a lot of plutonium. And actually, their, their health rates were a lot higher. Their mortality rates were lower than what was in the normal population. (86-7)

Response: These comments are noted. The comments do not provide new information and will not be evaluated further. No change was made to the EIS as a result of these comments.

E.2.18 Comments Concerning the Uranium Fuel Cycle and Waste Management

Comment: Section 6.2.4, Page 6-39, Line 24-26. It is stated that if the ACR-700 and IRIS were chosen, the transportation accident analysis would be performed at CP/COL. In the ESP application, Exelon applied for a site to be reserved for a future nuclear facility (See Administrative Section 1.1). As stated in the Environmental Report in Section 1.1.3, the selection of the reactor design is still under consideration and a set of bounding parameters was determined using a listing of reactor design-types listed. Therefore, the statement of selection of a particular reactor design for future analysis is not appropriate.

Nonetheless, it would be more accurate to state that the environmental impacts of LWR transportation accident analysis is SMALL and other-than-LWR fuel performance would be evaluated if the reactor design selected at CP/COL and environmental impacts greater than those evaluated at ESP. (141-119)

Response: As discussed in Section 6.2 of the EIS, none of the proposed light-water-cooled reactor (LWR) designs meets all the conditions in 10 CFR 51.52(a); therefore, a full description and detailed analysis are required for each LWR design. INEEL (2003) did not provide any spent fuel inventories for the ACR-700 and the IRIS designs; therefore, the staff could not perform a detailed analysis of transportation impacts from spent fuel for these reactor designs. If the ACR-700 or IRIS design is chosen by the applicant, a detailed analysis of transportation impacts from spent fuel would need to be performed at the CP or COL stage. However, as stated in Section 6.2.4, the staff concludes that the environmental impacts of transporting fuel and radioactive waste to and from advanced LWR designed facilities would be SMALL. No change was made to the EIS as a result of these comments.

Comment: Table G-5, Page G-14, Lines 17-18. The population density value at stops (30,000 persons/Km2) used in the RADTRAN run. The population density value at stops (30,000 persons/Km2) seems very high. Most of the routes from the selected cities utilize freeways with rest stops or check stations well away from residential areas (or, at least, at a favorable distance that dose rates from the cask would not be a contributing factor) and the density in and around truck stops would seem to be much less than that presented. This value should be re-evaluated and re-verified. (141-125)

Response: As discussed on page G-13 of the DIES, the 30,000 persons/km² value was taken from Sprung et al. (2000). This value is derived based on actual observations at truck stops and equates to about 9 people located within a ring from 1 to 10 m (3 to 30 ft) around the cask. This is used for close-proximity exposures. This ring is assumed to be surrounded by residents in a suburban area that has a lower population density. This is used for longer-distance exposures. Stop doses are the sum of the close-proximity and long-distance exposures. No change was made to the EIS as a result of the comment.

Comment: Section 6.0, Page 6-1, Line 11-15. It is stated that the transportation impacts of radioactive materials would have to be evaluated if a different reactor design was selected from those calculated in the EGC ESP. It should be stated that the transportation impacts of radioactive materials would have to be evaluated if the reactor design selected at CP/COL had environmental impacts greater than those evaluated in the EGC ESP. (141-96)

Comment: Section 6.2.1.2, Page 6-27, Line 32-34, Section 6.2.2.3, Page 6-36, Line 30-32. It is stated that the environmental impacts of other-than-LWR fuel performance needs to be assessed at CP/COL. It should be stated that the environmental impacts of other-than-LWR fuel performance would have to be evaluated if the reactor design selected at CP/COL had environmental impacts greater than those evaluated in the EGC ESP. (141-114)

Response: As stated in the EIS, the applicant did not use the PPE approach for evaluating the transportation impacts; therefore, if a different design is chosen at the CP/COL the transportation impacts for that design would need to be performed. This is the only way the staff will be able to determine if the impacts fall within those determined for the ESP plant designs. No change was made to the EIS as a result of these comments.

Comment: Fuel Cycle, Transportation, and Decommissioning, Section 6.1.1.5 Radioactive Effluents, page 6-12, paragraph 1. The information is this paragraph is misleading at best. The National Academy of Science has reviewed all studies through 1998 on low level exposures to radiation, with the results published in BEIR VI, on Health Effects of Exposure to Radon. The conclusion drawn for the studies was that the Linear No Threshold Theory was supported by the data from studies conducted world-wide to that point in time. These results were also concurred with by the National Council on Radiation Protection and Measurement (NCRP), as well as the International Commission on Radiological Protection (ICRP). Assertions that there is no unequivocal data is misleading. USNRC rules and regulations meet this viewpoint and are not in there merely for conservatism as implied by this statement. (172-48)

Comment: Fuel Cycle, Transportation, and Decommissioning, Section 6.2.1.1, Normal Conditions, pages 6-25, 6-26. The information is this paragraph is misleading at best. The

National Academy of Science has reviewed all studies through 1998 on low level exposures to radiation, with the results published in BEIR VI, on Health Effects of Exposure to Radon. The conclusion drawn for the studies was that the Linear No Threshold Theory was supported by the data from studies conducted world-wide to that point in time. These results were also concurred with by the National Council on Radiation Protection and Measurement (NCRP), as well as the International Commission on Radiological Protection (ICRP). Assertions that there is no unequivocal data is misleading. USNRC rules and regulations meet this viewpoint and are not in there merely for conservatism as implied by this statement. (172-57)

Comment: Fuel Cycle, Transportation, and Decommissioning, Section 6.2.2.1, Normal Conditions, pages 6-32, paragraph 3. The information is this paragraph is misleading at best. The National Academy of Science has reviewed all studies through 1998 on low level exposures to radiation, with the results published in BEIR VI, on Health Effects of Exposure to Radon. The conclusion drawn for the studies was that the Linear No Threshold Theory was supported by the data from studies conducted world-wide to that point in time. These results were also concurred with by the National Council on Radiation Protection and Measurement (NCRP), as well as the International Commission on Radiological Protection (ICRP). Assertions that there is no unequivocal data is misleading. USNRC rules and regulations meet this viewpoint and are not in there merely for conservatism as implied by this statement. (172-60)

Comment: Fuel Cycle, Transportation, and Decommissioning, Section 6.2.2.2, Accidents, page 6-36, paragraph 2. This paragraph is also misleading. The National Academy of Science has reviewed all studies through 1998 on low level exposures to radiation, with the results published in BEIR VI, on Health Effects of Exposure to Radon. The conclusion drawn for the studies was that the Linear No Threshold Theory was supported by the data from studies conducted world-wide to that point in time. These results were also concurred with by the National Council on Radiation Protection and Measurement (NCRP), as well as the International Commission on Radiological Protection (ICRP). Assertions that there is no unequivocal data is misleading. USNRC rules and regulations meet this viewpoint and are not in there merely for conservatism as implied by this statement. (172-61)

Response: *As stated in the EIS, the staff accepts the linear, no-threshold dose response model. In its recent report, the BEIR VII Committee of the National Research Council concluded that the current scientific evidence is consistent with the hypothesis that there is a linear, no-threshold dose-response relationship between exposure to ionizing radiation and the development of cancer in humans (National Research Council 2005). Having accepted this model, the staff does feel that this model is conservative when applied to workers and members of the public who are exposed to radiation from nuclear power plants. This is based on the fact that numerous epidemiological studies have not shown increased incidences of cancer at low*

Appendix E

doses. Some of these studies included the following: (1) the 1990 National Cancer Institute study of cancer mortality rates around 52 nuclear power plants, (2) the University of Pittsburgh study that found no link between radiation released during the 1979 accident at the Three Mile Island nuclear power station and cancer deaths among residents, and (3) the 2001 study performed by the Connecticut Academy of Sciences and Engineering that found no meaningful associated from exposures to radionuclides around the Haddam Neck nuclear power plant in Connecticut to the cancers studied. In addition, a position statement entitled "Radiation Risk in Perspective" by the Health Physics Society (revised August 2004) made the following points regarding radiological health effects:

- *Radiological health effects (primarily cancer) have been demonstrated in humans through epidemiological studies only at doses exceeding 0.05-0.1 Sv (5-10 rem) delivered at high dose rates. Below this dose, estimation of adverse effect remains speculative.*

- *Epidemiological studies have not demonstrated adverse health effects in individuals exposed to small doses (less than 0.1 Sv [10 rem] delivered in a period of many years).*

No change was made to the EIS as a result of these comments.

Comment: The draft EIS only considers the "no recycle" option for irradiated fuel management, which treats spent fuel as waste to be stored at a federal waste repository, and does not fully consider the possible reprocessing of spent nuclear fuel (EIS, pg. 6-6). Yet, the DoE has had significant setbacks in its attempt to attain a license for a federal repository for irradiated nuclear fuel at Yucca Mountain, and the federal policy banning the reprocessing of spent nuclear fuel far from intractable. In fact, the DoE was granted more than $67 million in fiscal year (FY) 2005 for the "Advanced fuel cycle initiative," a research and development program intended to provide technology to "recover the energy content in spent nuclear fuel," and it has requested $70 million from Congress for FY 2006 for the same program. This continued government interest in reprocessing, combined with the failure to establish a national repository for irradiated nuclear fuel, should compel the NRC to consider the impacts of spent fuel reprocessing in the final EIS. (150-11)(151-11)

Response: *Federal policy does not prohibit reprocessing; however, reprocessing is unlikely in the foreseeable future (NEPDG 2001). Table S–3 from 10 CFR 51.51 does include impacts from reprocessing. In Section 6.1.1 of this EIS, the contributions in Table S–3 for reprocessing, waste management, and transportation of wastes are maximized for either of the two fuel cycles (uranium only and no-recycle); that is, the cycle that results in the greater impact is used. As discussed in this EIS, 10 CFR 51.51(a) allows the applicant to use Table S–3 as the basis for evaluating the contribution of the environmental effects of the uranium fuel cycle that* *includes reprocessing. Section 6.1.1 was modified to indicate that Federal policy does not prohibit spent fuel reprocessing.*

Comment: Table 6-4, Page 6-25, Table G-1, Page G-5. Numerical values reported in Table 6-4 and Appendix G.
Most of the numerical values in Table 6-4 -page 6-25 - and Table G-1 - page G-5 - of the DEIS need to be re-calculated based on the information presented above and those reported in Appendix G of the DEIS. In addition text in the sections should be revised to reflect these any changes that would result from the re-calculation.

In Appendix G - page G.2 of the DEIS - which supports Table 6.4 it is stated that, "The surrogate AP1000 is a 1150-MW(e) advanced PWR power plant. The initial core load was estimated to be 84.5 MTU per reactor and annual reload requirements were estimated at 24.4 MTU/yr per reactor. The data in INEEL (2003) also indicated that the average uranium mass in an unirradiated surrogate AP1000 fuel assembly was 0.583 MTU and that 12 fuel assemblies per truck shipment would be transported. This resulted in about 14 truck shipments to supply the initial core and about 3.8 truck shipments per year to support refueling. For a site with two reactors, these estimates would be doubled."

If the staff then adds the initial load of 28 to refuel loadings (7.6 reloads per year x 39 years) to equal 324, not the 322 as stated in Table 6.4 and Table G-1 - line 7 - for the AP1000.

The same applies to the ACR-700, Appendix G - line 9 - page G-5 - which supports Table 6.4. It is stated that, "The AP-1000 is an advanced design Canada Deuterium Uranium (CANDU) reactor assumed to generate 731 MW(e). It was stated in INEEL (2003) that the initial core load for the ACR-700 included 61.3 MTU per reactor and the annual refueling requirements are 33.1 MTU/yr per reactor. Each fuel assembly contains 18 kg of uranium (INEEL 2003). This corresponds to 3406 fuel assemblies in the initial core loading and 1839 fuel assemblies per year for refueling. A range of truck shipment capacities was given in INEEL (2003) to be from 180 to 240 fuel assemblies per truck shipment. This equates to 15 to 19 truck shipments to supply the initial core load and from 7.7 to 10.2 annual refueling shipments. For a site with two reactors, these estimates would be doubled.

If the staff then added the initial loading of 19 to refuel loadings (10.2 annual reloads x 39 years) = 416, not the 628 as stated in Table 6.5.

The same applies to the IRIS reactor, where in appendix G - page G-2 - lines 22-29 - which supports Table 6.4. It is stated that, "The International Reactor Innovative and Secure (IRIS) design is a 335-MW(e) advanced PWR. It requires an initial core load of 48.67 MTU or 89 fuel assemblies per unit (546.9 kg of uranium per fuel assembly) (INEEL 2003). For refueling, the IRIS reactor was assumed to require an additional 6.26 MTU/yr of unirradiated fuel per reactor or approximately 40 unirradiated fuel assemblies every 3.5 years. INEEL (2003) indicates that a "typical" site may contain three reactors. Assuming each truck shipment carries eight fuel assemblies, the initial core load requires 28 truck shipments per three-reactor site and annual refueling requires an additional 4.3 truck shipments per year per three-reactor site.

If the staff adds the initial loading of 28 to refuel loadings (4.3 annual reloads x 39 years of operation) = 195, not the 201 as stated in Table 6.4.

On page G-2 of Appendix G - line 36 - it is stated that, "Annual average reload requirements are 510 fuel assemblies per reactor." However in Table 3.8-2 of the ER, this value is stated as 520 elements per reload.

It is unclear why the staff is only using values for the AP-1000 - one module site, as shown in Table 6-4/G-1.

Table 6.4 and G-1 of the DEIS should be revised to reflect the following:

Reactor Type	Number of Shipments/Site			Site Electric Generation MW(e)	Capacity Factor	Normalized Shipments per 1100 MW(e)
	Initial Core	Annual Reload	Total			
Reference LWR (WASH-1238)	18	6	252	1100	0.80	252
ABWR/ESBWR	30	6.1	268	1500 (1 reactor site)	0.95	165 (1 reactor site only)
AP1000	14	3.8	162 (1 module) or 324 (2 modules)	1150/module = 2300/ unit	0.95	131 (1 module site)
ACR-700	19	10.2	417 (1 module) or 834 (2 modules)	731/module = 1462/unit	0.90	279 (2 module site)
IRIS	28	4.3	196 (3 module site)	335/module = 1005/unit	0.96	178 (3 module site)
GT-MHR	51	20	831 (4 module site)	285/module = 1140/unit	0.88	729 (4 module site)
PBMR	44	20	824 (8 module site)	165/module = 1320/unit	0.95	113 (8 module site)

Since the numerical values in the last column of Table 6.4 - page 6-25, are used to calculate the values in the first column of Table 6.4 - page 6-26 of the DEIS. Conversely the numerical values in the last column of Table G-1 - page G-5, are used to calculate the values in the first column of Table G-3 - page G-8 of Appendix G, the tables should be changed to reflect the following:

Plant Type	Normalized Average Annual Shipments	Cumulative Annual Dose, Person-Sv/yr per 1100 MW(e)		
		Workers	Public-Onlookers	Public-Along Route
Reference LWR (WASH-1238)	6.3	1.10E-04	4.2E-04	1.0E-05
ABWR/ESBWR	4.1	7.43E-05	2.84E-04	6.76E-06
AP1000	3.3	5.88E-05	2.25E-04	5.35E-06
ACR-700	7.0	1.25E-04	4.79E-04	1.14E-05
IRIS	4.5	8.02E-05	3.06E-04	7.29E-06
GT-MHR	18.2	3.29E-04	1.25E-03	2.99E-05
PBMR	14.5	2.50E-04	9.6E-04	2.3E-05
10 CFR 51.52, Table S-4 Condition	<1 per day	4.00E-02	3.00E-02	3.00E-02

(141-134)

Response: For the AP1000, the difference between 322 shipments and 324 shipments is a rounding discrepancy. For the ACR-700, the calculations assumed the high end of the shipping cask capacities and assumed that there are two plants at the site. These assumptions resulted in 15 shipments for the initial load plus (7.7 per year x 39 yr) ≈ 314 total shipments. For two plants, the total would be approximately 628 shipments. For the IRIS reactor, the text on page G-2, line 27, states that "... the initial core load requires 34 truck shipments ...," not 28 as the commenter states. This explains the apparent six shipment discrepancy identified by the commenter. If the 34 shipment value is used, a total of 201 truck shipments are calculated, as indicated in Table 6-4. The discrepancy between the average reloading requirements for the GT-MHR (520 versus 510 fuel elements per year) is due to the selection of references. INEEL (2003) estimates that annual reloads will consist of 360 standard fuel elements, 60 control fuel elements, and 90 reserve shutdown fuel elements. The total is 510 fuel elements per year. The comment about the AP1000 one-module site arises from the fact that Appendix G is identical for all three ESP EISs. However, two of the ESP applicants assumed there would be one AP1000 per site and the other assumed two AP1000s per site. This difference is addressed by normalizing the number of shipments and impacts to a standardized reference electrical generation capacity, which effectively removes the assumed number of plants per site as a factor in the comparisons to 10 CFR 51.51, Table S-4. The conclusions are derived from comparisons involving the normalized values rather than the absolute values of the impacts. No change was made to the EIS as a result of the comment.

Comment: The draft EIS estimates that, for the reference reactor-year (a 1000-MW(e) LWR), 816,000 - metric tons (MT) of raw ore would be required to produce 900 MT of yellowcake for ultimate use as fuel after conversion, enrichment, and fabrication (EIS, § 6.1.2.4 and § 6.1.2.5).

Over time, as worldwide uranium ore supplies are depleted, requiring exploitation of less pure deposits of ore, would this ratio of ore to yellowcake increase? If so, would the environmental impacts of mining and milling become greater? (150-13)(151-13)

Response: If less pure ores are used, the ratio of raw ore to yellowcake would increase and the associated environmental impacts would increase proportionally. This also assumes that no new high-purity ore deposits are found and no fuel is reprocessed. The environmental impacts in the EIS were taken from Table S–3 of 10 CFR 51.51(a) which assumed conventional underground and strip mining of uranium ore. Two factors that will offset this increased impact are (1) the increased reliance on in situ leach mining for uranium, and (2) increased reliance on foreign sources for uranium. In situ leach mining has fewer environmental impacts compared to underground and strip mining of the ore because (1) the dusty ore-crushing process is not needed and (2) management of the extensive waste tailings that are generated is not needed. All steps in the in situ leach mining operation have the uranium in a less dispersible liquid form. In 2001 and 2002, the last years with reportable data, all the uranium produced in the United States was from in situ leaching operations (DOE 2003). This same report indicated that foreign-origin uranium accounted for 88 percent of the uranium purchases for U.S. civilian nuclear power plants in 2002. No change was made to the EIS as a result of these comments.

Comment: Fuel Cycle, Transportation, and Decommissioning, Section 6.1.1.6, Radioactive Wastes, page 6-13. Due to changes in the Yucca Mountain facility, changes in estimations of the waste to be transported for disposal may need to be re-evaluated to assure that the previous estimations still are applicable. (172-49)

Response: In Section 6.2.2 of the EIS, the staff estimated the environmental impacts of transport of spent fuel from the proposed advanced reactor designs on the proposed ESP site to a Federal waste repository assumed to be on the Yucca Mountain site. No change was made to the EIS as a result of the comment.

Comment: Fuel Cycle, Transportation, and Decommissioning, Section 6.1.1.5 Radioactive Effluents, page 6-10, last paragraph. For a compliance demonstration, using a postulated maximally exposed individual (MEI) with a modeled maximum anticipated exposure - would provide a better comparison than using a population dose model that is effectively averaged out over the entire populace of a given area. (172-46)

Comment: Fuel Cycle, Transportation, and Decommissioning, Section 6.1.1.5 Radioactive Effluents, page 6-11, paragraph 2. For a compliance demonstration, using a postulated maximally exposed individual (MEI) with a modeled maximum anticipated exposure would provide a better comparison than using a population dose model that is effectively averaged out over the entire populace of a given area. (172-47)

Response: Per the guidance in 10 CFR 51.51 and Section 5.7 of NUREG-1555 (NRC 2000), the staff relied on Table S–3 as a basis for the impact of uranium fuel-cycle impacts. Table S–3 only provides population dose estimates. No change was made to the EIS as a result of these comments.

Comment: A 1,000 megawatt reactor, the size of Clinton, generates 20 to 30 tons of high level radioactive waste per year. This waste contains byproducts of nuclear fission, which nature does not. They are man-made. One of these products is plutonium, the half life of which is 24,000 years. Plutonium is so deadly that less than one millionth of one gram is cariogenic. 24,240 years, which is the lethal life, is 10,000 generations. What on earth are we leaving our children? Are we even going to leave them an inhabitable earth? (94-6)(46-6)

Response: Plutonium formed in the fuel during reactor operation remains bound in the spent fuel and will be disposed of in a Federal repository. Radioactive material in the environment will be monitored from any new units on the Exelon ESP site as part of the Radiological Environmental Monitoring Program (REMP) similar to the program at the CPS. Results of REMP will be reported annually in the Annual Radiological Environmental Operating Report. Gaseous and liquid effluents from the plant will be monitored and reported annually in an Annual Effluent Monitoring Report. The EIS evaluated the impact of radiological emissions from the proposed reactor(s) on the Exelon ESP site in Section 5.9 of the EIS and found results to be within regulatory limits.

No change was made to the EIS as a result of these comments.

Comment: Question. On page 6-4 line 25 it seems to state that this reactor will have an effluent of 400,000 curies per year of Kr-85. I thought Kr-85 was a fission product and that it was captured in the fuel rod, not released as an effluent. Am I reading this right?? (183-1)

Comment: What concerns me most about this document is that on line 26, on page 6-4, it states that 400,000 curies, of the radioactive gas krypton 85 would be released every year by the new reactor, as is already probably being done by the current reactor. This means that a little bit of this highly radioactive gas is released into the air every day, or at least every week.

Now, curie is a measure of radiation, or actually, radioactivity. But how does 400,000 curies relate to that dot of uranium? One curie is the radioactive equivalent of 3 million of those dots, or those, I might say possibly lethal dots.

The question we need to ask is how much will this affect the health of the people and the animals downwind of the plant? Nobody knows. Governance made no studies, there's no statements to this. (73-2)

Comment: 6-4 shows that 240,000 metric-tons of tailings and 91,000 metric-tons of solids are generated making the nuclear fuel. In terms of radiation producing elements created, 18,000 Curies of Tritium and 400,000 Curies of Krypton-85 are produced per year. And yet this is listed in the effluents. Isn't this a tremendous amount of radioactive material to be vented into the surrounding atmosphere? (153-14)

Response: *Krypton-85 is a fission product that is released as a gaseous effluent. Section 6.1 of the EIS is referring to fuel-cycle facilities, and the tritium and krypton-85 releases are from reprocessing. Although current Federal energy policy does not prohibit reprocessing of spent fuel, additional work is needed before commercial reprocessing is begun (see Section 6.1.1 of the EIS). Krypton-85 releases from plant parameter envelope reactor design, as discussed in the applicant's ER, are estimated to be 3×10^{14} Bq (8200 Ci). No changes were made to the EIS as a result of these comments.*

Comment: Section 6.1.2, Page 6-16, Lines 20-25. "Exelon (2003) compared the impacts in Table S-3 LWR with those of the gas-cooled reactor designs. The comparison used an annual fuel loading as a starting point and then proceeded in reverse direction through the fuel cycle (i.e., fuel fabrication, enrichment, conversion, milling, mining, radioactive waste). Table 6-3 provides an estimate of the impacts for each phase of the uranium fuel cycle, assuming that the ESP site would host two GT-MHR units or one PBMR unit with the multiplier factors described above." The wording in the last sentence should be changed to read, "Table 6-3 provides an estimate of the impacts for each phase of the uranium fuel cycle, assuming that the ESP site would host two-four module GOT-MHR units or one-eight module PBMR unit with the multiplier factors described above." (141-100)

Response: *Section 6.1.2 of the EIS was revised to state, "Table 6-3 provides an estimate of the impacts for each phase of the uranium fuel cycle, assuming that the ESP site would host two four-module GT-MHR units or one eight-module PBMR unit."*

Comment: Section 6.2, Page 6-23. "Exelon used a sensitivity analysis to show that transportation impacts from advanced LWR designs would be bounded by the criteria identified in Table S-4 (Exelon 2003). Exelon referenced the related discussion and information in NUREG-1437, Addendum 1 (NRC 1999) to support its basis for exceeding 4 percent uranium-235 enrichment and 33,000 MWd/MTU. However, as discussed above, NUREG-1437, Addendum 1 applies to reactors that are listed in NUREG-1437, Appendix A and not to any other reactor designs. Exelon also used a sensitivity analysis to show that transportation impacts from the advanced gas-cooled reactor designs would be bounded by the criteria identified in Table S-4 (Exelon 2003); however, as discussed previously, this type of analysis does not adequately meet the requirements of 10 CFR 51.52." EGC disagrees with the staff's conclusion. 10 CFR 51.52 does not prohibit the use of sensitivity analyses to determine the environmental impacts of transportation of fuel from reactors not covered by the criteria in Section 51.52. (141-111)

Response: *Section 6.2 of the draft EIS states the following: "The environmental impacts of transportation of fuel and radioactive wastes to and from nuclear power facilities were resolved generically in 10 CFR 51.52, provided that the specific conditions in the rule are met; if not, then a full description and detailed analysis is required for initial licensing." As stated in Section 6.2, none of the proposed advanced LWR reactor designs meets all the conditions in 10 CFR 51.52; therefore, detailed analyses are needed. However, as stated in Section 6.2.4, the staff concludes that the environmental impacts of transporting fuel and radioactive waste to and from advanced LWRs would be SMALL. No change was made to the EIS as a result of this comment.*

Comment: Section 6.2, Page 6-22, Lines 27-28. "The ACR-700, ABWR, AP1000, and ESBWR designs exceed the 3800-MW(t) core thermal power-level limit." However in Appendix G - page G.4 - lines 20-22 - it is stated that, "As shown above, single unit ABWR and ESBWR plants exceed the 3800 MW(t) condition in 10 CFR 51.52 (a)(1). In addition, the twin reactor ACR-700 site exceeds the core thermal power condition." There is no information given about the AP1000. This sentence should read, "The ACR-700 (3,964 MW(t)/unit - 2 modules each producing 1,982 MW(t)), ABWR (single module unit producing 4,300 MW(t)), AP1000 (6,800 MW(t)/unit - 2 modules each producing 3,400 MW(t)), and ESBWR (single module unit producing 4,300 MW(t)) designs exceed the 3800-MW(t) core thermal power-level limit." (141-110)

Response: *Section 6.2 of the EIS was revised to read as specified in the comment: "The ACR-700 (3,964 MW(t)/unit - 2 modules each producing 1,982 MW(t)), ABWR (single-module unit producing 4,300 MW(t)), AP1000 (6,800 MW(t)/unit - 2 modules each producing 3,400 MW(t)), and ESBWR (single-module unit producing 4,300 MW(t)) designs exceed the 3800-MW(t) core thermal power-level limit."*

Comment: Section 6.2.2.1, Page 6-29, Lines 24-29. "The bounding cumulative doses to the exposed population given in Table S-4 are: * 0.04 person-Sv (4 person-rem) per reference reactor-year to transport workers * 0.03 person-Sv (3 person-rem) per reference reactor-year to general public (onlookers) * 0.03 person-Sv (3 person-rem) per reference reactor-year to general public (along route). Population doses to the crew and the onlookers for all the reactor types, including the reference reactor found in Table 6-7, exceed Table S-4 values."

Upon review of Table 6-7 - Routine (Incident-Free) Population Doses from Spent Fuel Transportation, Normalized to Reference LWR - page 6-31 - and Table G-7 - Routine (Incident-Free) Population Doses from Spent Fuel Transportation, Normalized to Reference LWR Net Electrical Generation - page G-18 - a lot of the reported values do not exceed Table S-4 criteria.

For example from Table 6-7 and G-7 for the Clinton Site (bold values indicate exceedance of Table S-4 values):

ABWR/ESBWR (person-rem)			Table S-4 Value (person-rem)		
Crew	Onlookers	Along Route	Crew	Onlookers	Along Route
2.9	**10**	0.18	4	3	3

AP-1000 (person-rem)			Table S-4 Value (person-rem)		
Crew	Onlookers	Along Route	Crew	Onlookers	Along Route
2.8	**9.7**	0.18	4	3	3

ACR-700 (person-rem)			Table S-4 Value (person-rem)		
Crew	Onlookers	Along Route	Crew	Onlookers	Along Route
6.4	**22**	0.41	4	3	3

IRIS (person-rem)			Table S-4 Value (person-rem)		
Crew	Onlookers	Along Route	Crew	Onlookers	Along Route
2.5	**8.5**	0.16	4	3	3

GT-MHR (person-rem)			Table S-4 Value (person-rem)		
Crew	Onlookers	Along Route	Crew	Onlookers	Along Route
2.4	**8.2**	0.15	4	3	3

PBMR (person-rem)			Table S-4 Value (person-rem)		
Crew	Onlookers	Along Route	Crew	Onlookers	Along Route
0.80	2.8	0.051	4	3	3

(141-116)

Response: *Section 6.2.2.1 of the EIS was revised to state the following: "Population doses to the crew for the ACR-700 and to onlookers for the ABWR, ESBWR, AP1000, ACR-700, IRIS, and GT-MHR exceed the Table S–4 values."*

Comment: Section 6.2.2.1, Page 6-29, Lines 1-4. "For purposes of this analysis, their design was assumed to be the same as those used for the existing LWRs. Spent fuel shipping cask designs for gas-cooled reactors will be evaluated at the CP or COL stage if the applicant references such designs." All casks designed for the shipment of spent fuel from advanced LWRs or Gas Cooled reactors will have to comply with the shipping requirements specified in 10 CFR 20 and therefore additional information regarding the design is not warranted. Therefore, the environmental impacts associated with Fuel Transportation are SMALL. (141-115)

Response: Section 6.2.4 states that the NRC staff believes the impact of transporting fuel and radioactive waste to and from gas-cooled reactors are likely to be small. However, gas-cooled reactor fuel performance and shipping cask design information is insufficient at this time to support a definitive conclusion. No change was made to the EIS as a result of the comment.

Comment: Fuel Cycle, Transportation, and Decommissioning, Section 6.1.1, Light-Water Reactors, page 6-7, paragraph 3. Please clarify whether the information cited in the table scaled all of the information columns as well as the potential of differing impacts that the scaling may cause. (172-44)

Response: Table 6-1 in the EIS is a reprint of Table S–3 from 10 CFR 51.51. The values in Table 6-1 are not the scaled values. Scaled values are presented and their impacts are discussed in Sections 6.1.1.1 - 6.1.1.9 of the EIS. No change was made to the EIS as a result of the comment.

Comment: Table 6-10, Page 6-36. Values in the second column - Annual Waste Volume m3/yr -For the AP-1000 and IRIS the value should be 110 (2 x 55) and 75 (3 x 25), respectively. (141-117)

Response: Table 6-10 of the EIS was revised to change the following entries in the "annual waste volume" column: (1) AP1000 was changed from 112 to 110 and (2) IRIS was changed from 74 to 75.

Comment: Table 6.3, Page 6-17, Line 19. "Category of Low-level waste from reactor decontamination and decommissioning (Ci per reference reactor-year) - data is not available." In Table 5.7.1 of the ER - Gas Cooled Fuel Cycle Impact Evaluation - the following information is listed. (141-103)

Response: Table 6-3 of the EIS was modified to include the low-level waste information for the pebble-bed modular reactor.

Comment: Table G-13, Page G-32, Line 3. Waste Generation Information. In the second column of Table G-13 - labeled DOE (2003) Waste Generation Information, there is no reference DOE (2003) listed in the reference table at the end of Appendix G. The reference should be INEEL (2003). (141-126)

Response: Table G-13 of the EIS was revised to change the DOE (2003) reference to the INEEL (2003) reference.

Comment: This is from, this flyer, called Radiation Nation, and this is what it says in the first paragraph. "The nuclear industry and its allies in Government, want to transfer nuclear, its nuclear waste problem to the American public. The industry is working behind the scenes to deregulate nuclear waste so that it can be recycled into household products and dumped into landfills." That is one of the most ludicrous statements I have ever read. I heard stuff like this when we were talking about siting a low level waste repository in this state. (160-3)(70-3)

Comment: Fuel Cycle, Transportation, and Decommissioning, Section 6.1.2 Gas-Cooled Reactors, pages 6-15, 6-16. Similar issue can be raised for this type of reactor as for the Light-Water Reactors. (172-52)

Response: The comments are noted. No change was made to the EIS as a result of these comments.

Comment: When I first decided to pursue a career in nuclear energy, my intention was to work towards the development of a solution toward the nuclear waste problem, which I -- because I don't believe the problem really exists...it became apparent to me that there were numerous technically sound, and scientifically valid options to deal with nuclear waste. The most important of which was recycling and reprocessing the spent fuel.

It is my opinion that the only reason that these options have not developed and implemented is political grandstanding by those who oppose anything nuclear, because the development of the implementation of a, as of the viable solution to the nuclear waste problem would remove the primary, their primary objection to nuclear energy in general. (75-2)

Comment: Fuel Cycle, Transportation, and Decommissioning, Section 6.1.1.6, Radioactive Wastes, page 6-14 paragraph 3. With this disposal location being problematic at this time, we concur with this estimation and expectation to meet any changes that are proposed for the Yucca Mountain facility. (172-50)

Comment: Fuel Cycle, Transportation, and Decommissioning, Section 6.2, Transportation of Radioactive Materials, page 6-23, paragraph 3 and following bullet points. The listing of the potential confounding factors to the evaluation is very helpful in getting a more realistic picture of the potential situations that may be encountered. (172-55)

Comment: Illinois has 14 nuclear reactors; three are not operating. We thus have more nuclear waste than any other state. (46-4)

Comment: Right here in Clinton and in reactors the size of Clinton, in the cooling pool sits spent fuel. The spent fuel is thousands and thousands of times more radioactive when it's taken out of the reactor core than when it went in. In the cooling pool there is the radioactive equivalent of one thousand Hiroshimas. (46-8)

Comment: The other thing about the fuel cycle, if the plans of the industry and the politicians they bought in their apparatchiks come to fruition and we have a nuclear economy, there's only

something on the order of 30 years worth of uranium. So that makes no sense unless there's a
plan afoot not fully communicated to build breeder reactors and go to plutonium. And I think
that's even more horrible. (55-5)

Comment: And spent nuclear fuel can hardly be considered waste when 95 percent of it can
be recycled as fuel for future reactors. (66-7)

Comment: I'd like to say that we have the technical expertise to deal with the nuclear waste.
The thing standing in the way are social and political issues. And the technical expertise is out
there, and processes are already developed. (77-4)

Comment: Their processing of spent uranium rods in the U.S. has proven uneconomical. It's
just cheaper to mine fresh uranium and enrich it. (83-1)

Comment: The issue of reprocessing was brought up, and yes, it is not economical now. But
with special nuclear material from, from, with weapons created from plutonium, we're buying
back our warheads, that's understandable. But eventually, the supply of uranium will run out.
And the political pressure of building another repository will make, will make it so reprocessing
will become politically economical, rather than just right, than just straight up costs I believe.
(86-4)

Comment: The cost of reprocessing spent nuclear fuel has already been paid for because
there is a small tax of less then 1/10 millionth of a cent per kilowatt hour that has generated
billions of dollars to pay for centralized storage and it will pay for the new waste as it is
produced. (93-5)

Comment: Illinois has 14 nuclear reactors; three are not operating. We thus have more
nuclear waste than any other state. (94-4)

Comment: Right here in Clinton and in reactors the size of Clinton, in the cooling pool sits
spent fuel. The spent fuel is thousands and thousands of times more radioactive when it's
taken out of the reactor core than when it went in. In the cooling pool there is the radioactive
equivalent of one thousand Hiroshimas. (94-8)

Response: *The comments are noted. The comments do not provide new information and will
not be evaluated further. No change was made to the EIS as a result of these comments.*

Comment: Section 6.1.2.2, Page 6-18, Lines 27-29. "To produce 120 MT of enriched U02 for
the 1000-MW(e) LWR-scaled model, the enrichment plant needs to produce about 156 MT of
UF6, which requires approximately 400 MT of SWUs (Exelon 2003). This is not a direct quote
from the ER. From Section 5.7.2.3.2 of the ER - Uranium Enrichment - the quote should have
been, "In order to produce the 40 MT of enriched UO2 for the reference LWR, the enrichment

plant needed to produce 52 MT of UF6, which required 127 MT of SWU (USNRC 1976)." The quote should have been followed by the Staff's sentence, "In order to produce 120 MT of enriched U02 for the 1000-MW(e) LWR-scaled model, the enrichment plant needs to produce about 156 MT of UF6, which requires approximately 400 MT of SWUs." (141-105)

Response: Section 6.1.2.2 of the EIS was modified to read, "In order to produce the 40 MT of enriched UO_2 for the reference LWR in WASH-1238 the enrichment plant needed to produce 52 MT of UF_6, which required 127 MT of SWU (NRC 1976). Therefore, to produce 120 MT of enriched UO_2 for the 1000-MW(e) LWR-scaled model, the enrichment plant needs to produce about 156 MT of UF_6, which requires approximately 400 MT of SWU."

Comment: Fuel Cycle, Transportation, and Decommissioning, Section 6.2.1.1, Normal Conditions, pages 6-25. MEI scenarios need to be provided for a better understanding of potential exposures. (172-56)

Comment: Fuel Cycle, Transportation, and Decommissioning, Section 6.2.2.1, Normal Conditions, pages 6-29, bulleted points. The population dose is provided, but no MEI data is provided to determine of the maximum exposures expected for each type of scenario evaluated. (172-58)

Response: Section 6.2.2.1 of the EIS was revised to include an analysis of the potential maximally exposed individuals (MEIs) during normal transportation of fuel and wastes. The analysis addressed maximum individual exposures to truck crews, vehicle inspectors, residents, individuals stuck in traffic next to a shipment, and persons at a service station where a shipment stops for refueling.

Comment: 6-3 shows that 118,000 metric-tons (2,200 pounds per metric-ton) of coal, 323,000 mega-Watt-hours of electricity and 135,000,000 cubic feet of natural gas are used to make the nuclear fuel. This is not a greenhouse gas-free process. (153-13)

Response: The energy requirements are based on the experience of the uranium industry during the 1970s (AEC 1974) when gaseous diffusion enrichment technology was used. Table S-3A in WASH-1248 (Environmental Survey of the Uranium Fuel Cycle) shows that 310,000 MW-hr of the 323,000 MW-hr electrical energy and 113,000 MT of the 118,000 MT coal equivalent total were due to the gaseous diffusion enrichment facility. Current enrichment technology in the United States will likely rely on the gaseous centrifuge technology that will use 90 percent less energy than the gaseous diffusion enrichment technology. No change was made to the EIS as a result of the comment.

Comment: The draft EIS lacks a consideration of the environmental and public health impacts resulting from military applications of depleted uranium, (DU), a byproduct of the enrichment process of the fuel cycle. Moreover, there is not a complete consideration of the impacts of managing this substance as a waste. There is no repository established for the permanent disposal of depleted uranium, but the impacts of such a hypothetical facility should be considered. (150-12)(151-12)

Response: The environmental and public health impacts resulting from military applications of depleted uranium and deposition of depleted uranium waste are beyond the scope of the EIS. No change was made to the EIS as a result of these comments.

Comment: Section 6.1.2, Page 6-16, Lines 6-11. "One of the other-than-LWRs considered by Exelon, the Gas Turbine-Modular Helium Reactor (GT-MHR), is a four-module, 2400-MW(t), nominal 1140-MW(e) unit assumed to operate at an annual capacity factor of 88 percent for a net electric output of 1032 MW(e). Therefore, the maximum number of GT-MHR units that could be sited at the Exelon ESP site and remain below the 2200-MW(e) total net electric output PPE for the site is two (i.e., 2 x 1032). This would result in a factor of 2.5 (i.e., 2064/800) for comparison with Table S-3 and LWRs." It should be noted that 1140 x .88 = 1003.2 therefore the paragraph should read: "One of the other-than-LWRs considered by Exelon, the Gas Turbine-Modular Helium Reactor (GT-MHR), is a four-module, 2400-MW(t), nominal 1140-MW(e) unit assumed to operate at an annual capacity factor of 88 percent for a net electric output of 1003 MW(e). Therefore, the maximum number of GT-MHR units that could be sited at the Exelon ESP site and remain below the 2200-MW(e) total net electric output PPE for the site is two (i.e., 2 x 1003). This would result in an approximate factor of 2.5 (i.e., 2006/800) for comparison with Table S-3 and LWRs." (141-99)

Response: The GT-MHR power rating was revised to 1003 MW(e) in Section 6.1.2 of the EIS.

Comment: Fuel Cycle, Transportation, and Decommissioning, Section 6.2, Transportation of Radioactive Materials, page 6-23, paragraph 2. The information refers to NUREG-1437, Addendum 1, and specific types of reactors that are covered in this evaluation of transportation issues. USNRC needs to clarify whether or not all of the proposed types of reactors that are reasonably expected to be considered for construction are covered under this evaluation. (172-54)

Response: The referenced sentence in Section 6.2 was revised to state, "However, the GEIS, Addendum 1, applies to reactors that are listed in the GEIS, Appendix A, which does not address advanced reactors."

Comment: Fuel Cycle, Transportation, and Decommissioning, Section 6.2, Transportation of Radioactive Materials, page 6-21, paragraph 2. The references with regard to transportation are from 1972 and 1975. Newer information and transportation requirements have been put in place by various Federal Agencies that would not be taken into consideration under this guidance. Re-evaluation of this aspect should be conducted to assure that these issues are appropriately addressed. (172-53)

Comment: Fuel Cycle, Transportation, and Decommissioning, Section 6.2.2.1, Normal Conditions, pages 6-30, paragraph 1. It is unclear if the evaluation provided has taken the new specifications for the transportation and disposal casks into consideration in this report. (172-59)

Response: *The NRC has conducted several transportation studies to evaluate the risk of transportation of radioactive material. NUREG-0170 (NRC 1977b), supported NRC's 10 CFR Part 71, "Packaging and Transportation of Radioactive Material" rulemaking. Based on this study, the Commission concluded that the transportation regulations are adequate to protect the public against unreasonable risks from the transport of radioactive materials, including spent fuel. The NRC sponsored another study in the 1980s entitled, "Shipping Container Response to Severe Highway and Railway Accident Conditions," NUREG/CR-4829 (Fischer et al. 1987), or the "Modal Study." Based on the results of this study, the NRC staff concluded that NUREG-0170 overestimated spent fuel accident risks by about a factor of three. In March 2000, the NRC initiated another spent fuel study, "Reexamination of Spent Fuel Shipment Risk Estimates," NUREG/CR-6672 (Sprung et al. 2000). This study focused on risks of a modern spent fuel transport campaign from reactor sites to possible interim storage sites and/or permanent geologic repositories. This study concluded that accident risks were much less than those estimated in NUREG-0170 and that more than 99 percent of transportation accidents are not severe enough to damage NRC-certified spent fuel casks. While very severe accidents could cause cask damage, the studies show that releases of material would be small and pose little risk to the local population/public. The most severe accidents might cause greater releases, but their likelihood is so remote that the NRC considers the risk to public health to be low.*

The NRC's regulations for the safe transportation of radioactive materials have evolved over the years, e.g., the 2004 revisions to achieve compatibility with International Atomic Energy Agency (IAEA) Transportation Safety Standards. However, the basic specifications for shipping containers have largely not changed. For instance, Type B shipping containers, such as those used for spent fuel, must withstand the effects of a 9-m (30-ft) drop test, puncture, fire, and immersion. Basic radiological dose rate limitations have not changed. While some of the details in the regulations have been changed, the NRC staff believes that the basic safety standards that determine the performance of shipping containers under normal and hypothetical accident conditions have been properly accounted for in the EIS. No change was made to the EIS as a result of these comments.

Comment: Why would an informed public support building additional nuclear reactors when we don't have a workable long term solution for the nuclear waste already in existence? (02-2)

Comment: This country's ability to manage nuclear waste in a responsible manner has never been demonstrated. In fact, THERE IS NO SAFE WAY TO DISPOSE OF NUCLEAR WASTE. (04-2)

Comment: Do not further deceive the public with claims that you have any handle on storage of waste or possible melt down. Please face the facts and the truth and stop making these things!!!!! (05-4)

Comment: Another grave concern I have is with nuclear waste transport and storage... The lifetime of high level radioactive waste is so long that it is unknown and will last for as infinite number of generations. (06-2)

Comment: There is still no solution to the waste problem; the proposed nuclear waste dump at Yucca Mountain is in a downward spiral and wouldn't be large enough to hold waste from a new reactor even if it did go forward. (09-6)

Comment: Nuclear waste products cannot be transported or disposed of in any remotely secure manner and must be stored for tens of thousands of years to be rendered safe. How would it be possible to find appropriate, unbreachable storage containers and an undisturbed location for such materials? How can we communicate the danger to our descendants eons in the distant future? How can we burden our future generations with such a perilous legacy? (10-2)

Comment: I am also appalled at the lack of foresight on the NRC's part when they judge environmental/health impacts of nuclear power as "small" and not significant. In exchange for a probable 30 years of functioning capacity for this proposed new nuclear plant, it will produce tons of radioactive waste that will remain dangerous for thousands of years (and which we have no safe way to dispose of). (109-3)

Comment: Calling this technology "clean" and "safe" is utterly ridiculous. Is a few years of increased energy generation worth thousands of years of contaminations of our environment? How can anyone claim to be protecting human health by advocating an energy source that produces dangerous waste that will persist for beyond all of our lifetimes? (109-4)

Comment: Additionally, the safe disposition of nuclear waste has not been resolved, particularly in light of the new questions raised about Yucca Mountain. (11-3)

Comment: The disposal of radioactive waste poses a threat to generations and is clearly not acceptable. (12-3)

Comment: Use your funds and energy to develop alternative sources of energy which do not put the population at risk and which do not produce waste products which are hazardous to all life and which are impossible to dispose of with long term safety. (13-3)

| **Comment:** Nuclear waste is a problem and will continue to be a very big problem. The site in Nevada has been determined unsafe and some research falsified concerning waste disposal and safety. (15-2)

| **Comment:** Nuclear power plants create nuclear waste, and our government -- as well as everybody else -- has no real answer to "What do we do with nuclear waste?" It is long past the time that we should have stopped destroying our world. We MUST stop now. No more nuclear waste, therefore no more nuclear plants. Quite simple! (19-2)

| **Comment:** Another unwarranted assumption of this report is that the radioactive waste will be successfully moved to a geologically secure site someplace else, such as the proposed Yucca Mountain repository in Nevada. This is very unlikely. The Yucca Mountain site is subject to earthquakes and water leakage, is tied up in the courts, and is opposed by the state of Nevada; the data used to justify the site may have been falsified by scientists. Even if it is eventually approved, it is likely to be full of radioactive waste long before the second Clinton plant is operational. Nothing in this report addresses what might happen if the Yucca Mountain repository or an alternative similar site for radioactive waste is not available; this is a serious oversight. The nuclear waste from the second Clinton plant may well have to stay on the Clinton site, perhaps in dry casks. I don't know how safe dry casks are on a year by year basis, but even if the risks are small, they are cumulative over time, year by year. I am absolutely certain that they will crumble, leak, and disintegrate within in the hundreds of thousands years that the highly concentrated radioactive waste containing plutonium-239 and other radioactive elements are deadly to humans, that is, to our descendants. (110-4)

| **Comment:** There is no safe means to store spent nuclear fuel. It is not acceptable to bury radioactive nuclear waste on site or anywhere. Doing so creates an enormous health risk for all living creatures (including people) that could contact the material for thousands of years. (148-4)

| **Comment:** The draft EIS fails to evaluate the environmental impacts and security threat of indefinitely storing the additional irradiated fuel that would be generated by the proposed additional nuclear unit onsite. Another nuclear unit at Clinton could create annually 20 to 30 metric tons of additional irradiated fuel to the site. Despite the NRC's Waste Confidence Decision, the only national repository site under consideration, Yucca Mountain in Nevada, is far from a done deal. Numerous scientific questions remain about whether the site can safely store waste, and, recently, a scandal has erupted over the possible falsification of scientific studies used to justify the geologic suitability of the site. The NRC's assumption that at deep repositories like Yucca Mountain "no [radioactive] release to the environment is expected" (EIS, pg. 6-13) is unfounded-the geologic integrity of this site is far from proven. Moreover, the Department of Energy (DoE) has not yet submitted its license application to the NRC, although the statutory deadline was more than two years ago. DoE was supposed to begin accepting waste in 1998 and is highly unlikely to meet its revised goal of accepting waste by 2012.

Further, Illinois law [220 ILCS 5/8-406(c)] prohibits the construction of a new nuclear power plant until the director of the Illinois Environmental Protection Agency finds that the U.S. government has identified and approved and demonstrable technology or, means for the disposal of high-level nuclear waste.

Even if Yucca Mountain is opened, the site cannot hold the high-level radioactive waste that will be generated by existing reactors after 2010. Therefore, in addition to the waste generated by existing reactors, waste created by a new nuclear unit at Clinton would also have to remain onsite for an indefinite period of time. The NRC recently approved an unprecedented 40-year license extension for the nuclear operator Dominion to store high-level nuclear waste on-site at its Surry nuclear plant near Williamsburg, Virginia, indicating that fuel can reasonably be expected to be stored at reactor sites for at least that long. The environmental impacts of, indefinite storage must be thoroughly evaluated in the final EIS. (150-10)(151-10)

Comment: 6-6 shows a figure that draws an object that represents the "proposed Federal Waste Repository." This is now known to be the Yucca Flats Repository in Nevada that is also known to be non-functional. So the spent nuclear fuel has no place to go. Where are you going to put it? (153-15)

Comment: The Draft EIS does not consider the impacts of the storage of high-level nuclear waste. Despite noting some concern that the proposed Yucca Mountain repository will not open in a timely fashion, the NRC Staff continues to rely on the Waste Confidence Rule ("WCR"), 10 C.F.R. 51.23, to conclude that any impacts from the storage of high-level waste would be "acceptable." (Draft EIS at 6-14). The Staff's discussion of this issue, however, is clearly inadequate. The WCR is based on the assumption that sufficient repository capacity will exist to store all waste created by nuclear plants. As described in Section II above, however, the Draft EIS significantly downplays the significant delays and safety concerns that raise serious questions about whether Yucca Mountain will ever open. More importantly, the possible construction of new nuclear power plants entirely undermines the WCR. As previously mentioned, the proposed Yucca Mountain facility does not even have the capacity to store all of the high-level wastes that will be created by existing nuclear power plants, much less new plants. Therefore, the NRC must consider the impacts of the storage of additional high level waste at the Clinton site. (170-13)

Comment: Section 6: This section does not address the current practice of onsite dry cask storage of spent nuclear fuel. Although technical and political solutions may yet be found to provide a disposal path for spent fuel, the EIS should identify the current impacts of the current practice of onsite storage, and evaluate the impacts of the proposed action. Onsite storage of spent fuel has significant public interest associated and should be specifically addressed in this EIS. (172-67)

Comment: The NRC's environmental impact statement also fails to evaluate the security threat of indefinitely storing onsite the additional nuclear waste that would be generated by the proposed new nuclear unit. Another nuclear reactor at Clinton could create 20 to 30 metric tons of high-level radioactive waste, annually. To date, there is no feasible solution to safely and permanently dispose of this waste, which must cool onsite for five years before it can be moved. (191-4)

Comment: Other important factors, such as waste, are ignored altogether. Given that the proposed Yucca Mountain repository is a long way from ever opening, and that even if it does it will not have the capacity to accept waste from a new reactor at Clinton, waste concerns must be taken into consideration. No analysis of whether the site is suitable for indefinite storage of high-level waste is included in the DEIS. (112-6)(113-6)(114-6)(115-6)(116-6)(117-6)(118-6) (119-6)(120-6)(121-6)(122-6)(123-6)(124-6)(125-6)(126-6)(127-6)(128-6)(129-6)(130-6)(131-6) (132-6)(133-6)(134-6)(135-6)(136-6)(137-6)(138-6)(139-6)(140-6)(142-6)(143-6)(144-6)(145-6) (146-6)(147-6)(149-6)(154-6)(155-6)(158-6)(159-6)(162-6)(163-6)(164-6)(165-6)(166-6)(167-6) (173-6)(174-6)(175-6)(176-6)(177-6)(178-6)(180-6)(181-6)(182-6)(185-6)(186-6)(187-6)(188-6) (189-6)(190-6)(192-6)(193-6)(194-6)

Comment: Since there is no safe way to store waste, it is idiocy to create more. (20-2)

Comment: (Because of weak regulations, bad science, and "pronuclear hanky-panky," NEIS is opposed to Yucca Mountain, Nevada, as the preferred high-level nuclear waste site being backed by the nuclear industry and the Bush-Cheney Administration.) IT IS JUST "GOOD OLE FASHIONED" COMMON SENSE TO SAY THAT BUILDING CLINTON-2 WITHOUT AN APPROVED NUCLEAR WASTE SITE IS LIKE BUILDING A NEW HOME WITHOUT A TOILET. (25-6)

Comment: But the radioactive waste, there isn't a long term plan. After decommissioning, who will take care of it? Chapter 6, as they referred to, in the EIS, addresses many factors, but not long term storage. Who takes care of it? Who pays for it? What is an acceptable risk?...No longer industry is responsible, the taxpayer is. Who will be responsible for this nuclear solid waste, when Exelon's done with Clinton, Illinois? Not Exelon. We will. The taxpayer. We will pay.

Not us, no, we will be gone. But our children, our children's children, and our children's children's children, and on and on, they will pay. It's a sacrifice I'm not willing to make with a nuclear power plant. Are we ready to sell out, to sell our soul to the highest bidder, because we want energy? (37-2)

Comment: We keep generating more and more nuclear waste and yet there is absolutely nowhere for it to go. Yucca Mountain is not safe. Dry casks are not safe. Cooling pools right here in Clinton are not safe. And I'm sure that nuclear engineers could differ with me. I am not a nuclear engineer. I'm a human being. (46-7)

Comment: A major problem that I see with the report is that I didn't see anything in there on how the radiation waste would be taken care of. It assumes that the Yucca Mountain will or something alternative to it, will take place. I just don't think that's a certainty. I think these wastes will remain in Clinton for over 100,000 or maybe more years with all the potential radiation leakage, drainage, water problems and so on. And for that reason I think the report is defective and should be rejected. (58-3)

Comment: I wanted to touch on another reason why a proposed nuclear power plant is a bad idea is that there is no way designs currently to permanently store the waste created by the plant. Nuclear power creates radioactive waste that must be stored for tens of thousands of years. Yet there is currently no repository for storing this waste. And we believe building a nuclear power plant without having a way to dispose of the waste is similar to building a house without a toilet. It just doesn't make sense.

The draft EIS tries to dodge this by relying on a waste confidence rule. It says there will be a site but the only site under consideration, Yucca Mountain, has been delayed for decades and it couldn't open till at least 2015, most likely, and most importantly, wouldn't even have capacity to store the waste from the existing plants much less from a new plant. So a whole new repository would be needed. And clearly there's no plans on the table for creating such a repository. (59-5)

Comment: In particular, the NRC draft environmental impact statement has trivialized the harmful environmental impacts of both the current and new nuclear waste generation with the proposed expansion of the Clinton site. The NRC staff has concluded that the environmental impacts of the radioactive waste is small and they interpret that as the affects are not detectable. In the same time, the EIS states that the staff acknowledges that there is uncertainty with respect to off site releases of radiation from Yucca Mountain, Nevada, should that be the site. And it's the only site under consideration right now before the Licensing Board. (67-1)

Comment: Yet how do you quantify uncertainty? Can you be just a little bit unsure? The NRC DEIS has, in fact, failed to quantify the uncertainty. More detail, what it's done is it's now relying upon the waste confidence decision. In fact, you've heard tonight that the Agency has said that they have confidence, that, in fact, they will develop a site by 2025, someplace, somewhere, somehow, that will hopefully comply with current health and safety standards and limits for peak radiation dose exposure to the U.S. populations.

With regard to this application, however, the NRC Atomic Safety and Licensing Board has already dismissed contentions with regard to the nuclear waste generated from this new facility basically using this same waste confidence decision. But where does the confidence come from? That's the question tonight. Or are the impacts much larger than the Nuclear Regulatory Commission is willing to disclose? In fact, it's our concern that to pass the environmental liability of nuclear waste on to succeeding generations that won't give one watt of electricity is more akin to revealing this confidence decision as a confidence game. Now we all know that a confidence game is, in fact, where the victim is defrauded after his or her trust has been won. And let me look at some of the events that raises the question about whether, in fact, this is a confidence decision or a confidence game. The Yucca Mountain safety standard; the Energy Department has for years been planning on designing for Yucca Mountain that would pass off as being safe for 10,000 years. But last year federal court threw out that standard for the mountain 90 miles northwest of Las Vegas. And the court, in fact, has deferred the recommendation back with regard to coming up with a standard that more appropriately addresses the hazards for hundreds of thousands of years.

Now, the EPA is working on that standard. But I can tell you that the nuclear industry is hard at work to push that conclusion back to 10,000 years. Now is that a confidence decision or a confidence game? Yucca Mountain capacity; according to DOE, in 2011 current reactors will produce 63,000 metric tons of highly radioactive waste across the country. Yucca Mountain's technical and legal limit, should it be licensed, will not even be open yet. And it will have surpassed that volume. By 2046, according to DOE figures, with the license extensions to Illinois' current reactor fleet, the state will be left with more than 5,000 tons of nuclear waste that would be in excess to Yucca Mountain. If an additional two units at Clinton are brought on line, we're talking about an additional 1,736 metric tons in excess to Yucca Mountain. (67-2)

Comment: We've heard all about the issue of waste being left out. I think that's insane, first of all. Waste is the, one of the primary drawbacks to these nuclear power plants. (68-3)

Comment: Building Clinton II, without an approved nuclear waste site, is like building a new house without a toilet. (71-7)

Comment: The Nuclear Waste Policy Act, to my knowledge, allows for only 70,000 tons of nuclear waste in a repository. Even if a repository begins accepting the annual waste load of 3,000 tons, in 2010, we'd have to find another place to store the nuclear waste by 2035. (83-3)

Comment: Common sense, I think, dictates where I have come to my conclusion. And the very simple common sense thing that I think has not really been articulated quite yet, is that over the course of the nuclear age, which is now over 60 years, we have yet as society, we have yet as a world community, to figure out what to do with the stuff. The stuff is a polite term, isn't it? I think with the best intentions, we've continued to license nuclear power plants and site them, thinking that we could push off into the future a satisfactory solution. But the solution has not materialized. 60 years on, no solution for long term storage. No solution for long term protection. I think Yucca Mountain has been the last best hope of our generation. I think many of are aware now that the science that has gone into Yucca Mountain is, to be polite, suspect. Yucca Mountain may not at all come on line. And since we can't figure out what to do with the stuff, in our generation, let us not license any more plants until we figure out what the heck to do. (89-3)

Comment: We keep generating more and more nuclear waste and yet there is absolutely nowhere for it to go. Yucca Mountain is not safe. Dry casks are not safe. Cooling pools right here in Clinton are not safe. And I'm sure that nuclear engineers could differ with me. I am not a nuclear engineer. I'm a human being. (94-7)

Response: The NRC's Waste Confidence Rule, found in 10 CFR 51.23, states:

The Commission has made a generic determination that, if necessary, spent fuel generated in any reactor can be stored safely and without significant environmental impacts for at least 30 years beyond the licensed life for operation (which may include the term of a revised or renewed license) of that reactor at its spent fuel storage basin or at either onsite or offsite independent spent fuel storage installations. Further, the Commission believes there is reasonable assurance that at least one mined geologic repository will be available within the first quarter of the twenty-first century, and sufficient repository capacity will be available within 30 years beyond the licensed life for operation of any reactor to dispose of the commercial high-level waste and spent fuel originating in such reactor and generated up to that time.

In its Statement of Considerations for the 1990 update of the Waste Confidence Rule (55 FR 38472), the Commission addressed the impacts of the disposal of spent fuel discharged from the current fleet of reactors operating under existing and renewed licenses and from a new generation of operating reactors. Therefore, the current rule covers new reactors and can be used in the staff's review of an ESP application. The rule was last reviewed by the Commission in 1999 when it reaffirmed the findings in the rule (64 FR 68005, dated December 6, 1999). Furthermore, the Atomic Safety and Licensing Board presiding over the proceeding on the Exelon ESP application affirmed that the Waste Confidence Rule and its subsequent amendments resolve issues associated with long-term disposal of high-level waste as they relate to future reactors (Exelon Generation Company, LLC, Early Site Permit for Clinton ESP Site, L.P.-04-17, 60 NRC 229, 246-47 (2004)). No change was made to the EIS as a result of these comments.

Comment: Table 6-5, Page 6-26, Table G-3, Page G-8. Numerical values presented for Radiological Impacts of Transporting Unirradiated Fuel to Advanced Reactor Sites. There is not enough information presented to be able to verify the numerical values presented in Table 6-5 (page 6-26) or Table G-3 (page G-8) of the DEIS - Radiological Impacts of Transporting Unirradiated Fuel to Advanced Reactor Sites. (141-113)

Response: The RADTRAN 5 computer code was used to calculate the values in Table 6-5 and Table G-3 of the EIS. Table G-2 of the EIS provides RADTRAN 5 input parameters used in calculating doses from unirradiated fuel shipments. No change was made to the EIS as a result of the comment.

Comment: Section 2.2.3 ends with the Governor's "Opportunity Returns" goals that specify "investing in renewable energy and the environment." There's a limited supply of uranium, which is expected to run out in a few decades, so nuclear power is clearly not renewable. (169-15)(179-15)

Response: The staff agrees that nuclear power is not a renewable energy source. No change was made to the EIS as a result of these comments.

Comment: Section 6.1.2.7, Page 6-20, Lines 22-25. "Exelon expects that low-level waste impact from decontamination and decommissioning will be comparable to or less than that of the reference LWR (Exelon 2003). On this basis, the staff concludes that the environmental impacts from solid low-level radioactive waste generated during decontamination and decommissioning for gas-cooled reactors would likely be small, but these impacts will need to be assessed again at the CP or COL stage." EGC disagrees that further assessment at CP/COL is warranted. Based on the analyses presented in, "Early Site Permit Environmental Report Sections and Supporting Documentation. Engineering Design File, Number 3747, INEEL, Idaho Falls, Idaho" it is believed that there exists enough information to conclude that the impacts are comparable and the associated impacts are SMALL. The Decommissioning and Decontamination Impacts would have to be evaluated at the CP/COL stage if the reactor design selected has environmental impacts greater than that evaluated at ESP. (141-107)

Response: The staff bases this decision on the fact that Table 5.7-1 of the ER does not provide a low-level waste estimate from reactor decontamination and decommissioning for the GT-MHR and the estimate for the PBMR is 22,000 Ci per reactor year, compared to only 1500 Ci per reactor year for the reference LWR. There isn't conclusive evidence that impacts from the gas-cooled reactor designs would be bounded by the reference LWR. No change was made to the EIS as a result of the comment.

Comment: Section 6.1.2.1, Page 6-18, Line 11-14. "By comparison with the fuel fabrication impacts for LWR technologies, the staff concludes that the environmental impacts from producing gas-cooled reactor fuel likely would be small, but these impacts will need to be assessed at the CP or COL stage, when the staff will consider the environmental data that is available on a large-scale, fuel fabrication facility for gas-cooled reactors." The analyses in, "Early Site Permit Environmental Report Sections and Supporting Documentation. Engineering

Design File, Number 3747, INEEL, Idaho Falls, Idaho" provided a basis for estimating impacts that were bounded by Table S-3 values. It would be more appropriate to state that the fuel fabrication impacts would have to be evaluated if the reactor design selected at CP/COL had environmental impacts greater than those evaluated at the ESP. (141-104)

Response: The staff believes that environmental impacts from a larger-scale fuel fabrication facility for gas-cooled reactor fuel would need to be evaluated to make a final determination on acceptability. No change was made to the EIS as a result of the comment.

Comment: Fuel Cycle, Transportation, and Decommissioning, Section 6.1.1, Light-Water Reactors, page 6-7, paragraph 4. While the information provided gives a range of expected values, the probability of exceeding these estimates needs to be included. (172-45)

Response: The staff believes that the probability of exceeding environmental impacts in Table S–3 would be small. Section 6.1.1 provides several reasons why the staff feels that impacts in Table S–3 will bound current impacts (e.g., fuel management improvements, elimination of restriction on foreign uranium, use of gas centrifuge enrichment technology, and increased reliance on in-situ leach mining activities). No change was made to the EIS as a result of the comment.

Comment: Section 6.1.1, Page 6-7, Line 15-21. Bounding PPE power rating of 6,800 MW(t) (assuming two AP1000 units) with an associated plant capacity factor of 0.95 from the PPE table. The staff used the bounding PPE power rating of 6,800 MW(t) (assuming two AP1000 units) with an associated plant capacity factor of 0.95 from the PPE table in order to estimate impacts from the LWRs for comparison to Table S-3 values. It appears that a ratio of the Net MW(e) for the bounding advanced reactor to that of the reference reactor was used to calculate the impacts in Table S-3. As stated in Section 6.1.1, page 6-7, lines 15-21 "The fuel cycle impacts in Table S-3 are based on a reference 1000-MW(e) LWR operating at an annual capacity factor of 80 percent for a net electric output of 800 MW(e). In the following review and evaluation of the environmental impacts of the fuel cycle, the staff considered the capacity factor in the PPE of 95 percent with a total net electric output of 2200 MW(e) for a new nuclear unit at the ESP site (Exelon 2003); this is approximately three times the impact values in Table S-3 (see Table 6-1). Throughout this chapter, this will be referred to as the 1000-MW(e) LWR-scaled model,- reflecting 2200 MW(e) for the site." It is agreed that the ESP bounding LWR reactor may have approximately three times the net electrical generation capacity but it is overly conservative to state that, based on this methodology, the new breed of reactors would yield three times the impact values presented in Table S-3. Additional supporting information is needed to justify the use of this methodology. (141-97)

Response: The staff believes that this scaling approach is a valid method for estimating the bounding environmental impacts from the uranium fuel cycle. The staff recognized that this approach is conservative and included examples of this conservatism in the EIS. Section 6.1.1 provides these examples, including a discussion of (1) improved fuel management, which results in higher performance and reduced fuel and enrichment requirements, and (2) use of foreign uranium sources. Section 6.1.1 also references Section 6.2 of NUREG-1437 (NRC 1996) for a further discussion of recent changes to the fuel cycle. No change was made to the EIS as a result of the comment.

Comment: Appendix S, Table S-3. Discussion of multiplier approach based on a ratio of the electrical rating for the proposed gas-cooled plants versus the reference LWR. The staff has adopted a multiplier approach based on a ratio of the electrical rating for the proposed gas-cooled plants versus the reference LWR i.e., 2.5 for the GOT-MHR and 1.5 for the PBMR and then has applied these factors to the impacts cited in Table S-3 of 10 CFR 51.51(b). The NRC also utilized this same approach when estimating the impacts from the proposed advanced LWR reactor designs. Although a very simplified approach to estimating the impacts from these types of advanced reactors where there is currently very little information available, it is a very conservative method. Please provide additional supporting information to justify the use of this methodology. (141-133)

Comment: Section 6.1.2.1, Page 6-16, Lines 29-33. "The quantity of UO_2 required for reactor fuel is a key parameter. The more UO_2 required, the greater the environmental impacts (i.e., more energy, greater emissions, and increased water usage). The 1000-MW(e) LWR-scaled model described in Section 6.1.1 would require the equivalent of 120 MT of enriched UO_2 annually. This compares to 14.3 to 15.3 MT of enriched UO_2 annually for the gas-cooled reactor technologies." In Section 5.7.2.3.1 of the ER - Fuel Fabrication/Operations - it is stated that, "The reference LWR required 35 MTU of new fuel on an annual basis. This is equivalent to 40 MT of enriched UO_2, the annual output needed from the fuel fabrication plant. In comparison, the normalized annual fuel needs for the new gas-cooled reactor technologies ranged from 4.3 MTU to 5.3 MTU, approximately 88 percent to 85 percent lower than the reference plant." If the staff is going to use the multipliers of 2.5 for the GT-MHR and 1.5 for the PBMR in the DEIS when comparing to the 1000 MW(e) reference reactor, then the values for the required MT of enriched UO_2 should be 2.5 X 4.3 = 10.75 and 1.5 x 5.3 = 7.95 and not 14.3 and 15.3 listed in the DEIS. (141-101)

Comment: Section 6.1.2.6, Page 6-20, Lines 3-10. "Table S-3 (see Table 6-1) of 10 CFR 51.51 (a) states that there are 3.4 x 1014 Bq (9100 Ci) of low-level waste generated annually from operation of the reference LWR; operation of the1000-MW(e) LWR-scaled model would result in 1 x 1015 Bq (27,300 Ci) of low-level waste annually. Gas-cooled reactor technologies are projected to generate 3.6 x 1012 Bq to 1 x 1014 Bq (98 to 2750 Ci) of low-level waste scaled annually, far below the amounts generated by the reference LWR (Exelon 2003)." However in Table 5.7.1 of the ER - Gas Cooled Fuel Cycle Impact Evaluation - the following information is listed: TABLE 5.7-1 Gas-Cooled Fuel Cycle Impact Evaluation Reactor Technology Facility/Activity Reference LWR (Single unit) (~1,000 MWe) 80% Capacity GOT-MHR (4 Modules) (2,400 MWt total) (~1,140 MWe total) 88% Capacity PBMR (8 Modules)

(3,200 MWt total) (~1,320 MWe total) 95% Capacity Solid Radioactive Waste Annual LLW from reactor operations Ci 9,100 1,100 Ci; 98 m3 65.4 Ci; 800 drums.

In addition, the staff has not consistently applied their 2.5 (GOT-MHR) and 1.5 (PBMR) multiplier approach used throughout the other sections of the document. For example if we use the 2.5 (GOT-MHR) and 1.5 (PBMR) multiplier approach then the sentence should read, "Table S-3 (see Table 6-1) of 10 CFR 51.51 (a) states that there are 3.4 x 1014 Bq (9,100 Ci) of low-level waste generated annually from operation of the reference LWR; operation of the1000-MW(e) LWR-scaled model would result in 1 x 1015 Bq (27,300 Ci) of low-level waste annually. Gas-cooled reactor technologies are projected to generate 5.1 x 1014 Bq to 8.5 x 1014 Bq (1.5 x 9,100 = 13,650 to 2.5 x 9,100 = 22,750 Ci) of low-level waste scaled annually, far below the amounts generated by the reference LWR." (141-106)

Response: The staff believes that this scaling approach is a valid method for estimating the bounding environmental impacts from the uranium fuel cycle. The staff recognized that this approach is conservative and included examples of this conservatism in the EIS. Section 6.1.1 provides these examples, including a discussion of (1) improved fuel management which results in higher performance and reduced fuel and enrichment requirements, and (2) use of foreign uranium sources. Section 6.1.1 also references Section 6.2 of NUREG-1437 (NRC 1996) for a further discussion of recent changes to the fuel cycle.

The staff revised the method used to calculate values in Table 6-3 of the EIS. The number of four-module GT-MHR units (with a net power rating of 1003 MW(e) that could be placed on the ESP site and remain within the PPE net power rating for the site of 2200 MW(e) is two. The number of eight-module PBMR units (with a net power rating for 1254 MW(e) that could be placed on the ESP site and remain within the PPE net power rating for the site of 2200 MW(e) is one. These scaling factors were multiplied times the appropriate values in Table 5.7-1 of the ER to estimate fuel-cycle impacts from the gas-cooled reactor designs. For example, the enriched UO_2 MT estimate for the fuel fabrication plant operations in Table 6-3 of the EIS would equal 12.2 MT for the GT-MHR and 9.5 MT for the PBMR. Table 6-3 of the EIS was revised to include the values calculated using the new scaling factors.

Comment: My question is about solid waste. In the entire impact statement there are 29 lines on radioactive waste management. There are 33 lines on transportation of radioactive waste. This is a big issue. Is there more in there that I missed because it seems like there's not much discussion of the problem that nuclear energy causes? (37-1)

Response: The staff believes the discussion of radioactive waste in the EIS is adequate. The EIS contains the following information on radioactive waste: (1) Section 3.2.3 provides a discussion of the radioactive waste management system for the ESP site, (2) Section 6.1.1.6 provides a discussion of the impacts of radioactive waste disposal from the uranium fuel cycle operations for advanced light-water reactor designs, (3) Section 6.1.2.6 provides a discussion of the impacts of radioactive waste disposal from the uranium fuel cycle operations for advanced gas-cooled reactor designs, (4) Section 6.1.2.7 provides a discussion of the impacts of solid low-level radioactive waste generation from decontamination and decommissioning activities for advanced gas-cooled reactor designs, and (5) Sections 6.2.2 and 6.2.3 provides a discussion of the environmental effects of transporting spent fuel and radioactive waste, respectively, from the proposed ESP site. Section 6.1.1.6 of the EIS also refers the reader to Section 6.2 of NUREG-1437 (NRC 1996) for a more detailed description of radioactive waste generation, storage, and disposal from power reactors. No change was made to the EIS as a result of the comment.

Comment: Section 6.1.1.5, Page 6-10, Lines 21-37. Discussion of radioactive effluents estimated to be released to the environment from waste management activities. Using these data, the staff has calculated the 100-year environmental dose commitment to the U.S. population from the LWR-supporting fuel cycle for one year of operation of the 1000-MW(e) LWR-scaled model. This calculation estimates that the overall whole body gaseous dose commitment to the U.S. population from the fuel cycle (excluding reactor releases and the dose commitments due to radon-222 and technetium-99) would be approximately 12 person-Sv (1200 person-rem) per year of operation of the 1000-MW(e) LWR-scaled model; this reference reactor-year is scaled to reflect the total electric power rating for the site for a year. The additional whole body dose commitment to the U.S. population from radioactive liquid effluents due to all fuel cycle operations other than reactor operation would be approximately 6 person-Sv (600 person-rem) per year of operation of the 1000-MW(e) LWR-scaled model. Thus, the estimated 100-year environmental dose commitment to the U.S. population from radioactive gaseous and liquid releases due to these portions of the fuel cycle is approximately 18 person-Sv (1800 person-rem) to the whole body per reference reactor-year. Intuitively, these figures appear to be improper. However, the DEIS has not presented sufficient information to substantiate the gaseous and liquid effluent doses as to how the values were calculated in order to come to the same conclusion as the staff in that there are gaseous and liquid effluent doses of 1,200 person rem and 600 person-rem. (141-98)

Response: *The staff derived the 12 person-Sv (1200 person-rem) and 6 person-Sv (600 person-rem) population annual doses for the gaseous and liquid pathways, respectively from Section 6.2.2.1 of NUREG-1437 (NRC 1996). NUREG-1437 estimated the annual population dose estimate from gaseous releases of the fuel cycle to be 400 man-rem. The annual population dose estimate from liquid releases of the fuel cycle was 200 man-rem. These estimates were based on the 1000-MW(e) LWR (800 MW(e) net). The staff applied a scaling factor of 3 to the NUREG-1437 values to account for the 2200 MW(e) net power rating for the ESP site. Section 6.1.1.5 of the EIS was revised to provide a discussion on how the person-Sv estimates were derived from NUREG-1437 and the use of a scaling factor of 3.*

Comment: The Draft EIS fails to adequately consider impacts relating to the nuclear fuel cycle, waste storage, and safety. (170-4)

Response: The staff in its EIS includes an entire chapter on the uranium fuel cycle, solid waste management, and transportation impacts. These impacts are evaluated in accordance with NRC regulations in 10 CFR 51.51. No change were made to the EIS as a result of the comment.

Comment: Table 6.3, Page 6-17, Line 10. Numerical values for the PBMR. The numerical value for the PBMR is incorrect and should be 569. It is stated in Table 5.7.1 of the ER - Gas Cooled Fuel Cycle Impact Evaluation - under the UF6 Production category that the annual UF6 (MT) required would be 379 before normalization therefore 1.5 x 379 = 569 and not 659. (141-102)

Response: The staff revised the method used to calculate values in Table 6-3 in the final EIS. The number of eight-module PBMR units that could be placed on the ESP site and remain within the 2200 MW(e) net power rating for the site is 1. For UF_6 production, the annual UF_6 MT in Table 5.7-1 of the ER for the PBMR of 379 MT was multiplied by one to obtain an estimated impact of 379 MT of UF_6.

Comment: Section 6.2, Page 6-21, Lines 16-18. "Non-radiological impacts during accident conditions were estimated as one fatal injury per reference reactor-year and one nonfatal injury in 10 reference reactor-years." The staff has misquoted the information in Table S-4 of 10 CFR 51.52 and the sentence should read, "Non-radiological impacts during accident conditions were estimated as one fatal injury per 100 reference reactor-years and one nonfatal injury in 10 reference reactor-years." (141-108)

Response: The staff revised Section 6.2 of the EIS to read, "Nonradiological impacts during accident conditions were estimated as 1 fatal injury per 100 reference reactor years and 1 nonfatal injury in 10 reference reactor years."

Comment: Appendix G, Page G-4, Lines 28-30. "The average enrichments for the other advanced LWR fuels exceed the 4 percent uranium-235 by weight condition in 10 CFR 51.52(a)(2)." There is insufficient information in the DEIS to support this statement. A reference should be provided for this statement. (141-124)

Response: The staff used the average enrichment values in INEEL (2003). Section G.1.2.2 was revised to reference the INEEL document.

Comment: Table 6.4, Page 6-25. Values reported for each of the proposed reactors that denote site electric generation capacity MW(e) and capacity factors.

Reactor Type	# of Units	MW(t)	MW(e)	Eff.	MW(e) Net
ABWR	1	3926 total	1500	.95	1425
AP1000	2	3400/unit = 6800 total	1150/module = 2300/unit	.95	2185
ESBWR	1	4000 total	1500	.95	1425
ACR-700	2	1983/unit = 3966 total	731/module = 1462/unit	.90	1316
IRIS	3	1000/unit = 3000 total	335/module = 1005/unit	.96	965
GT-MHR	4	600/unit = 2400 total	285/module = 1140/unit	.88	1003
PBMR	8	400/unit = 3200 total	165/module = 1320/unit	.95	1254

From the ER, SSAR, and PPE table some of the values could be verified specifically the ones for the Gas Cooled Reactors. However information regarding MW(e) for the LWRs was not listed in the PPE Table and therefore not verifiable in the DEIS in Table 6.4. From reviewing the table and sections of the DEIS the following table presents the staff's values for each of the proposed reactors. A basis for these values should be provided as they form the basis for the staff's conclusions.

In addition, values for plant capacity factors specified in the DEIS differ marginally from those reported in the PPE table.

Plant Capacity Factors		
Reactor Type	Table 6.4 of the DEIS	PPE Table
ABWR	.95	.92
AP1000	.95	.93
ESBWR	.95	.92
ACR-700	.90	.93
IRIS	.96	.95
GT-MHR	.88	.96
PBMR	.95	.95

(141-112)

Comment: Appendix G Page G-4, Line 6. Discussion of thermal rating of the ABWR. There is some disparity about the thermal rating of the ABWR which in Appendix G of the DEIS (page G-4, line 6), the value is stated as 4,300 MW(t) but in Section 1.3.1 of the SSAR - Advanced Boiling Water reactor - the thermal rating is reported as 3,926 MW(t). In addition there is some disparity about the thermal rating for the ACR-700 which in Appendix G of the

DEIS (page G-4, line 12), is stated as 1982 MW(t)/reactor x two reactors per site = 3964 MW(t) per site. However in section in Section 1.3.5 of the SSAR - Advanced CANDU Reactor - the thermal rating is reported as 3,983 MW(t). (141-123)

Response: The staff used the power ratings in INEEL (2003). No change was made to the EIS as a result of these comments.

Comment: Section 6.2, Page 6-22, Lines 12-16. "Five of the designs are LWRs and include the ACR-700 (3964 MW(t)/unit); the ABWR (4300 MW(t)/unit; the AP1000 (6800 MW (t)/unit); the ESBWR (4000 MW(t)/unit), and the IRIS (3000 MW(t)/unit). For the ACR-700 and AP1000 reactor designs, two reactors make up a unit. For the IRIS design, three reactors (modules) make up a unit. For the remaining LWR designs, one reactor makes up a unit."

In Section 3.2 of the ER - Reactor Power Conversion System - it is stated that, "The bounding parameters indicate that the proposed reactor(s) could generate up to 6,800-MW core thermal power. In general, the ABWR (one unit) is rated at 3,926 MWt, the AP1000 (two units) is rated at 6,800 MWt, the IRIS (three units) is rated at 3,000 MWt, the GT-MHR (four modules) is rated at 2,400 MWt, the PMBR (eight modules) is rated at 3,200 MWt, the ESBWR (one unit) is rated at 4,000 MWt, and the ACR-700 (two units) is rated at 3,966 MWt." In Section 3.8 of the ER - Transportation of Radioactive Materials it is stated that, "The standard configuration for these reactor technologies (assumed in this analysis) is as follows. The ABWR is a single unit, 4,300 MWt, nominal 1,500 MWe boiling water reactor. The ESBWR is a single unit, 4,000 MWt, nominal 1,390 MWe boiling water reactor. The AP1000 is a single unit, 3,400 MWt, nominal 1,117-1,150 MWe pressurized water reactor. The IRIS is a three module pressurized water reactor configuration for a total of 3,000 MWt and nominal 1,005 MWe, and the ACR-700 is a twin unit, 3,964 MWt, nominal 1,462 MWe, LWR with a heavy water moderator."

The PPE table states the following values for MWt:

MWt for Proposed Reactors		
Reactor Type	Lines 12-14 - page 6-22 of the DEIS	PPE Table
ABWR	4300/unit	**3926** (1 module)
AP1000	6800/unit (2 modules)	6800 (2 modules)
ESBWR	4000/unit	4000 (1 module)
ACR-700	**3964**/unit (2 modules)	**3966** (2 modules)
IRIS	**3000**/unit (3 modules)	**3006** (3 modules)
GT-MHR	2400/unit (4 modules)	2400 (4 modules)
PBMR	3200/unit (8 modules)	3200 (8 modules)

There is a difference in the MW(t) for the ACR-700, IRIS and the ABWR. (141-109)

Response: *The staff used the power ratings in INEEL (2003). The power level differences between the PPE table and the values the staff used in the EIS are not significant for the ACR-700 (3964 vs 3966 MW (t)) and the IRIS (3000 vs 3006 MW(t)). No changes was made for the ACR-700 and IRIS in the EIS.*

The applicant's ER is not consistent on the power level for the ABWR. In Section 3.8.1 (p. 3.8-2) of the ER and Section 5.7.1 (p. 5.7-1) of the ER, the applicant refers to the uprated power level of 4300 MW(t). The 3926 MW(t) value was used in the radiological analysis in Section 5.4 of the ER. The staff added a statement similar to the one in Section 5.7.1 of the ER to Section 6.1.1 of the EIS which states "Note that for this analysis (transportation and fuel cycle), the ABWR is conservatively presumed to be the uprated design (4300 MW(t)) while other evaluations within this ESP application are based on the certified design configuration (3926 MW(t))."

Comment: Moreover, the environmental impact statement does not adequately consider the possibility and consequences of severe accident scenarios resulting from the transportation of spent nuclear fuel. (191-5)

Comment: This section and the accompanying Appendix G of the draft EIS do not give adequate weight and consideration to the possibility and consequences of severe accident scenarios resulting from the transportation of spent nuclear fuel. The possibility of extreme accidents, while slight, exists, as evidenced by recent incidents such as the Baltimore train tunnel fire of 2001 and the more recent accident in Graniteville, South Carolina in January, where a violent train crash and release of chlorine killed nine people, sent, hundreds to the hospital, and required thousands to evacuate their homes. (150-14)(151-14)

Response: *The transportation impact analysis in Section 6.2 and Appendix G analyzed the full spectrum of transportation accidents, from minor fender-benders to severe collisions and fires. Detailed supporting studies for the accident frequencies, conditional probabilities, and releases from potential spent fuel transportation accidents formed the basis for the analysis of transportation accidents in the EIS.*

The NRC has sponsored studies to analyze the consequences of specific accident scenarios on rail and truck transportation casks carrying spent fuel. For example, the NRC undertook an investigation of a July 2001 accident that involved a freight train carrying hazardous materials that derailed and caught fire while passing through the Howard Street railroad tunnel in downtown Baltimore, MD, to determine the possible regulatory implications of this particular event for the transportation of spent fuel by railroad. NRC assembled a team of experts from the National Institute of Standards (NIST), Center for Nuclear Waste Regulatory Analyses (CNWRA), and the Pacific Northwest National Laboratory (PNNL) to determine the thermal conditions that existed in the Howard Street tunnel fire and to analyze the effects of this fire on various spent fuel transportation cask designs. The staff concluded that the spent fuel transportation casks analyzed would withstand a fire with thermal conditions similar to those that existed in the Baltimore tunnel fire event. No release of radioactive materials would result from exposure of the casks analyzed to such an event.

No change was made to the EIS as a result of these comments.

Comment: Now that the federal government has addressed the waste storage at Yucca Mountain, the most critical open issue when Unit One was built is solved. (106-5)

Response: *The comment provides general information acknowledging that radioactive waste can be disposed of safely. No change was made to the EIS as a result of the comment.*

E.2.19 Comments Concerning Postulated Accidents

Comment: Table 5-12, Page 5-74. Population Dose from Water Ingestion values. There was not enough information given in Section 5 text to duplicate the values in the last column of Table 5-12 - Population Dose from Water Ingestion (page 5-74), so the values could not be verified. Information or reference to information should be provided. (141-90)

Response: *In the first paragraph on the surface water pathway, the EIS states that the MACCS2 code evaluates the ingestion of contaminated water. The water ingestion dose risks in the last column of Tables 5-11 and 5-12 are the product of the water ingestion pathway dose computed by MACCS2 and the core damage frequency. No change was made to the EIS as a result of the comment.*

Comment: The recent National Academy of Science report on nuclear fuel pool storage hazards found the storage pools at the Dresden and Lasalle plants (downwind of Chicago) are particularly vulnerable, meaning a radioactive fire with blowing smoke are realistic risks: http://www.nap.edu/books/0309096472/html. (169-2)(179-2)

Response: *Searches of the National Academy of Sciences (NAS) report at the internet location cited above for the words Dresden, LaSalle, and Chicago did not find any instances in which the words appeared.*

The NRC believes that the NAS study reinforces the validity of recent NRC studies, which indicate that spent fuel storage systems are safe and secure, and of NRC action to improve safety and security of such systems. In the aftermath of the events of September 11, 2001, the NRC has taken a number of steps to reduce the risks associated with spent fuel storage. Among these steps are a February 2002 order requiring "…licensees to develop specific guidance and strategies to maintain or restore SFP [spent fuel pool] cooling capabilities using existing or readily available resources (equipment and personnel)…" and a July 2004 letter advising "…licensees to implement additional "spent fuel pool mitigative measures." Should

Exelon apply for a CP or COL, it will be required to meet the applicable regulations concerning spent fuel pools.

The comment provides no new information and no change was made to the EIS as a result of the comment.

Comment: In its analysis of the potential consequences of "design basis" accidents, Exelon used the characteristics of two particular reactor designs, assuming the impacts of such accidents would bound those of other possible reactor designs (EIS, pg. 5-66). For its analysis of "severe" accidents, Exelon evaluates the consequences for the current generation reactors- not of the kind that it would build at the CPS (EIS, pg. 5-66)-and the NRC only considers two reactor designs it considers bounding in its evaluation of potential hazards from a serious accident (EIS, pg. 5-69). How can the NRC reasonably judge accident consequences when several of the potential reactor designs proffered by Exelon have never been deployed? (150-6)(151-6)

Response: *The ABWR and AP1000 reactors are reactors that could be included in an application for a new nuclear plant should a construction permit or combined license application be submitted for the Clinton site. These reactors have well-defined source terms for postulated accidents. The other reactor designs include engineered safety systems that are expected to keep potential releases in the event of an accident below those for the ABWR and AP1000 designs. The staff concludes that the site is acceptable for advanced light water reactors based on the analyses presented here. While the staff considers it likely that the site would be acceptable for other types of advanced reactors based on the enhanced safety of those designs, the staff did not reach a conclusion with respect to those designs. The comments provide no new information. No change was made to the EIS as a result of these comments.*

Comment: Section 3.2, Page 3-4, Line 22. Discussion of radiological consequences. The DEIS indicates that the radiological consequences was based on the certified ABWR with an uprated power level of 4300 MW. The megawatt rating used in the EGC ESP application was for 3926 MWt. EGC did not use the uprated value for these analyses. (141-49)

Response: *The comment is correct. Section 3.2 was revised to state that the radiological consequences of ABWR design basis accidents were based on the certified ABWR design without a power uprate.*

Comment: Table 5-11, Page 5-72, Line 8. Values listed in Population Dose category. In the above-mentioned table, the value of 3.80E-10 under the Population Dose category is incorrect and should be 3.80E-09. (141-89)

Response: *The comment is correct. The value in Table 5-11 has been changed.*

Comment: Station Operation Impacts, Section 5.10, Environmental Impacts of Postulated Accidents, page 5-61, paragraph 5. The statements here are misleading. There are some studies that have been peer reviewed that provide data that contradicts this assumption, i.e., the BEIR VI report and similar studies. This information needs to be included in the FEIS to provide an overview that is not skewed to minimize potential issues of radiation exposures. (172-40)

Response: The comment provides no new information. The statement in the EIS is consistent with the findings reported in the BEIR VII report (National Research Council 2006). No change was made to the EIS as a result of the comment.

Comment: Station Operation Impacts, Section 5.10.1, Design Basis Accidents, page 5-66. The information provided in this DEIS may not provide an adequate basis to make the assertions and assumptions that this information adequately bound potential accident impacts. Further review that will be conducted at the construction permit (CP) and combined construction permit-operating license (COL) should provide a better information base to make this type of determination. (172-41)

Response: The design basis accident analyses for the ABWR and AP1000 reactors presented in the EIS were made as part of the design certification process for these reactors. These analyses are based on detailed information about the designs and represent a range of accidents. Should a CP or COL application be submitted, the applicant would be required to demonstrate that impacts of DBAs remain within the bounds of this analysis. The comment does not provide new information and no change was made to the EIS as a result of the comment.

Comment: Table 5-11: The discussion of this table should specify the dose or risk criteria on which the Land Requiring Decontamination values were based. (172-66)

Response: Table 5-11 in the EIS was revised to include the criteria used to estimate the land area requiring decontamination.

Comment: Table 5-13. Comparison of Environmental Risks for ABWR. There was not enough information given in Section 5.0 text to verify the numerical values for early and latent cancer risks in the last column of Table 5-13 - Comparison of Environmental Risks for ABWR or a Surrogate AP1000 at the Exelon ESP site with Risks for Five Sites Evaluated in NUREG 1150 (page 5-75). Information or reference to information should be provided. (141-92)

Response: The MACCS2 code computes average individual early and latent cancers for each accident class. The EIS text has been revised to state that these values are computed by the MACCS2 code.

Comment: Section 5.10.1, Page 5-66, Line 17-19. It is stated that the environmental impacts of design basis accidents for gas-cooled reactors have not been evaluated and would need to be evaluated at CP/COL. It is more accurate to state that the design basis accidents for LWRs have been determined to be SMALL. Assuming that EGC selects a design that did not form the basis of its PPE, it may be necessary, during the review at CP/COL, to determine whether the impacts associated with design basis accidents for the selected reactor design are bounded by the ESP. (141-85)

Response: The staff believes that the statement made in the EIS is correct. However, the staff has clarified its conclusions in Section 5.10.1

Comment: Section 5.10.2, Page 5-70, Line 17-18. Page 5-71, Line 15-17. Page 5-76, Line 25-26. Page 5-77, Line 9-10. It is stated that the environmental impacts of severe accidents for gas-cooled reactors have not been evaluated and would need to be evaluated at CP/COL. It is more accurate to state that the severe accidents for LWRs have been determined to be SMALL. The NRC requested additional information asking EGC to justify the generic conclusion that was used in the ER Section 7.2.2. EGC responded, on 7/23/04, by providing example evaluations that were used to justify the generic evaluation used in the ER. These evaluations concluded that the consequences due to severe accidents at the CPS site listed in NUREG-1437 remain valid for the purposes of evaluating the environmental impacts of severe accidents at the EGC ESP site. As such, EGC believes that, during the review at CP/COL, a determination will be made as to whether the impacts associated with severe basis accidents for the selected reactor design are bounded by the ESP. (141-87)

Response: The staff considers the statements in the EIS to be correct. The staff does not believe that the generic evaluation of severe accident consequence assessments for gas-cooled reactors is adequate to serve as a bounding analysis. However, the staff has clarified its conclusions in Section 5.10.2

Comment: Section 5.10, Page 5-60, Line 26-28. It is stated that the consequences of Severe Accidents are based on the ABWR and AP1000. As stated in the ER, the GEIS provides the basis for the environmental impacts of severe accidents. Examples of this generic rationale were provided in response several RAIs using the ABWR and AP1000. Exelon is requesting an assessment of impact based on the GEIS supported by the examples given in response to RAIs. (141-84)

Response: The staff's position is that an assessment based on site-specific information, a state-of-the-art computer code (MACCS2), and design-specific source-term information is appropriate for an ESP application review. The comment provides no new information. No change was made to the EIS as a result of the comment.

Comment: And any danger from an accident shouldn't be any worse from a double reactor as from the single one. (24-3)

Comment: And lastly there was a mention of nuclear explosion. And I noticed all T.V. shows on nuclear energy open with a mushroom cloud. That's nonsense. We have national labs like Los Alamos and Livermore who spent big bucks designing things that will explode, okay? It takes that much expertise and effort. Nuclear plants don't do that. The fuel is a ceramic, a little bit like floor tile in your bathroom. It's kept pretty much chemically inert and it sits in cans. And when it's removed from the reactor, it's placed in other cans and stored. (62-3)

Response: *These comments generally support the postulated nuclear plant. The comments do not provide new information and no change was made to the EIS as a result of these comments.*

Comment: Please reconsider building a possible disaster so close to Chicago. (05-2)

Comment: There is one overriding reason why Clinton II should not be built, and indeed why Clinton I should be decommissioned, with all deliberate speed. Namely, if anything really serious should ever go wrong, the resulting devastation would go beyond what most people can imagine. Ah, but our design is so modern, so technologically advanced, that nothing could ever go wrong, Exelon might say. So thought the builders of the Titanic. And indeed, the builders of the World Trade Center. (100-1)

Comment: There is a risk of a potentially serious accident, which would endanger most of Central Illinois, let alone areas east of Illinois. (148-3)

Comment: Champaign-Urbana is due east of the current facility, not even mentioning all of the people and communities in the path of an accident. Please do not build another NUCLEAR POWER PLANT. (15-3)

Comment: Illiopolis PVC plant explosion last year. Turns out when the plant exploded, power was lost, which shut off the water pumps, so fighting a major fire became extremely difficult. In addition, the emergency warning sirens also didn't work since there wasn't any backup power. The winds shifted several times, blowing a toxic cloud of dioxin in multiple directions. There were also rains that pulled the toxins down to surrounding lands and homes. A nuclear accident or attack would be similar, but the consequences would be even more devastating. (169-1)(179-1)

Comment: There could be an accident. There could be. There already have been; Chernobyl and Three Mile Island. There have already been countless near misses. So an accident could happen here to you and to me. And I disagree completely with the idea that the environmental risk is small. It is potentially catastrophic. (46-3)

Comment: I think it's very clear that one of the differences of opinion on, from different sides is how important a problem or catastrophe with the nuclear explosion or nuclear radiation is. And those of you who are willing to live with the risks, that is your choice. To me, those risks are considerable and would be catastrophic if there was an accident. The report says the probabilities are small. I think the risks are very high. (58-2)

Comment: I think that what's happening is that we're getting lost and sort of daydreaming. I think we have to look at the history of this technology and look at all the accidents that have occurred for the last 50 years starting with Char River in Canada, Browns Ferry, Indian Point, Three Mile Island. Many accidents in Russia. There have been so many accidents all over the world, in Brazil, with waste. Now it would take me about 20 minutes to list all the world-wide cast nuclear accidents... And now you want to create another power plant? Are you immune to the fact that this state could have an accident? (65-1)

Comment: One is about Chicago and why people should come from there because we're not local. If you, if you look at the blast maps, in the worst case scenarios, we are, in Chicago, totally at risk from either an explosion or a meltdown. And that's current data that, that is publicly available. As Dr. Caldicott said, many times tonight it's been mentioned that it's a nuclear accident. Her point is that an accident is something that surprises you because you didn't know about it. So there are no more nuclear accidents from nuclear generators, because we know the consequences. It is a silent bomb. (81-1)

Comment: There is one overriding reason why Clinton II should not be built, and indeed why Clinton I should be decommissioned, with all deliberate speed. Namely, if anything really serious should ever go wrong, the resulting devastation would go beyond what most people can imagine. Ah, but our design is so modern, so technologically advanced, that nothing could ever go wrong, Exelon might say. So thought the builders of the Titanic. And indeed, the builders of the World Trade Center. (90-1)

Comment: There could be an accident. There could be. There already have been; Chernobyl and Three Mile Island. There have already been countless near misses. So an accident could happen here to you and to me. And I disagree completely with the idea that the environmental risk is small. It is potentially catastrophic. (94-3)

Response: *These comments, which refer to nuclear accidents and their consequences, are opposed to the postulated nuclear plant. Such accidents are discussed in Section 5.10. The comments provide no new information. No change was made to the EIS as a result of these comments.*

E.2.20 Comments Concerning Alternatives and Alternative Sites

Comment: Alternative energy from renewable sources is the solution. Not yet another nuclear power plant. Wind farms to harness wind energy have no harmful waste. They can even share land with agricultural use, because the turbines only take up 5% of the needed land. Given the agriculture that central Illinois depends on, wind energy has great potential to help meet our energy demands and is less costly, in every sense of the word. (109-6)

Comment: Sec. 8.2.3.1 Wind The examination of the potential of wind capacity both in Illinois and in the region in which a merchant plant would sell power is seriously flawed and understated, so much so as to be a worthless conclusion. (161-5)

Response: At any given time, the probability that a nuclear plant will be generating at or near its rated value is high (typically greater than 90 percent); the probability that a wind turbine will be generating at or near its rated value is much lower (typically less than 30 percent). In addition, the outages of a nuclear plant are generally scheduled to coincide with periods when the demand is low. Periods of low wind cannot be scheduled and may not correspond to periods of low demand. No change was made to the EIS as a result of these comments.

Comment: Section 8.2.3.2 fails to address ground-source heat pumps or earth tubes that extract free energy from the earth to heat/cool homes at a tiny fraction of the cost of using electricity or burning fossil fuels. (169-20)(179-20)

Response: Section 8.2.3.2 of the EIS discusses geothermal sources of energy that might be viable as sources of baseload power. This resource does not exist in the region of interest. Therefore the alternative was dismissed as not being viable. The comment provides no information that indicates that the ground-source heat pumps mentioned in the comments are a viable source of baseload power. The comments were not evaluated further. No change was made to the EIS as a result of these comments.

Comment: I'd like to know why you didn't consider coal gasification technology, which makes coal a relatively clean fuel source since it doesn't burn coal? And Illinois has a lot of coal. (41-1)

Response: Coal gasification technology was not considered by either Exelon or the NRC staff. However, the staff considered coal gasification technology as an alternative to license renewal of Browns Ferry Nuclear Plant, Units 1, 2, and 3. In that review, land-use impacts were found to moderate to be large. Waste and air quality impacts were found to be moderate. Other impacts ranged from small to large depending on the site. These impacts were similar to the impacts that the staff found for natural gas combined-cycle generation, except in the area of waste, where the impacts were moderate for coal gasification and small for natural gas combined-cycle generation. The comment provided no new information. No change was made to the EIS as a result of the comment.

Comment: Section 8.2.3.1 fails to mention that Illinois could easily get 15 percent of its energy from wind, which is much more scalable than nuclear. Wind is base load energy because once it's operating the fuel is free, so it's always used first. (169-19)(179-19)

Response: *Exelon and the NRC staff agree that there is viable wind energy resource in Illinois. This resource is discussed in Section 8.2.3.1 of the EIS. However, wind energy is not suitable for use as baseload power because the resource is intermittent. Capacity related to wind energy is an annual average generation value. It is not the same as the probability that a wind turbine will be generating at or near its rated value. The comment does not provide any new information and was not considered further. No change was made to the EIS as a result of the comment.*

Comment: The Draft EIS also improperly rejects clean energy alternatives to new nuclear power. Wind, solar, natural gas, and "clean coal" generation, both individually and in combination, along with energy efficiency, are reasonable alternatives for satisfying whatever future energy needs that would be met by the Clinton 2 nuclear power plant. Such alternatives would be not only environmentally preferable to and safer than new nuclear power, but would also cost less and bring important economic development benefits to Illinois. None of the reasons that the Draft EIS presents for ejecting clean energy alternatives withstand scrutiny. First, the Draft EIS claims that wind, solar, and other alternatives are not reasonable alternatives to new nuclear power because they do not generate baseload power. (Draft EIS at 8-17, 8-18). The Draft EIS acknowledges, wind, solar, and other energy sources can contribute to a combination of alternatives that can serve the purpose of creating baseload power. (Draft EIS at 8-21, 8-22). Therefore, wind and solar power should not be rejected as reasonable alternatives to new nuclear power. (170-8)

Comment: Finally, the Draft EIS rejects a combination of clean energy alternatives on the ground that any combination would purportedly not be environmentally preferable to new nuclear power. (Draft EIS at 8-21, 8-22). In reality, however, the Draft EIS's own analysis demonstrates that many more resources would be impacted by nuclear power than by clean energy alternatives. The Draft EIS concludes that nuclear power would have land use, air quality, thermal, aesthetic, water use and quality, human health, accident, ecological, and waste management impacts. (DEIS at 5-80 to 5-82, Table 5-15). By contrast, the only impacts that wind power would have are fairly minor impacts regarding land use, bird deaths, aesthetics, and noise. (Draft EIS at 8-17). Certainly an energy source that only has land use, bird deaths, aesthetic and noise impacts should be considered environmentally preferable to an energy source that impacts at least 10 resources including human health and air and water quality. Similarly, the only major impact from natural gas generation identified by the Draft EIS is air quality impacts. (DEIS at 8-23). In reality, however, a combination of alternatives that uses a proper amount of wind and solar power would significantly reduce those air quality impacts. (DEIS at 8-13). In addition, the other impacts of natural gas are minor, the Draft EIS acknowledges that human health impacts from natural gas are "not expected . . . [to] be detectable," (DEIS at 8-13), and the NRC Staff have not claimed that natural gas presents the type of accident risks that nuclear power does. As with wind, it is arbitrary and capricious to suggest that an energy source that presents human health and accident risks is environmentally preferable to a clean energy alternative that does not. Certainly, those energy sources in combination, along with energy efficiency efforts, could not be considered to have greater environmental impacts than new nuclear power. Therefore, the NRC Staff must reconsider its rejection of clean energy alternatives, and engage in the rigorous and objective analysis of such alternatives that is required by NEPA but not found in the Draft EIS. (170-10)

Response: Exelon and the NRC staff considered wind energy in combination with conventional methods of generation. None of the other renewable energy alternatives was considered to be a viable source of power on a large enough scale to be examined in detail. Even when the staff assigned no adverse environmental effects to wind energy, the impacts of the conventional generation needed to establish an adequate baseload capacity were sufficient that the combination of alternatives was not found to be environmentally preferable to new nuclear generation. The comments provide no new information. No change was made to the EIS as a result of these comments.

Comment: Nor did NRC do a proper analysis of the ability of a combination of renewable energy technologies to meet any power needs. (112-2)(113-2)(114-2)(115-2)(116-2)(117-2) (118-2)(119-2)(120-2)(121-2)(122-2)(123-2)(124-2)(125-2)(126-2)(127-2)(128-2)(129-2)(130-2) (131-2)(132-2)(133-2)(134-2)(135-2)(136-2)(137-2)(138-2)(139-2)(140-2)(142-2)(143-2)(144-2) (145-2)(146-2)(147-2)(149-2)(154-2)(155-2)(158-2)(159-2)(162-2)(163-2)(164-2)(165-2)(166-2) (167-2)(173-2)(174-2)(175-2)(176-2)(177-2)(178-2)(180-2)(181-2)(182-2)(185-2)(186-2)(187-2) (188-2)(189-2)(190-2)(192-2)(193-2)(194-2)

Response: Exelon and the NRC staff considered wind energy in combination with conventional methods of generation. None of the other renewable energy alternatives was considered to be a viable source of power on a large enough scale to be examined in detail. Even when the staff assigned no adverse environmental effects to wind energy, the impacts of the conventional generation needed to establish an adequate baseload capacity were sufficient that the combination of alternatives was not found to be environmentally preferable to new nuclear generation. The comments provide no new information and will not be evaluated further. No change was made to the EIS as a result of these comments.

Comment: 8.2.2 Alternatives Requiring New Generating Capacity (Introductory paragraph). The premises put forth by Exelon and endorsed by NRC staff in the introductory statements of this section are flawed and unnecessarily restrictive. They do not reflect adequately realistic possibilities in the area of renewables. Further, what is "technically reasonable and commercially viable" is totally in the hands of those with a clear agenda to build more nuclear plants, making this and the following statements and analysis a self-fulfilling prophecy, not an objective analysis of the realistic possibilities. NEIS staff has examined their premises, and find them erroneous and without adequate justification. We do not concur with their initial premise, and therefore do not accept the erroneous conclusions emanating from them. (161-4)

Response: Exelon applied for a early site permit for a site for generation of baseload power. It would not be appropriate for the NRC staff to alter the applicant's business objective. Therefore, Exelon and NRC evaluated existing power generation technologies on the basis of their ability to provide baseload power. Conventional and alternative renewable generation technologies were considered. Renewable technologies were found not to be viable as sources of baseload power. The comment questions the criteria used in the evaluation of alternative energy sources but does not provide any new information. No change was made to the EIS as a result of the comment.

Comment: I was wondering, you were discussing about the other alternative sites. And you said that none of them were preferable to this site. What makes this site in particular preferable to all the other sites? (40-1)

Response: Exelon chose the preferred site for business reasons. Exelon, and the NRC in its independent review in the EIS, undertook a site-by-site comparison of alternative sites with the proposed site (Clinton Power Station [CPS]) to determine if there were any alternative sites environmentally preferable to the proposed site. Not all possible alternative sites were considered, just a "reasonable" subset of possible alternatives. The review process involved the two-part sequential test outlined in NUREG-1555 (NRC 2000). The first stage of the review used reconnaissance-level information to determine whether there were environmentally preferable sites among the alternatives. If environmentally preferable sites were identified, the second stage of the review considered economics, technology, and institutional factors for the environmentally preferred sites to see if any of these sites was obviously superior to the proposed site at CPS. None of the alternative sites proved to be obviously superior to the ESP site at CPS. Just because an alternative site is not obviously superior to the preferred site does not mean that the alternative site cannot be considered for future nuclear development. No change was made to the EIS as a result of the comment.

Comment: The first is that the draft EIS fails to give a reasonable and objective analysis of alternatives. The draft EIS essentially defers blindly to first to Exelon's stated purpose of creating new base load power...And we believe that alternatives such as wind and solar power, energy efficiency in combination with natural gas and clean coal technology is a more sensible and preferable way to meet our future energy needs in new nuclear power. In particular, there would be four major benefits. First, wind, solar and energy efficiencies have very little to no environmental impacts, which in contrast [to alternative energy sources] nuclear power creates significant human health, radiation, land use, air and water quality impacts from the mining and enrichment of uranium, the operation of the plant, the transportation of nuclear waste and then the storage of high level nuclear waste for tens of thousands of years. The draft EIS, unfortunately, down plays or entirely ignores these impacts. (59-1)

Comment: As for alternatives, it is unacceptable to allow Exelon to define the project goals so narrowly that only nuclear power can achieve them; for instance, requiring that any alternative be constructed in the immediate vicinity of the proposed ESP site and provide baseload power unfairly precludes consideration of less polluting and less dangerous energy sources such as wind. If sufficiently distributed geographically, and combined with other forms of renewable energy generating technologies and conservation/efficiency measures, there are economic alternatives to nuclear power that can meet our energy needs without falling victim to the intermittency problem cited in the DEIS. An analysis by the Union of Concerned Scientists found that Illinois has the technical potential to generate up to eight times its current electricity needs through renewable sources; NRC should examine UCS's methodology and perhaps modify its conclusion that renewable energy resources are incapable of providing reliable power. (112-5)(113-5)(114-5)(115-5)(116-5)(117-5)(118-5)(119-5)(120-5)(121-5)(122-5)(123-5) (124-5)(125-5)(126-5)(127-5)(128-5)(129-5)(130-5)(131-5)(132-5)(133-5)(134-5)(135-5)(136-5) (137-5)(138-5)(139-5)(140-5)(142-5)(143-5)(144-5)(145-5)(146-5)(147-5)(149-5)(154-5)(155-5) (158-5)(159-5)(162-5)(163-5)(164-5)(165-5)(166-5)(167-5)(173-5)(174-5)(175-5)(176-5)(177-5) (178-5)(180-5)(181-5)(182-5)(185-5)(186-5)(187-5)(188-5)(189-5)(190-5)(192-5)(193-5)(194-5)

Response: Exelon has not defined its project so narrowly that meaningful examination of alternatives is precluded. Exelon has applied for an early site permit for a merchant power plant to produce baseload power. It is not appropriate for NRC to define an applicant's business objectives. The Exelon and NRC staff have considered alternative sites within Illinois. Exelon and the NRC staff have also considered alternative means of generation. Chapter 8 of the EIS considers both conventional, non-nuclear means of generation and alternative "clean"-energy alternatives. None of the clean-energy alternatives were found to be capable of supplying baseload power. Although, Illinois has a commercially viable wind resource, the State of Illinois is not large enough for the wind resource to be counted on for baseload power. Wind power was also considered in combination with conventional energy generation and was determined not to be environmentally preferable to a new nuclear unit. These comments did not provide any new information and will not be evaluated further. No change was made to the EIS as a result of these comments.

Comment: Section 8.2.3.8 Fuel Cells The examination of the potential for fuel cells is much too cursory and quickly dismissed. Again it suffers from the same self-fulfilling prophecy thinking that characterizes much of the renewables section. (161-6)

Response: Fuel cells have not been developed to the point where they can be considered a viable alternative for baseload power on the scale that would be required to match the generating capacity of the postulated nuclear power plants. Larger fuel cells are envisioned, but they do not exist at this time. The comment provides no new information and was not considered further. No change was made to the EIS as a result of the comment.

Comment: Section 8.2.3.10 Combination of Alternatives This section also suffers from significant lack of detail, and willingness on the part of either Exelon or NRC staff to think outside of the box. What may be "acceptable" to Exelon for analysis as potential "partnering" combinations of alternatives may not be acceptable to the State of Illinois as part of the Renewables Portfolio Standards, or to local people who may have other energy assets that are being ignored. (161-7)

Comment: 8.2.3.10 fails to provide each alternative with estimated megawatts of production or savings, along with costs of each compared to nuclear. The Governor proposed 3,000 MW of wind because it's realistic, so only using 60 MW is unfairly arbitrary. (169-21)(179-21)

Response: In Section 8.2.3.10, the staff considered a combination of alternatives that it considered to be viable and had the least environmental impact of potentially viable combinations for the production of baseload power. Even ascribing no adverse impacts to wind energy, the impacts of the combination are sufficient to indicate that it is not environmentally preferable to the postulated nuclear plant. Other combinations, while possible, would have greater impacts or would not be viable for production of baseload power. The comments present no new information and were not considered further. No change was made to the EIS as a result of these comments.

Comment: My first concern I want to state here is as far as alternative energy sources that the NRC has looked at here, I would say the most important one that we have in this country is conservation, which is not much talked about and it's not actually a technology. (47-3)

Response: In Section 9.2.1.1 of its ER, Exelon presented an assessment of the viability of conservation measures as an alternative to the postulated nuclear plant. This assessment concluded that conservation is not a viable alternative. The staff has reviewed this assessment and concurs in the conclusion. The comment did not provide new information. No change was made to the EIS as a result of the comment.

Comment: What happened to safe conservation alterative energy, micro-power? We must look to the future, for the future. (37-3)

Response: In Section 9.2.1.1 of its ER, Exelon presented an assessment of the viability of conservation measures as an alternative to the postulated nuclear plant. This assessment concluded that conservation is not a viable alternative. The staff has reviewed this assessment and concurs in the conclusion. Section 8.0 of the EIS addresses those alternative methods of power production that potentially have the ability to provide baseload power on a scale consistent with the power production postulated in the early site permit application. The energy source mentioned in the comment does not have this ability. This comment does not provide new information. No change was made to the EIS as a result of the comment.

Comment: NEPA requires a detailed statement on "the relationship between local short-term uses of man's environment and the maintenance and enhancement of long-term productivity." Productivity is the key word for energy because we want to maximize the amount of energy we squeeze out of every source to save money. Conservation is always the cheapest, followed by energy efficiency, then producing electricity, where wind has become very cost effective. (169-23)(179-23)

Response: *In Section 9.2.1.1 of its ER, Exelon presented an assessment of the viability of conservation measures as an alternative to the postulated nuclear plant. This assessment concluded that conservation is not a viable alternative. The staff has reviewed this assessment and concurs in the conclusion. However, the crux of these comments is related to discussion of "the relationship between the local short-term uses of the environment and the maintenance and enhancement of long-term productivity." Section 10.3 of this EIS briefly discusses this relationship. Full assessment of the relationship is deferred to the CP or COL stage because the EIS for an early site permit is not required to contain the evaluation of benefits needed to complete the assessment (10 CFR 51.18). The comment contains no new information. No change was made to the EIS as a result of these comments.*

Comment: Instead of a thorough evaluation, these issues receive only brief, perfunctory attention in Chapter 10 of the draft EIS. For example, only a half-page is devoted to energy conservation as an alternative, which Exelon considers unreasonable, an assessment that the NRC staff appears to agree with (EIS, § 8.2.1.1). (150-8)(151-8)

Comment: By the way, it is noticeable in the Clinton DEIS that the only reference for the section about improved energy productivity ("8.2.1.1 Energy Conservation," p. 8-3) is a 2003 report by Exelon. But There is NO SPECIFIC COST COMPARISON (IN TERMS OF KILOWATT-HOUR) between the electricity saved by proven utility DSM programs and the projected electric generation from the Clinton Unit 2 reactor. And what about utility DSM programs in California (which has a "deregulated electricity market")? (171-7)

Comment: Earlier was mentioned the testimony of physicist and Energy Consultant Amory B. Louins with the Rocky Mountain Institute before the Illinois Commerce commission about replacing the electrical output from Commonwealth Edison's Braidwood nuclear station with energy-efficient technologies and techniques. A part of Louins' analysis was an examination of ComEd's (then) energy efficiency program. The conclusion was that, for a utility the size of ComEd (now an Exelon company), this energy efficiency program was the weakest in the nation. Apparently, this regressive trend still continues to day with Exelon. This valid point is NOT mentioned in the Clinton DEIS. (171-9)

Comment: THE MORAL IS: If an environmentally progressive state like California has room for improvement when it comes to its electric utility energy efficiency programs, then where does that leave Illinois (with Exelon, the nation's "nuclear giant")? It looks like the Clinton EIS needs to be redone in this respect! (171-10)

Response: In Section 9.2.1.1 of its ER, Exelon presented an assessment of the viability of conservation measures as an alternative to the postulated nuclear plant. This assessment concluded that conservation is not a viable alternative. The staff has reviewed this assessment and concurs in the conclusion. The comments did not provide new information and will not be evaluated further. No change was made to the EIS as a result of these comments.

Comment: It is stated that, "Exelon did not consider nuclear power plants license renewal in its ER." NRC staff concurs with this position, stating that license renewal does not add additional generating capacity. This has already been proven wrong historically. The power up-rate process HAS already resulted in added capacity from existing plants. Using the staff and Exelon assumption that the plant at the Clinton site would be a merchant plant, it must therefore take into account the possibility of added capacity coming from other already existing reactors within the region in which the merchant plant would compete. To fail to take this into account demonstrates the lack of thoroughness on the part of Exelon and NRC staff in developing and analyzing scenarios based on demonstrated historic and potential industry operation. This erodes confidence in their conclusions significantly. NEIS staff has examined their conclusions, and find them erroneous and without adequate justification. We do not concur with their findings. (161-3)

Response: Renewal of an operating license does not, in itself, increase generating capacity, but as pointed out in the comment, power uprates do increase the generating capacity. The text of EIS Section 8.2.1.3 has been revised to discuss power uprates.

Comment: In section 1.4, NRC falsely claims that NEPA does not require a detailed statement about alternatives to the proposed action. The application is to build a nuclear reactor on this site, so reasonably foreseeable future actions and alternatives must be considered. Furthermore, alternative sites in other states were not considered. (169-12)(179-12)

Comment: Federal law does require a consideration of alternative energy sources, but the NRC's review of renewable energy as an alternative source of power saying that such - - sources are not "environmentally preferable" to nuclear power, despite acknowledging that Illinois has the untapped potential to produce as much electricity from wind as from nine additional nuclear reactors. (191-7)

Response: Section 1.4 does not state that a detailed statement about alternatives is not required by NEPA. In fact, Section 1.4 starts with a statement that NEPA requires such a discussion. Section 1.4 goes on to state that the Commission has determined that a discussion of energy alternatives is not required for an ESP, which is correct. However, not withstanding the Commission determination, Exelon included a discussion of energy alternatives in its ER. Therefore, the staff conducted an evaluation of energy alternatives. The discussion of alternatives is found in Section 8.0 of this EIS.

The staff based its dismissal of wind energy on the basis of the intermittency of the wind resource; wind energy is not sufficiently reliable to serve as baseload power.

Commission guidance allows applicants to define a region of interest for evaluation of alternative sites. Exelon chose the State of Illinois as its region of interest. The NRC staff reviewed this choice and considers it to be reasonable.

The comments provide no new information. No change was made to the EIS as a result of these comments.

Comment: So I want to talk about alternatives... We have actual visions for what you can do in your community. It involves wind. There's a wind farm going in in Arrowsmith that's not too far that will employ many people from this community. You have wind resources here in DeWitt County. There's also other resources available. And I recommend that you look into that. (32-3)

Comment: In a discussion on energy alternatives, it's been said that wind is not viable because of its intermittent nature. Zion and Three Mile Island turned out not to be particularly reliable. The life time capacity for all operating nuclear plants are far short of hundred percent, even though those statistics allow and discount the off time for refueling, which is an extended period of time. The Dresden Plant had to shut down when there were cracks in the turbine. They came very close to turbine missile scenario, which is a worse case accident that's unresolved and there's no solution for that at this time. The LaSalle Plant is operating much under its designed capacity yet along its EPU capacity because of a piece of sheet metal that got loose and I guess is in an unknown location and they're running at reduced power, I understand, until the next fueling cycle. So it turns out the Commonwealth Edison's reactors and reactors in general are much less reliable than we're led to believe. So I actually feel that wind combined with the ability to wheel power across states is a much more reliable source of energy than nuclear reactors. When we have a calm day it lasts a day not several years, such as in the case of the Clinton shut down, actually, right? There was an extended period of several years. (55-6)

Comment: Alternative energy sources can be great for the economy. Wind turbines are a cash crop for farmers as they can place them in the middle of their farms and they're perfectly capable with growing crops right around them. We just believe that the draft EIS fails to objectively analyze these alternatives. (59-4)

Comment: We believe wind is a viable alternative. There are ways to do it, as was alluded, spreading it out geographically, that can contribute to the stability and regularity of that wind producing needed power. But wind is, is a resource that not everyone is blessed with. And you guys here, you've got it. And I believe that you should take advantage of this opportunity...You know, we can't, we can't make your decisions for you. All we can do is let you know what the options are and what our views are. And, and I would say that wind is one great option for folks here in Clinton. It will bring in tremendous investments. I think more of those investments will stay here in Clinton, if you go with wind versus a nuclear plant...You can, you can have more of that money stay right here in Clinton with wind, than you can with a nuclear plant. (68-6)

Comment: I've worked in areas associated with the development of wind power, so I'm familiar with the benefits of wind and I fully support the implementation of wind power in Illinois as well. (75-5)

Comment: Wind Power is unreliable, disrupts local wildlife with its widespread distribution and interferes with the weather patterns. (93-3)

Response: Section 8.2.3.1 of the EIS describes the wind energy resource in Illinois and environmental impacts that have been ascribed to wind energy generation. The environmental impacts associated with wind energy generation are small but frequently overstated. The resource has sufficient energy to be viable on an annual basis, but it is an intermittent resource such that the likelihood of wind power availability is not sufficient for wind power to be viable as a baseload power. Consequently, the staff has determined that wind power is not a viable alternative to the postulated nuclear plants. These comments did not provide any new information. No change was made to the EIS as a result of these comments.

Comment: One of the big inherent advantages that SMUD's solar photovoltaics (PVs) have over Rancho Seco is that solar electricity is incapable of having such a qualitatively – severe accident. This safety point of solar PVs vs. Nuclear reactors is NOT mentioned in the Clinton DEIS section on solar-thermal power and photovoltaic cells (p.8-18). (171-2)

Comment: While the Clinton DEIS discusses the environmental (air quality) impacts of coal and natural gas ("8.2.2.1 Coal-Fired Generation," and "8.2.2.2 Natural-Gas-Fired Generation," pp.8-6 to 8-8, 8-11 to 8-13), there is NO SUCH COMPARISON between the air pollutants emitted from these methods of fossil fuel generation and the emissions avoided by using PV electricity. (171-4)

Comment: First off, there is no mention of a "cofunding scheme" for Exelon which is comparable to what SMUD used for the PV1 solar-electric power plant. Secondly, the Clinton DEIS excludes the positive environmental and reliability advantages of grid-connected, utility-scale PV systems which would help justify Exelon's investment in them. (171-5)

Comment: It is interesting to note that the Clinton DEIS does NOT include such a calculated value for Illinoisan on-peak solar photovoltaic throughout Exelon's entire service territory ("8.2.3.4 Solar Thermal Power and Photovoltaic Cells," p. 8-18). This kind of quantitative analysis should provide "ammunition" in favor of the expansion of solar electricity under Illinois own REPS that is a top priority in Springfield for NEIS and other environmental groups during the current legislative session. (171-6)

Response: Section 8.2.3.4 describes the solar resource for the Clinton ESP site. By its nature, the solar resource is intermittent (solar thermal and photovoltaic systems do not generate power at night). In addition, these systems have a large land requirement and high generation costs. As a result, the staff determined that they are not viable options for generating baseload power. Having determined that solar thermal and photovoltaic systems are not viable for generation of baseload power, there is no point in evaluating environmental impacts of the systems. The comments provide no new information. No change was made to the EIS as a result of these comments.

Comment: In the on-line version, it references the dry cooling tower but I don't see it in here. Could you explain what that option is. I don't know anything about this stuff but does that have an impact then on the amount of radiation released? (33-1)

Response: Section 8.3 of the DEIS describes cooling tower alternatives to using Clinton Lake for the primary cooling system. Two of these alternatives involve the use of dry cooling towers. These towers would cool without evaporative loss of water from Clinton Lake. All of the cooling systems considered, including dry cooling, would be isolated from coolant passing through the postulated reactors. None of these cooling systems would be a pathway for release of radioactive effluents under normal operation. No change was made to the EIS as a result of the comment.

Comment: We need all forms of non-polluting energy supplies in this country. We will need renewable energy sources such as wind and solar. But these are not enough to meet all of our future energy needs. Coal is plentiful and cheap and will no doubt play a part in our energy future. (66-4)

Comment: It is an absolute travesty to waste a finite resource such as natural gas to create electricity. The impact of limited gas supplies and increasing demands has made it financially impossible for low income families to heat their homes. As a farmer, my fertilizer costs have doubled as the price of precious natural gas has increased. (105-6)

Comment: It is an absolute travesty to waste a finite resource such as natural gas to create electricity. The impact of limited gas supplies and increasing demands has made it financially impossible for low income families to heat their homes. As a farmer, my fertilizer costs have doubled as the price of precious natural gas has increased. (29-6)

| **Comment:** It's an absolute travesty to waste a finite resource such as natural gas to create electricity. The impact of limited gas supplies and increasing demands has made it financially impossible for low income families to heat their homes. And as a farmer, my fertilizer costs have doubled as the price of the precious natural gas has increased. (42-5)

| **_Response:_** _The comments are noted. No change was made to the EIS text as a result of these comments._

| **Comment:** Section 8, Environmental Impacts of the Alternatives, addresses air emissions from coalfired generation and other fossil-based alternatives, but fails to include carbon dioxide. The benefits of the nuclear alternative with regard to voluntary US programs to reduce greenhouse gas emissions are similarly overlooked. Estimates of tons of carbon dioxide emissions avoided by the nuclear alternative should be included in the Final EIS. (172-62)

| **_Response:_** _The comment makes a valid point related to carbon dioxide emissions. The EIS text was revised to include a discussion of carbon dioxide emissions for nuclear (Section 5.2.2), coal-fired (Section 8.2.2.1), and natural gas fired generation (Section 8.2.2.2)._

| **Comment:** Regarding these NEPA requirements, of particular concern to Public Citizen is the deficient consideration of renewable energy sources draft EIS. While addressing renewable energy sources as an alternative, the draft EIS does not give a fair and thorough consideration of the potential of clean, sustainable energy, and it relies far too heavily on the faulty evaluations performed by Exelon (see EIS, § 8.2.3). Public Citizen and others have successfully intervened in the licensing proceeding for the Clinton ESP on the grounds that Exelon's application "does not provide the basis for the rigorous exploration and objective evaluation of all reasonable alternatives to the ESP that is required NEPA."

| The evaluation of alternatives to the proposed action in the EIS fails to achieve the requirements of 40 C.F.R. 1502.14, which compels agencies, inter alia, to "devote substantial treatment to each alternative considered in detail." While the draft EIS gives fair attention to alternative sites for a new reactor, it gives only scant attention to renewable energy alternatives, despite the conservative admission that Illinois has at least 9000 MW(e) of wind power potential (EIS, pg. 8-17). The draft EIS overstates the impacts of clean energy alternatives and understates the impacts of nuclear power, wrongly concluding that a new nuclear unit at the CPS would be "environmentally preferable" to a combination of clean energy generation alternatives such as wind, solar, and biomass, and even suggesting that a new nuclear unit is preferable in the areas of "air resources, ecological resources, water resources, and aesthetics" (EIS, § 8.2.4). (150-9)(151-9)

| **Comment:** When the issue of the wind alternative to a new nuclear generator was raised, the initial comment from the one member of the NRC team who was the "expert" on the subject (his name escapes me) was that the Draft EIS had not concluded one way or another on the viability

of wind as an alternative energy source. It seemed that as the discussion proceeded and the wind alternative was again raised, he became more agitated until he personally concluded that "you'd have to cover the whole state of Illinois with windmills!", implying that wind is in fact NOT a viable choice. There seemed to me to be two very serious problems with his comments: First, he was implying that the electric generating capacity of the one new reactor at Clinton would equal the generating capacity of wind-farms covering the whole state--and that seems grossly inaccurate, no? Second, and most importantly, he seemed--on the spot, and personally, with no scientific data--to be drawing a conclusion in defense of nuclear power and against the wind alternative, a conclusion that he himself had admitted earlier in the meeting the commission report (EIS) had not addressed one way or the other. (152-2)

Comment: The scenarios and positions of Exelon, and those of NRC staff analyzing the Impacts of Renewables are selectively narrow and self-serving; and totally incomplete. They further fail to examine and therefore take into account very realistic scenarios of aggressive growth in the renewable energy sector in the next 15 years, the period during which new reactors might be contemplated for the Clinton site. One such realistic factor is the intention of the State of Illinois to institute a renewable energy portfolio standard as soon as 2006. This change on the law would require utilities to achieve real, on the ground targeted additions to capacity coming exclusively from the renewables sector. Benchmarks and timelines are already publicly available for analysis. It should also be noted that in Illinois, the bulk of such capacity is likely to come from expansion of wind energy. The State notes that 3,119 MW of installed wind capacity has been proposed for Illinois - roughly three times the size of the single reactor proposed for the Clinton site. Further, the entire capacity projected for the Midwest amounts to 11,759 MW of proposed installed wind capacity alone - nearly 12 times the size of the proposed Clinton reactor. Addition of such large amounts of State mandated renewables capacity totally eviscerates the proposed need for the Clinton site reactor, either as a baseload generator providing for in-state power, or as a merchant plant selling to region about to experience a further glut of power. The facts that both Exelon and NRC staff failed to take this into account calls into serious question their assumptions and methodology used for reaching their conclusions. NEIS staff has examined their conclusions, and find them erroneous and without adequate justification. We do not concur with their findings. (161-2)

Comment: At the public hearing, we were told that the comparison of alternatives to nuclear at Exelon's ESP was based on the present situation. Yet, a nuclear reactor isn't expected to operate earlier than 2014, possibly later, making the rationale arbitrary. The comparison should be based on when the alternative is both realistic and most cost effective. Detailed tables that project both megawatt potential and future costs are needed for the proposed nuclear plant and for alternatives including biofuels, biomass, biogas, mechancial solar, methane hydrates, micro water turbines, natural gas pipeline from Alaska, energy efficient products and appliances, insulation, wind power, ground source energy, combined heat and power, micro CHP by Honda & Toyota, solar metal alloys, coal gasification, fuel cells, time-based metering, DSM, wave/tide turbines, more accurate or higher mileage standards and hybrid vehicles freeing energy for electricity generation, and higher energy prices decreasing consumption and improving efficiency. (169-17)(179-17)

Comment: The Draft EIS fails to "rigorously explore and objectively evaluate" better, lower-cost, safer and environmentally preferable clean energy and energy efficiency alternatives to new nuclear power. 40 C.F.R. 1502.14(a). (170-2)

Comment: In the Draft EIS, the NRC Staff has failed to comply with its duty under NEPA to "rigorously explore and objectively evaluate all reasonable alternatives" to the granting of the ESP. 40 C.F.R. 1502.14(a). The Draft EIS's purported analysis of alternative energy sources is flawed because it: (1) assumes, but does not analyze, a need for power, (2) uses an improperly constrained purpose of creating baseload power to reject reasonable alternatives to new nuclear power, and (3) improperly concludes that clean energy alternatives are environmentally preferable and cheaper than new nuclear power. Because of these shortfalls, the Draft EIS improperly rejects better, lower-cost, safer, and environmentally preferable energy efficiency, renewable energy resource, distributed generation, and "clean coal" resource alternatives to the siting of a new nuclear power plant at the Clinton ESP site. (170-5)

Comment: Roughly 207 inherent benefits of "distributed resources" (decentralized sources of electricity like solar photovoltaic, wind turbines, fuel cells, microturbines, cogeneration, and energy efficiency) can make them up to 10 times more valuable than previously thought by improving system planning, utility construction and operation, and service quality, and by avoiding societal costs. How unfortunate that ALL of these 207 inherent benefits (and how they can be so valuable to Exelon's distribution system) are OMITTED from the Clinton DEIS. (171-12)

Comment: Some concerns about the environmental impact statements, just one specific example, the fuel cells, it was essentially described as a non-viable source. Actually, this applies not just the fuel cells, but all of the alternative options that are available, described as a non-viable economically. But that was based on the present value, but not the future. That's an arbitrary and capricious decision, that was made in that decision, in that process. There's no reason to reasonably expect that that shouldn't be five years out, in terms of building the plants, if not 10 to 15, 20, for evaluating whether it's going to be economically viable. I want to also go into some of the alternatives. We've already heard about wind. But just to mention, Bloomington Normal has a plan for 400 megawatts of wind. That's one of the two ends of the major transmission line coming from Clinton. So that pretty well takes care of it, on that end. Some other alternates, in the geothermal section, there was, there was no mention whatsoever of getting energy from the ground. Not from the traditional geothermal that you find out west, where it's really hot springs, but the ground source heat pumps where you can actually get the heating and the cooling for your home, right from the ground beneath your feet. It's free, it doesn't cost anything other than a little of energy to run a pump, to either circulate water, or to circulate air if you're using an earth tube. Now, you can use an earth tube if you actually have a well enough insulated house, that, and the insulation is going to be far cheaper than actually building a nuclear power plant. So, when you're looking at ways to save, conservation, it's always the cheapest, fastest, healthiest, safest. It can be done like that. Energy efficiency is next. I mean, in terms of the payoff, the air conditioning. Everybody wants to have that cool, everybody wants, everybody wants to have their fridge with the cold beer, you can do that, but if you get a more efficient fridge, then you don't need as much electricity generated in the first place. Some other alternatives that were NOT considered. Micro-water turbines, micro-natural gas turbines, combined, I don't recall combined heat and power, pulling energy off the waste heat. Methane hydrates. Mechanical solar as opposed to photovoltaic, bio gas from algae sources and such. Was mentioned before, coal gasification. Since we're talking 15, 20 years down the line, most likely. And then, in the combination section of putting all of those alternatives together, it wasn't, and the whole problem with the alternatives was that there were no specific numbers. You couldn't see what are the potential megawatts available from this, that or the other source, in the EIS. Those should be addressed. (80-1)

Response: The foregoing comments question the staff's analysis of alternative energy sources and conclusion that none of the alternatives considered is environmentally preferable to the postulated nuclear plant. There is a statement that the environmental impacts of the clean energy alternatives is overstated, and the absence of a discussion of need for power is mentioned.

In the case of an EIS related to an early site permit, NRC regulations defer the consideration of need for power until an application for a construction permit (CP) or combined license (COL) is received. This is reasonable because an early site permit is not an application to build a nuclear power plant.

In its evaluation of energy alternatives, the staff considered whether the technology was commercially available on a scale to provide the capacity needed to match the capacity of the postulated nuclear plant and whether the technology together with the resource could provide baseload power, i.e., power on demand. Wind power is the most mature of the clean-energy technologies but fails to meet the requirements for baseload power. The other clean-energy alternatives fail one or more of the tests for viability.

The staff considered wind energy in combination with natural gas and biomass generation, purchased power, and demand-side management. The only adverse environmental impacts assigned to the combination were those associated with natural gas generation; no adverse impacts were assigned to the clean-energy portion of the combination. Thus, the staff did not overstate the impacts of clean-energy alternatives in this evaluation. In fact, the staff understated the impacts. Nevertheless, as long as enough natural gas generation was included in the combination to make the combination viable for baseload power, the adverse impacts of the natural gas generation were sufficient to conclude that the impacts of the combination were not environmentally preferable to those of the postulated nuclear plant.

These comments did not provide any new information and were not evaluated further. No change was made to the EIS as a result of these comments.

Comment: Section 8.2.1.1 says NRC reviewed Exelon's assumptions and analysis to reach a conclusion without sharing any of those same numbers or data in the EIS to allow the public to reach its own conclusions. (169-18)(179-18)

Response: The report referred to in these comments is the environmental report (ER) submitted with the early site permit application. That ER is publicly available at the NRC web site (http://www.nrc.gov/reactors/new-licensing/esp/clinton.html). The comment contains no new information and was not considered further. No change was made to the EIS as a result of these comments.

Comment: Section 8.6.8, Radiological Impacts of Normal Operations, characterizes health impacts as small because dose would be "small compared to the population dose from natural background." Natural background radiation doses generally exceed acceptable regulatory criteria for exposure to the public from nuclear power plants, and comparison to background is not useful in this case. Comparison of anticipated doses should be made against regulatory limits for radioactive emissions and the principle of ALARA to determine whether doses are "small." (172-63)

Response: The staff considered doses to the maximally exposed individual (MEI) as well as population doses in making the determination that radiation doses and resultant health impacts from a new nuclear unit's operations would be SMALL at all of the alternative sites. Section 8.6.8 of the EIS states that even with differences in pathways, atmospheric and water dispersion factors, and population, doses estimated to the MEI for the alternative sites would be expected to be well within the Appendix I design objectives. No change was made to the EIS as a result of the comment.

Comment: You said you did an environmental impact, a study of alternative sources. So are you saying that the potential environmental impact of this plant is not any greater than say a windmill going berserk or a coal plant? Is that what you're saying? There's no greater environmental impact, potential environmental impact of this plant than alternatives? (39-1)

Response: The staff considered the safety of two types of new nuclear reactors in Section 5.10. The consequences of design-basis accidents for the postulated reactors would be within regulatory limits. Therefore, the staff considers the environmental impacts of design basis accidents to be small. The probabilities of severe accidents are extremely small. As a result, the probability-weighted consequences (risks) of severe accidents are considered small. The risks associated with severe accidents for the ABWR and AP1000 reactor designs considered in the EIS are much smaller than risks set forth in the Commission's safety goals. Therefore, the staff also considers the environmental consequences of severe accidents to be small. The staff has not evaluated the environmental consequences of accidents associated with alternative energy sources, but considers them to be small. In making its comparisons among alternatives, the staff does not distinguish impacts within a significance level. Thus, the staff did not use environmental impacts of accidents in evaluation of either alternative energy sources or alternative sites. The comment does not provide new information. No change was made to the EIS as a result of the comment.

Comment: Which alternative energy generating technologies were deemed to be economically viable? (36-1)

Response: *The staff considers conventional generating technologies using coal and natural gas to be economically viable methods of generating large-capacity baseload power. Wind generation technology is economically viable for generating energy, but it is not viable for generating baseload power because of the intermittent nature of the resource. Several other technologies discussed in Section 8.2.3 may also be economically viable for small-capacity applications but were judged not to be viable for large-capacity baseload generation. The comment did not provide new information. No change was made to the EIS as a result of the comment.*

Comment: We will learn to live with the energy we can make without polluting our planet's future. (05-3)

Comment: Please consider wind, solar energy instead. (06-3)

Comment: We have the technology for safe and sustainable energy production right now. We have only to develop and use it. Illinois already has more nuclear reactors than any other state and has tremendous untapped renewable energy potential. (09-3)

Comment: I would like to think that with the best interests of the environment--most especially human health--in mind, the EIS would look much more closely at alternatives to nuclear power that are renewable, have zero emissions, and do not create radioactive waste that we and our kids and their kids will have to contend with. (152-4)

Comment: Instead of building another nuclear plant effort should be re-directed to developing large scale solar and wind sources. (168-2)

Comment: Part of my remarks are in regards to "8.0 Environmental Impacts of the Alternatives" (starting on p. 333). Probably the USNRC thinks that this is beyond the specific scope of the DEIS, but the NEGATING EFFECT of the proposed Clinton Unit 2 reactor on the state's "political environment" to actively promote truly sustainable appropriate renewable-electric technologies has to be considered. (25-3)

Comment: I think what has just been recently said about hydrogen really emphasizes the point of the no new nukes movement and moving away from nuclear power...All you're going to do is switch it with water. Water's not, not renewable, not a renewable resource. (35-2)

Comment: A combination of alternatives is better for the reliability. Rather than having a single source it will shut down, like the Clinton plant did, if you have wind farm, solar, natural gas and energy efficiency distributed throughout the state, it's better for reliability. (59-3)

Response: These comments are generally opposed to nuclear energy and to the postulated nuclear plant. They do not provide new information and were not considered further. No change was made to the EIS as a result of these comments.

Comment: It [a second reactor] beats the alternative of imported oil. (24-5)

Comment: I heard at a press conference earlier today passionate and involved people extol the virtues of diverse, renewable, sustainable energy supplies. In many cases I applaud the spirit of their intent. I look forward to more reliance on solar, wind, geothermal and other renewable energy sources. But we need both. Nuclear can continue to reduce emissions by fueling a conversion to a hydrogen economy. Renewables can't do that. (61-1)

Comment: Also bear in mind that when you're comparing alternative energy sources, a thousand megawatts of wind power does not equal or replace a thousand megawatts of nuclear power. There is an issue of capacity factor. How often nuclear energy is available versus how often the wind energy is available. The wind capacity factor best worldwide is about 35 percent. And nuclear on average right now runs about 92 percent. (61-5)

Comment: There are just a couple of factual errors that I wanted to clear up, one of which was addressed by the immediately preceding speaker, plant available compared with wind. And basically over the last several years, across the United States nuclear plant availabilities have been about 90 percent. So 90 percent of the time these plants have been operating and operating more or less at full power. Obviously there are exceptions. There are plants that have to shut down to refuel and occasionally there need to be some repairs made. But 90 percent is a pretty fair number. (62-1)

Comment: Wind certainly can play a part [in generating large base load power] you know, solar not so much, hydro certainly helps out already. But the only options for increasing significant demand is nuclear and fossil fuels. And when you compare the statistically small risks of nuclear power, compared to the very real risks and consequences of fossil fuel, such as asthma and the, the many deaths of lung cancer. These are very real, known consequences of coal. And that needs to be considered. (78-3)

Response: *These comments are generally supportive of nuclear energy and to the postulated nuclear plant. They do not provide new information and were not considered further. No change was made to the EIS as a result of these comments.*

Comment: It would be good to replace this capacity [21,000 megawatts of actual non-renewable generation capability in Illinois] with renewable sources, such as wind, solar and biomass sources, if that could be possible. According to the Department of Energy's wind map for Illinois, about 9,000 megawatts of wind power capacity exists in Illinois, and that's including both the good and the not so good sources. Solar power can help to supplement this amount, but will require significant capital expenditure, and I doubt we'll be able to make up the 12,000 megawatt gap, on its own. Now, biomass energy is a promising source for an agricultural state like Illinois. Crops such as Miscanthus Giganteus might be able to, to help with that. However, to obtain large amounts of energy, would require us to divert large amounts of our valuable farmland, away from producing food crops for ourselves, our nation and our world. Now, maybe, just maybe, if we put solar panels on all of our buildings, if we tapped all of the wind power which nature provides to us, and if we used our less valuable tracts of farmland to grow -- fuels, we could reach the 21,000 megawatts of actual electrical generating capacity, that we had in 2003, from non-renewable sources. And that would be great. We need a diversified energy portfolio. (79-1)

Response: *This comment supports combinations of renewable energy sources. It does not provide new information and was not considered further. No change was made to the EIS as a result of the comment.*

Comment: Means of renewable power generation need to be encouraged. (12-5)

Comment: Instead of building new reactors, we need to rid our country of nuclear power altogether. We must look to renewable, sustainable energy, wind turbines and solar energy for example. (10-6)

Comment: If Illinois and Gov. Blagojevich are serious about meaningfully expanding the role and market share for renewable energy resources in the state, then there is simply no room for one or two new 1000 Mw nuclear plants. (102-2)

Comment: Furthermore, as a state, we should be moving towards sustainable energy sources, instead of continuing to rely on harmful energy such as nuclear power. Not only are renewable resources safer for our communities and the earth, they are becoming cheaper and easier to access. For example, it will take seven to 10 years to build a nuclear power plant; it would take two years to build a wind farm that could produce the same amount of energy. This energy would be produced cleanly and safely. I encourage Exelon and other power companies in Illinois to begin investing in sustainable energy sources such as wind and solar power. (104-2)

Comment: It's time to pursue alternate forms of energy and divert our attention to that goal. (11-4)

Comment: Furthermore, as a state, we should be moving towards sustainable energy sources, instead of continuing to rely on harmful energy such as nuclear power. Not only are renewable resources safer for our communities and the earth, they are becoming cheaper and easier to access. For example, it will take seven to 10 years to build a nuclear power plant; it would take two years to build a wind farm that could produce the same amount of energy. This energy would be produced cleanly and safely. I encourage Exelon and other power companies in Illinois to begin investing in sustainable energy sources such as wind and solar power. (17-2)

Comment: Illinois has plenty of solar, wind, and biomass, [For example, Chicago itself is third in the country for solar usability (the "Windy City" has 80% of sunlight that is in Florida). The political commitment to sustainably maximize our use of what "homegrown" renewables we have by passing a state-wide "Renewable Energy Portfolio Standard" (REPS) supercedes Excelon's self-serving promotion of Clinton Unit 2 as part of the national "nuclear relapse." (171-3)

Comment: I think instead that renewable sources of energy should be encouraged. (21-3)

Comment: Focus our energies (and money and manpower) in alternative energy resources!! (23-2)

Comment: And I think it's up to the people in Illinois to really support economical and sustainable energy development. To demand that Exelon create wind farms. If it's not feasible, of course, I mean, of course, it's not going to be feasible where they put the plant. I mean, anyone can say that. So you find a place where it is feasible. And if not, then you go to an area in DeWitt, in Clinton where you can develop wind, solar, biomass, all of these. (35-5)

Comment: Like, please save us, Exelon, right? Exelon's great for the community. But I don't think, I don't think it has to be this way. And I think that it's up to people in Illinois, the residents of Clinton to really demand that their leaders, that the corporations are held accountable and start developing sustainable development. (35-7)

Comment: Our state should be moving towards wind energy or another renewable source of energy as opposed to harmful nuclear energy. (44-2)

Comment: Instead, I am in favor of passing a state-wide renewable energy portfolio standard, this year in Springfield. (71-3)

Comment: Secondly, Exelon makes no bones about opposing a state-wide renewable energy portfolio standard. This would set realistic goals to ramp up our use of renewably generated electricity, requiring three percent of all electricity by 2007 to come from renewable sources, solar, wind and biomass, and then 10 percent by 2012. Such a renewable energy portfolio standard, should be passed in Springfield first, before Exelon's proposed second Clinton reactor is even considered. (71-8)

Comment: If we would invest in renewables, and in this diversified portfolio of options that we have in the same way, for the next 50 years, we would certainly get to that sustainable future in energy that we all want. (81-2)

Comment: I believe that the current energy policy, endorsed by the current administration is a step in the wrong direction. Instead of subsidizing nuclear power plants, why not invest and explore renewable energy sources. Why not explore and create renewable energy sources, rather than creating nuclear waste. (83-2)

Comment: I did want to mention, that the Government does invest quite a bit in renewable energy research. They do more in renewable energy research than they do nuclear energy research. And when that develops, I do hope that it becomes a part of the energy mix. (87-4)

Comment: And the time has finally come, for us as a society, to wrap it up. To cut our losses. To do our very best to stuff the genie back in the bottle, if indeed that's at all possible, and to move on to other forms of energy. (89-2)

Response: These comments, while generally supporting alternative, renewable energy sources, deal primarily with energy policy. The NRC is not involved in establishing energy policy. Rather it regulates the nuclear industry to protect the public health and safety within existing policy. These comments provide no new information and were not considered further. No change was made to the EIS as a result of these comments.

Comment: You start with water. You add energy and you split it into hydrogen and oxygen. You put those into a fuel cell. What comes out? Energy and water. How perfectly renewable is that? (64-1)

Response: This comment addresses fuel cells. The staff considers fuel cells in Section 8.2.3.8. On the basis of the cost and capacity of current fuel cells, the staff concluded that they are not viable alternatives to the postulated nuclear plant. The comment provides no new information. No change was made to the EIS as a result of the comment.

E.2.21 Comments Concerning the Site Redress Plan

Comment: Site preparation (Section 1.1.2) refers to 10 CFR 50.10(e)(1), which says activities permitted under an ESP include part (v): "the construction of structures, systems and components which do not prevent or mitigate the consequences of postulated accidents that could cause undue risk to the health and safety of the public." The last part of the sentence was left out of the draft EIS. This carefully worded NRC code would allow construction of the entire reactor, so long as nothing radioactive is installed. (169-9)(179-9)

Response: The ESP does not authorize construction or operation of a nuclear power plant. An early site permit is a Commission approval of a site or sites for one or more nuclear power facilities. However, as discussed in Section 4.11 of this EIS, certain site-preparation activities and preliminary construction activities are allowed provided that a site redress plan is submitted by the applicant and the final ESP EIS concludes that the activities will not result in any significant adverse environmental impacts that cannot be addressed.

The filing of an application for an ESP is a process that is separate from the filing of an application for a construction permit (CP) and operating license (OL) or a combined operating license (COL) for such a facility. The ESP application makes it possible to evaluate and resolve safety and environmental issues related to siting before the applicant makes large commitments of resources. If the ESP is approved, the applicant can "bank" the site for up to 20 years for future reactor siting. If an ESP holder decides to pursue construction of a nuclear power plant beyond any approved limited activities identified in Section 4.11 of this EIS, it must obtain a CP or a COL, the issuance of which would be a major Federal action requiring preparation of an EIS under 10 CFR 51.20 that, among other things, would address the benefits of the proposed action, such as the need for power and cost of power. No change was made to the EIS as a result of the comment.

E.2.22 Comments Concerning Editorial Issues

Comment: Appendix K-14, Line 2. Statement regarding downstream release. The statement considered by the NRC regarding maintaining the 5 cfs discharge minimum could not be found on page 5.2-6 or the rest of Section 5.2. (141-129)

Response: Although the wording is different, the wording on page 5.2-6 is considered to be the same commitment. No change was made to the EIS as a result of the comment.

Comment: Section 3.2, Page 3-2, Line 25. Appendix Table of PPE values. The DEIS refers to the Appendix for the complete set of PPE values. The Table in the Appendix is not current and a new updated Table should be provided or those values that have changed should be listed. (141-47)

Comment: Appendix J, Table J-1, Various sections, Table J-1. Table J-1 appears to be based on an earlier version of the SSAR, in that some elements have since been included in PPE Table 1.4-1 in the SSAR, but are missing from Table J-1 in the DEIS. Specifically, Table J-1 is missing PPE Elements 2.1 "Air Temperatures", 3.1 "Ambient Air Requirements", 4.

"Containment Heat Removal System", and 14, "HVAC Systems". These revisions were previously made in response to NRC RAI 2.3.1-8. (141-131)

Comment: Appendix J, Table J-2, PPE Element 1.22 "Snow Load". Table J-2 site characteristic value snow load is 35 lb/ft2. ER Section 2.3.1.2.3 and SSAR Section 2.7.3.3, both entitled "Heavy Snow and Severe Glaze Storms", indicate that the snow load is 40 lb/ft^2, a point that was clarified in several RAI responses to NRC (i.e., NRC RAI's 2.3.1-5, 2.3.1-6, and 2.3.1-10). (141-132)

Response: Appendix J was revised to include the most current PPE values.

Comment: Section 1.5, Page 1-7, Line 15 NRC's use of the word 'reactor'. In the ESP application, Exelon applied for a site to be reserved for a future nuclear facility (See Administrative Section 1.1). As stated in the Environmental Report in Section 1.1.3, the selection of the reactor design is still under consideration and a set of bounding parameters was determined using the reactor design-types listed. In the Site Safety Analysis Report, Section 1.2.3, Proposed Development, EGC describes where the EGC ESP facility will be located and that the facility may consist of a single reactor or multiple reactors (or modules) of the same reactor type. The use the term 'unit' implies that the EGC ESP would be restricted to a single reactor of the same design. Since the EGC ESP application was based on a set of bounding parameters and not on a single reactor design, the term 'unit' should be changed to 'facility' throughout the Draft Environmental Impact Statement. (141-1)

Comment: General Comment NRC's use of the word 'unit'. In the ESP application, Exelon applied for a site to be reserved for a future nuclear facility (See Administrative Section 1.1). As stated in the Environmental Report in Section 1.1.3, the selection of the reactor design is still under consideration and a set of bounding parameters was determined using the reactor design-types listed. In the Site Safety Analysis Report, Section 1.2.3, Proposed Development, EGC describes where the EGC ESP facility will be located and that the facility may consist of a single reactor or multiple reactors (or modules) of the same reactor type. The use the term 'unit' implies that the EGC ESP would be restricted to a single reactor of the same design. Since the EGC ESP application was based on a set of bounding parameters and not on a single reactor design, the term 'unit' should be changed to 'facility' throughout the Draft Environmental Impact Statement. (141-2)

Comment: Section 1.5, Page 1-7, Line 15. NRC's use of the word 'reactor'. In the ESP application, Exelon applied for a site to be reserved for a future nuclear facility (See Administrative Section 1.1). As stated in the Environmental Report in Section 1.1.3, the selection of the reactor design is still under consideration and a set of bounding parameters was determined using the reactor design-types listed. In the Site Safety Analysis Report, Section 1.2.3, Proposed Development, EGC describes where the EGC ESP facility will be located and that the facility may consist of a single reactor or multiple reactors (or modules) of the same reactor type. The use the term 'reactor' implies that the EGC ESP would be restricted to a single reactor of the same reactor design. Since the EGC ESP application was based on a set of bounding parameters and not on a single reactor design term 'reactor' should be changed to 'facility' throughout the Draft Environmental Impact Statement. (141-3)

Comment: Section 3.1, Page 3-1, Line 33. It is stated that the multiple units would be grouped into one operating unit. Due to the nature of the ESP, it would more accurate to state that the multiple units could be grouped into one operating unit. There is no requirement to place these multiple units into one operating unit, e.g., 2-AP1000s. (141-41)

Response: Section 3.0 of the EIS defines how the staff has used the terminology of "unit" relating to the Exelon ESP site. It does not exclude multiple reactors at the ESP site. No change was made to the EIS as a result of these comments.

Comment: Section 6.3, Page 6-39. "At the end of the operating life of a power reactor, the NRC regulations require that the facility undergo decommissioning. Decommissioning is the removal of a facility safely from service and the reduction of residual radioactivity to a level that permits termination of the NRC license. The regulations governing decommissioning of power reactors are found in 10 CFR 50.82, 50.75, and 50.82." 10 CFR 50.82 is repeated twice in the last sentence. (141-118)

Response: Section 6.3 of the EIS was revised to delete one of the 10 CFR 50.82 entries.

Comment: Section 4.0, Page 4-1, Lines 10-13. Discussion of Site Redress Plan. "The site redress plan allows for specific site-preparation activities to be conducted with approval of an ESP. The activities evaluated for the Exelon ESP site are those permitted by Title 10 of the Code of Federal Regulations (CFR), 50.10(a)(1) and 52.25(a). In the event that the ESP is approved and Exelon conducts site preparation activities but does not build the new nuclear unit, Exelon would be required to implement its site redress plan." The correct 10 CFR callout should be 10 CFR 50.10(e)(1). (141-53)

Response: The callout was corrected to read 10 CFR 50.10(e)(1).

Comment: Section 3.2, Page 3-4, Line 11. It is stated that the selected reactor would be bounded by values. It is more accurate to state that the selected reactor(s) would be bounded by values. (141-48)

Response: The comment is noted and the change was made in the EIS.

Comment: Section 5.9.3.1, second paragraph: The Appendix I thyroid dose design objective is incorrectly stated as 15 rem/yr. The correct value is 15 mrem/yr. (172-64)

Response: The commenter is correct. The text in Section 5.9.3.1 was changed accordingly.

Comment: The first thing that bothers me about this 650 page document is that there is no index. Today, it is a trivial matter for anyone using word processors like Microsoft Word and others to add an index to any document. Why wouldn't they put in an index of major topics in such an important document? (153-1)

Comment: Introduction, Section 1.5 Compliance and Consultations, page 1-7. A listing of the specific contacts for consultation would be helpful to provide other sources of information or clarification on specific points within this document that may fall in other areas of expertise in other Agencies. (172-7)

Comment: Station Operation Impacts, Section 5.9.2.1, Liquid Effluent Pathway, page 5-5 1, paragraph 1. Provide the summary of the data from the cited tables. (172-32)

Response: The comments are noted. No change was made to the EIS as a result of these comments.

Comment: It is also unclear whether the Presidential Memorandum of June 1, 1998, requiring all Federal documents to be written in a plain language format has been assessed for this document. Please clarify this point. (172-51)

Response: Every attempt has been made to implement the plain language policy of the NRC for the purpose of this document; however, an EIS is by its very nature a technical document that requires the use of specific technical terms and discussions. No change was made to the EIS as a result of this comment.

Comment: Section 5.10.2, Page 5-71, Line 4, 10. It is stated that Exelon owns Clinton Lake. AmerGen Energy Company, LLC owns Clinton Lake. (141-88)

Response: The EIS was changed to indicate that AmerGen Energy Company, LLC, owns Clinton Lake.

Comment: Section 5.10.2, Page 5-66, Line 40. Reference in Section 5.10. In the above-mentioned section, the reference Exelon 2004b does not exist in Section 5.0 and should be changed to Exelon 2004. (141-86)

Response: The reference was changed as indicated.

Comment: Appendix K-12, Line 6. Statement regarding construction impacts. The statement "A NOI will be filed with the federal and state agencies to receive authorization for land disturbance under the general storm water permit. A SWPPP will also be prepared in accordance with the requirements of the general permit. A NOT will be filed with the IEPA upon completion of construction and stabilization of the disturbed areas" is made in ER Section 4.2.1.2.2. (141-127)

Comment: Appendix K-13, Lines 5 and 6. Statement regarding storm water sediment. Statement is found in ER Sections 4.6.3.7.2.2 and 4.2.2, rather than 4.6.3.7.2.1 and 4.2.1.2.2 as identified. (141-128)

Response: The section and page number for this statement were modified.

Comment: Section 2.1, Page 2-1, Line 23-24. "between the cities of Lincoln and Urbana-Champaign". Site location of the ER lists the city as "Champaign-Urbana". (141-5)

Response: The text was changed to read Champaign-Urbana.

Comment: Affected Environment, Section 2.6.3.4, Chemical Monitoring, page 2-23. The requirements of the current permit should be stated and not just cited. (172-13)

Response: There are no current permits for the proposed nuclear unit. No change was made to the EIS as a result of the comment.

Comment: Appendix K, Page K-14, Line 6. Statement regarding makeup water for the normal (non-safety) plant operations. The second sentence of the statement needs to be revised. "The makeup water for the normal (non-safety) plant operations will be taken up through a new intake structure located next to the CPS intake structure on the northern basin of Clinton Lake, should read, "The makeup water... located 65 feet south of the CPS... of Clinton Lake". In addition, the EIS Statement Sections for this item needs to be revised. It is currently identified as "Lance," which is not a Section of the Statement. (141-130)

Response: These corrections were made in the EIS.

Comment: Please define the term "best management practices," which occurs throughout the draft EIS. (150-22)

Comment: Please define the term "best management practices," which occurs throughout the draft EIS. (151-22)

Response: These suggestions were implemented in the EIS.

Comment: Section 4.12, page 4-44. Statement regarding construction impacts. Add a sentence to Section 4.12 of the DEIS (as in 5.12): "The impact column designates negligible and beneficial impacts as SMALL." (141-64)

Response: This sentence was added in Section 4.12 of the EIS.

E.2.23 Comments Concerning the Safety Review for the ESP

Comment: But as the weather becomes increasingly unstable, we get much greater highs and much greater lows and much more of the severe events such as the tornados. So, you know, I would very much hate to see one of those hit the earthen dam and lose coolant or something like that. (55-3)

Response: Nuclear power plants are extremely robust structures that are designed to survive hurricane and tornado strikes. Should an extreme weather event cause a nuclear power plant to be shut down (i.e., reactor is shut down as a hurricane is approaching, rather than the reactor being shut down by the hurricane), the reactor can be maintained in a safe condition by the reactor's ultimate heat sink. Ultimate heat sinks are designed to withstand extreme weather events such as hurricanes and tornadoes. The likelihood of the maximum wind speed in a tornado exceeding the design wind speed for the ultimate heat sink is typically less than 1 in 10 million years. There is no evidence that the frequency of the most violent tornadoes is increasing. No change was made to the EIS as a result of the comment.

Comment: Affected Environment, Section 2.4, Geology, pages 2-16, 2-17. The location of the New Madrid fault relative to the proposed Exelon ESP site should be included as a point of information and evaluation relative to the potential of earthquake and structure requirements to meet this potential need. (172-10)

Comment: On page 2.4, of the, this environmental impact study, I would like to recommend that the NRC look more closely at the two major geological faults that run through this area, the Wabash fault, and more seriously, the New Madrid fault. The New Madrid fault is the location, the fault location with the greatest earthquake that there's ever been in North America, was on the New Madrid fault. These are not solid faults. They're active. And the most serious result, again, another risk factor and you've heard a lot of them here, for the people that live here in Clinton, and for the people that live around the area, including Chicago, for a major meltdown. What would happen if you had a, an earthquake, the only thing holding up Clinton Lake is an earthen damn. That damn would liquify, and the lake would retract. And then the, there would be no more water for the reactors. This issue has not been properly addressed in this environmental fact study, or in the current running of the Clinton reactor there. (72-2)

Response: The geology of the Exelon site is described only briefly in the EIS. Section 2.5 of the Safety Evaluation Report (NRC 2006) contains a description of the geology of the site. The discussion includes a detailed description of the seismic characteristics of the region including the New Madrid and Wabash faults. The comments provide no new information. No changes were made to the EIS as a result of these comments.

E.2.24 Comments Concerning Safeguards and Security

Comment: Also, transportation of spent nuclear materials is not secure, the materials might be used to make dirty bombs, and the reactors themselves remain terrorist targets. (04-3)

Comment: And security standards at nuclear plants are downright pathetic. (09-9)

Comment: Nuclear plants are vulnerable to...terrorist attack. (10-4)

Comment: I feel that the long-term safety and security issues far outweigh the supposed benefit of a new facility. (18-2)

Comment: I have witnessed first hand the level of security and professionalism involved with the work at the Clinton Power Plant. I live 5 miles from the facility and have no fear about the safety of the community. We live in an era where all need to be concerned about worldwide activities. (28-2)

Comment: On April 7th, 2005, the National Academy of Science disclosed that the Nation's reactors are vulnerable to terrorist attack on these fuel pools. We are sure that the folks here, that work at Clinton, know that we have about several hundred metric tons of irradiated fuel stored on the top of the reactor building, that is, would be vulnerable to an attack. (67-4)

Comment: The security of the waste, the inability to store it safely, as, as Paul said. The FBI is currently investigating staff as the USGS, for falsifying data on the, on the safety of -- (68-4)

Comment: The second reactor will be just another tempting target for terrorists. (71-1)

Comment: The National Academy Science report that came out talking about how LaSalle and Dresden are extremely vulnerable in particular. Apparently, here in Clinton, the fuel pool is sitting on top of your, sitting on top of your reactor. A medium long range mortar has a range of five to eight miles. It would be very hard, very difficult, it fires about 30 rounds per minute. It would be very hard, very difficult to neutralize that. And we don't know whether it's actually safe from that sort of an attack, because that information is not publicly available. (80-3)

Comment: Nuclear security enhancements since September 11th. And I commend them for that, they recognize, the Department of Homeland Security recognizes, President Bush and the Congress recognize, that nuclear power plants are a prime target for terrorism. But what are we doing here in Clinton? We're not mitigating the threat, we're increasing the threat. We are going to make the attractiveness of a nuclear facility more enhanced for terrorists. They're going to want to come to Clinton more than ever to, to wreak havoc on us here in this community, throughout the state. For anybody who thinks this won't happen in Clinton, tell it to the folks in Oklahoma City who would never have thought that they would be the subject of the worst terrorist attack 10 years ago today. (89-4)

Comment: I have witnessed first hand the level of security and professionalism involved with the work at the Clinton Power Plant. I live 5 miles from the facility and have no fear about the safety of the community. We live in an era where all need to be concerned about worldwide activities. (99-2)

Comment: The safety, security and wisdom of building new nuclear plants near the world's busiest airport in a post 9-11 world is also a consideration that must be taken seriously, according to Kraft. NRC officials consider airline crashes into reactors as "unrealistic scenarios," and would not allow this as a topic for discussion or analysis in the current ESP process. Yet, interviews of captured al Qaeda operatives revealed that reactors are indeed potential terrorist targets. "O'Hare Field, the world's busiest airport, is less than 27 minutes of normal-appearing flight time away from every Illinois reactor; and the new Airbus A-380 aircraft - 500+ tons, carrying nearly 300,000 liters of fuel - begins flights to O'Hare in 2006. Illinois cannot afford to disregard the potential for serious crash and burn events - accidental or otherwise" Kraft asserts. "NRC's glib disregard of these facts, and refusal to permit their discussion at this hearing demonstrate that "NRC" surely means "Never Really Concerned" about public safety. Yet this agency will decide whether Exelon's proposed Clinton-2 nuke is 'safe enough,'" he observes. (102-4)

Comment: With the terrorist risk in our country, a new reactor becomes a new target. Too risky. (148-2)

Comment: Nuclear power plants have known vulnerabilities to terrorist attack and sabotage. According to the 9/11 Commission Report, the infamous terrorist organization al Qaeda specifically discussed targeting U.S. nuclear plants. Fuel storage pools, dry storage facilities, and reactor control rooms are not designed to withstand the type attack that occurred on September 11, 2001. The U.S. Government Accountability Office (GAO) concluded in recent testimony before the U.S. Senate that cargo and general aviation airfields are more vulnerable to security breaches than commercial airports. Ignoring the threat because it is "highly speculative" does not make the threat go away, and indicates one shortfall of using an exclusively risk-based approach. One possible security measure to protect the reactor from assault by aircraft is to place a reactor below ground level. Therefore, an analysis in the draft EIS of the suitability of the site to place the reactor containment below-grade level should be done, which would require an in-depth analysis of geological and hydrological conditions at the site. (150-19)(151-19)

| **Comment:** Upgrades [to roads] may be a possibility, but then again, access to the plant for emergency measures will be limited by lane closures and repairs making it a more attractive target for either foreign or domestic terrorism. (157-5)

| **Comment:** New York has already been attacked, Los Angeles may be next, but Chicago is the third largest city with vulnerable fuel rod storage nearby. Both Chicago and nuclear facilities were discovered to be terrorist targets. National attention is already highly focused on nuclear issues with Iran and North Korea, as well as dirty bombs. (169-3)(179-3)

| **Comment:** We need to face up to the reality that nuclear power plants are not safe in an age of terrorism. (169-6)(179-6)

| **Comment:** The Draft EIS fails to adequately consider the safety risks that would exist at the proposed Clinton 2 nuclear plant. Although the Draft EIS discusses the impacts of various postulated accidents, the document does not discuss the likelihood or impacts that would result if there were to be a terrorist act at the Clinton plant. This omission occurs at a time heightened security concerns and, apparently, real vulnerability of nuclear plants to infiltration and attack. In fact, the National Academy of Sciences recently concluded that "there are currently no requirements in place to defend against the kind of larger-scale, premeditated, skillful attacks that were carried out on September 11, 2001, whether or not a commercial aircraft is involved." This is an issue that needs to be fully evaluated before any new nuclear power plants are sited. (170-14)

| **Comment:** NRC also fails to consider the security implications of expanding the Clinton nuclear site. It is well known that nuclear plants are considered prime terrorist targets. However, the Clinton plant, like all Exelon-owned plants, is guarded by the private security firm Wackenhut. Wackenhut also has a contract to test security at all the country's nuclear plants, posing a tremendous conflict of interest. Without an unbiased system for testing security, the actual level of preparation by guards is unknowable, and this gap in our knowledge should be enough to preclude further reactor construction. If the NRC will not remove Wackenhut from testing duty, it should take this conflict of interest-and the security questions it raises-into account. (112-7)(113-7)(114-7)(115-7)(116-7)(117-7)(118-7)(119-7)(120-7)(121-7)(122-7) (123-7)(124-7) (125-7)(126-7)(127-7)(128-7)(129-7)(130-7)(131-7)(132-7)(133-7)(134-7) (135-7)(136-7)(137-7) (138-7)(139-7)(140-7)(142-7)(143-7)(144-7)(145-7)(146-7)(147-7) (149-7)(154-7)(155-7)(158-7) (159-7)(162-7)(163-7)(164-7)(165-7)(166-7)(167-7)(173-7) (174-7)(175-7)(176-7)(177-7)(178-7) (180-7)(181-7)(182-7)(185-7)(186-7)(187-7)(188-7) (189-7)(190-7)(192-7)(193-7)(194-7)

| *Response: The NRC is devoting substantial time and attention to terrorism-related matters, including coordination with the Department of Homeland Security. As part of its mission to protect public health and safety and the common defense and security pursuant to the Atomic Energy Act, the NRC staff is conducting vulnerability assessments for the domestic utilization of radioactive material. In the time since the horrific events of September 2001, the NRC has*

identified the need for license holders to implement compensatory measures and has issued several orders to license holders imposing enhanced security requirements. Finally, the NRC has taken actions to ensure that applicants and license holders maintain vigilance and a high degree of security awareness. Consequently, the NRC will continue to consider measures to prevent and mitigate the consequences of acts of terrorism in fulfilling its safety mission.

Major NRC actions include the following:

- *Ordering plant owners to sharply increase physical security programs to defend against a more challenging adversarial threat*

- *Requiring more restrictive site access controls for all personnel*

- *Enhancing communication and liaison with the Intelligence Community*

- *Improving communication among military surveillance activities, NRC, and its licensees to prepare power plants and to effect safe shutdown, should it be necessary*

- *Ordering plant owners to improve their capability to respond to events involving explosions or fires*

- *Enhancing readiness of security organizations by strengthening training and qualifications programs for plant security forces*

- *Requiring vehicle checks at greater stand-off distances*

- *Enhancing force-on-force exercises to provide a more realistic test of plant capabilities to defend against an adversary force*

- *Improving liaison with Federal, State, and local agencies responsible for protection of the national critical infrastructure through integrated response training*

- *Working with national experts to predict the realistic consequences of terrorist attacks on nuclear facilities, including one from a large commercial aircraft. For the facilities analyzed, the results confirm that the likelihood of both damaging the reactor core and releasing radioactive material that could affect public health and safety is low.*

No changes were made to the EIS as a result of these comments.

E.2.25 Comments Concerning Emergency Preparedness

Comment: Exelon states in the ESP, "access to the site is limited primarily by Illinois Route 54." If access in either direction were limited via Route 54, then Emergency Responders would be unable to access the plant. Clinton I relies heavily on the volunteer emergency responders in the outlying communities. Even the closest community would have to take time to reach the plant, clear security, and reach the scene. To have both an operating plant and construction of a new plant being done concurrently would be of much more than a small or medium risk. It simply places too much of a burden on local law and emergency responders. (157-3)

Comment: And I remember very clearly that our school [which was a mile away from the Zion nuclear plant] was not very well prepared at all for a nuclear accident, despite the fact that mine had one of the worst safety records in the country. You know, recently, Vermont Yankee tried to, tried to do a drill, back in December. And the emergency management officials were astounded by the failure of half the busses did not show up. This left thousands of students stranded, waiting for the busses to come. Now, what about all the other reactors, where we haven't done any kind of drill preparation. This, this needs to be taken seriously. And, to make matters worse, a lot of Exelon reactors don't have backup power systems on their emergency sirens. So, if there's a power failure in the event of an accident, when these systems are needed the most, no one's going to be able to, no one's going help. I mean, you're relying on police with bullhorns trying to evacuate a community. That's insufficient. The people of Clinton deserve better. (85-2)

Comment: The local governments have also received a side benefit from this planning. DeWitt County has a state-of-the-art emergency operations center. The plan in place, as well as the warning sirens, can be used for disasters other than nuclear. The training received by the emergency personnel, as well as the drills, has proved to be a huge asset to the community. (106-7)

Response: *Emergency preparedness is a safety issue that is addressed in the Safety Evaluation Report (SER). The SER is available on the NRC's website at www.nrc.gov/reactors/new-licensing/esp/clinton.html. The NRC evaluates emergency plans for nuclear power reactors to determine whether there is a reasonable assurance that adequate protective measures can and will be taken in the event of a radiological emergency. The Commission must determine, in consultation with the Federal Emergency Management Agency (FEMA), whether the information submitted by the applicant shows there is no significant impediment to the development of emergency plans. This determination was made in the Exelon ESP SER. No change was made to the EIS as a result of these comments.*

E.2.26 Comments Concerning Decommissioning

Comment: Section 6.3, Page 6-40, Line 12. It is stated that if an other-than-LWR design were chosen, the decommissioning impacts analysis would be performed at CP/COL. It would be more accurate to state that the environmental impacts of LWR decommissioning analysis is SMALL and other-than-LWR decommissioning impacts would be evaluated if the reactor design selected at CP/COL and environmental impacts greater than those evaluated at ESP. (141-120)

Response: The comment is noted. The staff believes that the text is correct as stated. No change was made to the EIS as a result of the comment.

E.2.27 Comments Concerning the Cost of Power

Comment: I don't want to subsidize another Nuclear plant in IL. We have enough to subsidize as is. We (tax payers) build the plants, subsidize the plants, pay for the clean ups created by the plants...etc created by the plants. This is a bad idea at the environmental and financial levels. It would make much $ for very few at such a cost for the other 99% of us. That's immoral. (07-1)

Comment: Nuclear power continues to rely heavily on taxpayer subsidies because it is so expensive, and draft language in the energy bill in the current Congress indicates billions more dollars could be on the way. (09-5)

Comment: The flat refusal of insurance companies to touch nuclear power. The judgment of the professionals, whose basis it is to assess risks, has been 100 percent consistent from the beginning. Nuclear power is too risky for us to touch, they say. That is an objective judgment we would do well to heed today. (100-3)

Comment: We recognize that new power plants, of any kind, must be competitive in the market place. Operators must be able to supply power reliably and affordably. (101-3)

Comment: NRC should save the public and Exelon time and money, and deny the Early Site Permit Exelon is requesting for these nukes,...Illinois is in the process of enacting legislation that would mandate utilities to increase their production of electricity from renewable sources such as wind and biomass. "Enacting this 'renewables portfolio standard' legislation would provide even more environmentally friendly electricity at more competitive prices as increased market share reduces the costs for the renewables," Kraft explains. "Conversely, building more large nuclear plants will deliberately sabotage these future plans for renewables," he warns. (102-3)

Comment: The construction of a new reactor could be done more efficiently at the Clinton Power Station than other plants since quite a bit a preparation for the second unit was done when the first unit was built many years ago. The excavation and foundation work done years ago would save millions of dollars. (106-3)

Comment: Also nuclear power relies much too heavily on taxpayer subsidies for construction; and in our current economy, there simply is no money for it. (11-2)

Comment: The plant was built initially with cost overruns of over 4 billion dollars that were ultimately paid by consumers. That will happen again! (12-2)

Comment: Terminating the process early will save everyone - the public, NRC, even Exelon - significant amounts of time and money, both better spent on a sustainable and renewable energy future. (161-11)

Comment: If we end government subsidies, including reactor insurance the market refuses to bear, and create a level playing field for investors, then decentralized energy efficiency and renewable energy can readily compete and save taxpayers a lot of money. (169-5)(179-5)

Comment: This allows a huge initial investment into the plant, creating a potentially very costly white elephant if the final construction and operating license were not approved. Approving the ESP approves a new nuclear reactor that will be under tremendous pressure to start operating so it can repay investors for the portion not subsidized by government. (169-10)(179-10)

Comment: Another Governor goal is "encouraging investment and opening markets." (169-16)(179-16)

Comment: The Draft EIS is also erroneous in suggesting that wind, solar, and other alternatives should be rejected because they are "too expensive" or not "economical." In fact, the U.S. Department of Energy's Energy Information Administration's 2005 Annual Energy Outlook ("AEO 2005") projects that wind power would cost only 4.5 to 6 cents per kWh. By contrast, nuclear power is projected to cost 6.8 cents per kWh, leading the AEO 2005 to state that new nuclear power is "not likely to be economical.'" Similarly, a recent Massachusetts Institute of Technology study projected that new nuclear power would cost 6.7 cents per kWh. The NRC Staff has not explained how wind, solar, and other energy sources can be rejected as too expensive, when the U.S. DOE's own projects show that new nuclear power is more costly and not likely to be economical. (170-9)

Comment: The first reactor at the Clinton Power Plant cost consumers a whopping extra $4 billion, mostly because of bad planning, design, management, and decision-making. (21-1)

Comment: We bought into the first Clinton plant with promises that will probably never be fulfilled, yet we not only paid for it. (22-2)

Comment: Being the facility and lake were originally intended for a double-plant, it should certainly be more efficient and cost-effective to utilize the area. (24-2)

Comment: Recent studies by the University of Chicago and others have found that nuclear generation can be an efficient and cost-competitive option to coal and natural gas generators. In the face of volatile oil and gas supplies, this is an opportunity to reduce our reliance on foreign energy sources. Natural gas, in particular, is an important source of fertilizer for agriculture and increasing use of natural gas for generating electricity is placing a financial burden on farmers through higher fertilizer prices. (26-3)

Comment: We need to keep things in perspective. We in Illinois did not experience drastic increases in our electric rates or shortages of power because of the utility companies invested in nuclear power 30 years ago. We cannot take the attitude of the NIMBY groups without running the risk of making Illinois another California. (42-9)

Comment: Nuclear electricity also has one of the lowest productions cost per kilowatt hour. (48-3)

Comment: And life cycle emission analysis show that per kilowatt hour, the impact of nuclear electricity is among the lowest of any electricity generation, including wind and solar. (48-7)

Comment: On the issue of cost, wind, natural gas and energy efficiency efforts can meet future energy needs at a cost of approximately three to six cents per kilowatt hour. Credible estimates of nuclear power cost are much higher and history has shown are often underestimated. And, in fact, the U.S. Department of Energy, its most recent energy outlook states that new nuclear power plants, quote, are not expected to be economical. It's pretty clear that the government itself considers new nuclear power not economical. (59-2)

Comment: Nuclear is significantly lower in cost than many of the alternatives. You can support a cleaner environment and limit your growing power bills. Production cost for nuclear including paying construction, operation, decommissioning and waste disposal costs and still it is cheaper than coal, natural gas or wind power, none of which includes their full life cycle costs. (61-4)

Comment: Says that if nuclear reactors prove too expensive to operate and too costly to shut down, we could have an economic recipe for a nuclear disaster like Three Mile Island. (65-3)

Comment: I also like that nuclear power is not susceptible to fluctuations in natural resource pricing resulting from frequent unrest in certain regions of the world. (75-6)

Comment: The nuclear industry itself can't get insurance because the risk is too high. They are asking the public to do the insuring. It seems to me self, self-evident that if the nuclear industry itself won't insure itself, that that's the bottom line. That they know the risks are too great, and they refuse to insure their own plants. They want the public and the taxpayers to, to pay for it. (81-3)

Comment: We recognize that new power plants, of any kind, must be competitive in the market place. Operators must be able to supply power reliably and affordably. (84-3)

Comment: One is the nuclear industry doesn't have insurance. Well, it is true there is the Price Anderson Act, and that only covers liability insurance. And the nuclear industry does pay quite a good premium for this Act's insurance. However, this in no way pays for any of the, this only pays for the off site health losses and in no way pays for any onsite damages. So, the nuclear industry is not getting subsidized that way. (86-3)

Comment: And we need a source of energy that is economical. And quite frankly, nuclear energy is a reliable and economical source of energy, merely by the fact that people are looking at it. Exelon may be looking, in the future, at building a plant. And they wouldn't be doing that if it wasn't good business. So, we can trust that. (87-2)

Comment: In a time a rising oil and natural gas prices, and greenhouse concerns about burning coal, I can, as an economist, appreciate the potential appeal of nuclear energy. But as someone who spend decades researching nuclear economics, I am skeptical about the technology's economic viability. From the earliest days of the Atomic Energy Commission, people have underestimated the economic implications of nuclear power's unique hazards. They have also fought to shift the cost responding to these hazards, from private to public shoulders. In a post 9/11 world, this underestimation, and mis-accounting seems especially unwise. The last 40 or so nuclear plants, completed in the 1990's, generated electricity at more than four times the real inflation adjusted cost predicted in the 1960's. The dominant reason for this, was the underestimation of the cost of containing nuclear hazards. I expect these unpleasant surprises to continue. And the proposed plan to cost more than predicted. (88-1)

Comment: I urge you to include liability for terrorist related accidents, in the full costs of nuclear power, otherwise we can't have a fair competition, between nuclear energy and alternative energy options. The market cost of such liability, could be proxied for, by requiring private sector insurance coverage, and waiving Price Anderson.

Without including the implications of terrorist hazards, we will bias technology development in potentially dangerous ways. (88-3)

Comment: If we let market forces ride, nuclear power would be an absolute non-starter. Price Anderson is one, as the professor pointed out. We have a very peculiar thing that's going on here. We have lost track of capitalism in this country. We have privatized the reward and socialized the risk. What kind of deal is that?... If this plant is licensed, let's not palm off the responsibility on the taxpayers and the rate payers, let's make Exelon and any other nuclear utility in this country, put an escrow up front, multi-millions of dollars. Probably hundreds of millions of dollars, to ensure that the waste will be disposed of properly. No more free rides. Let capitalism raid and nuclear energy will be out of business altogether. (89-5)

Comment: The flat refusal of insurance companies to touch nuclear power. The judgment of the professionals, whose basis it is to assess risks, has been 100 percent consistent from the beginning. Nuclear power is too risky for us to touch, they say. That is an objective judgment we would do well to heed today. (90-3)

Response: The regulations under 10 CFR 52.18 specify that the EIS prepared for an ESP need not include an assessment of the benefits (for example, need for power of the proposed action). Cost of power is part of the assessment of the need for power. These issues will be reviewed at the combined operating license (COL) stage because they were not reviewed at the ESP stage. The Atomic Energy Act prohibits the NRC from promoting nuclear power in any manner including rebates and incentives. No change was made to the EIS as a result of these comments.

E.2.28 Comments Concerning the Need for Power

Comment: I believe we need to develop our nuclear generating capacity. It is vital to our national energy policy and our security. We must reduce our reliance on fossil fuel in order to provide a energy source for our future. (01-1)

Comment: Our position is that the building of the next generation of nuclear power plants is very important, to provide the electricity that will be needed in the year 2020. (101-1)

Comment: We need dependable base load plants to prevent brown outs during the peak demand periods. (105-8)

Comment: According to NRC regulations at 10 CFR 52.17(a)(2), the need for power does not have to be addressed in the ESP process. But an evaluation of the need for power and who benefits is crucial to determining whether the ESP application should be considered at all. In fact, the first question that should be asked is whether residents of Illinois will receive any of the benefit of a new nuclear unit. Much of the electric power produced by Clinton will be fed into the PJM interconnection. PJM is the largest regional transmission organization (RTO) in the U.S. It coordinates the movement of electricity in all or parts of Delaware, Illinois, Indiana, Kentucky, Maryland, Michigan, New Jersey, Ohio, Pennsylvania, Tennessee, Virginia, West Virginia and the District of Columbia. The final EIS should include an analysis of the exportation of electricity

Appendix E

generated by the new nuclear unit at Clinton to other states where electricity prices are higher and revenues will be greater for Exelon. (150-20)(151-20)

Comment: The purpose and need for the proposed action (Section 1.3) does not address why we need more nuclear power. (169-11)(179-11)

Comment: In section 10.3, it says "The benefit is the production of electricity." This fails to address why we even need more energy. (169-22)(179-22)

Comment: The discussion of the Draft EIS starts off on the wrong foot by failing to analyze whether there is any need for the power that would be produced by Exelon's proposed Clinton 2 nuclear power plant. Instead, the NRC Staff has accepted Exelon's stated purpose that the Clinton 2 project is intended to create baseload power, and then refused to consider whether such power is needed. According to the NRC Staff, 10 C.F.R. 52.17 and 52.18 precludes the consideration of the need for power in determining whether or not to grant an ESP to Exelon. (Draft EIS at 8-15) It is also important to note that the Draft EIS appears to demonstrate that there is not a need for the baseload power that Exelon is seeking to produce here. In particular, as the Draft EIS notes, Illinois is a net exporter of power. (Draft EIS at 8-4). The NRC Staff, therefore, must explain how it can accept Exelon's stated need for baseload power, and reject alternatives for purportedly failing to meet that need, when the need itself appears to not exist. (170-6)

Comment: Aggressive "demand-side management" for companies around Illinois to save as much energy as possible should be considered FIRST BEFORE the proposed Clinton Unit 2 nuclear reactor is even considered. (171-11)

Comment: We are concerned about the lack of a documented justification for the purpose and need for the proposed action. According to the DEIS, the environmental report submitted by the applicant is not required to include a discussion of the need for power. Despite this legal exclusion, it seems that an increased need for power must be the single most important reason driving the applicant's proposed action. Without a discussion of the need for power, the purpose and need section of the DEIS is deficient, frustrating the NEPA process. Therefore, we urge the USNRC to include a discussion of the need for power in later environmental documentation, in spite of the legal exclusion. (172-3)

Comment: The draft Environmental Impact Statement (DEIS) for Exelon's Early Site Permit (ESP) application is incomplete for a variety of reasons. First, it avoids consideration of important siting factors. Specifically, the need for power in the central Illinois region was not examined. (112-1)(113-1)(114-1)(115-1)(116-1)(117-1)(118-1)(119-1)(120-1)(121-1)(122-1) (123-1)(124-1)(125-1)(126-1)(127-1)(128-1)(129-1)(130-1)(131-1)(132-1)(133-1) (134-1)(135-1) (136-1)(137-1)(138-1)(139-1) (140-1)(142-1)(143-1)(144-1)(145-1)(146-1)(147-1)(149-1)(154-1) (155-1)(158-1)(159-1)(162-1)(163-1)(164-1)(165-1)(166-1)(167-1)(173-1)(174-1)(175-1)(176-1)

NUREG-1815   E-188   July 2006

(177-1)(178-1)(180-1)(181-1)(182-1)(185-1)(186-1)(187-1)(188-1)(189-1)(190-1)(192-1)(193-1)
(194-1)

Comment: The United States Environmental Protection Agency (EPA) found NRC's approach flawed, claiming that "since it ignores the justification for the power plant addition in the early stage of project development...[it] biases the subsequent energy alternative analysis toward nuclear power." According to the U.S. Department of Energy, Illinois already exports approximately 18% of the electricity generated in the state; additional generating capacity is unwarranted. (112-4)(113-4)(114-4)(115-4)(116-4)(117-4)(118-4)(119-4)(120-4)(121-4)(122-4) (123-4)(124-4)(125-4)(126-4)(127-4)(128-4)(129-4)(130-4)(131-4)(132-4)(133-4)(134-4)(135-4) (136-4)(137-4)(138-4)(139-4)(140-4)(142-4)(143-4)(144-4)(145-4)(146-4)(147-4)(149-4)(154-4) (155-4)(158-4)(159-4)(162-4)(163-4)(164-4)(165-4)(166-4)(167-4)(173-4)(174-4)(175-4)(176-4) (177-4)(178-4)(180-4)(181-4)(182-4)(185-4)(186-4)(187-4)(188-4)(189-4)(190-4)(192-4)(193-4) (194-4)

Comment: NRC regulations do require consideration of the need for the plant and a detailed consideration of need is absent from the agency's impact statement. (191-6)

Comment: In time we'll need the increased power out-put. (24-4)

Comment: Additional electric generation capacity is needed to meet the growing demand for electrical power in the United States. The Exelon proposal to construct and operate a new nuclear power generating facility would supply new electrical power to a wide range of consumers. (26-2)

Comment: We must have adequate, stable, and low cost power to provide for the jobs of the future in Illinois...We must now plan for the future as our jobs rely upon having good power supplies. (28-4)

Comment: We need dependable base load plants to prevent brown outs during the peak demand periods. (29-8)

Comment: We need a dependable baseload plant to prevent brownouts during peak demand periods. (42-7)

Comment: Currently, nuclear power provides one fifth of our nation's electricity and about 50 percent of Illinois's electricity. (48-2)

Comment: When the last hearing was made at the Public Library in Clinton, I simply asked the question are there more people in the country? Is the demand for electric power still there? Has it increased? The answer to both of those questions is yes. The population has increased and the demand for power has increased. This is an opportunity to fulfill that need for power. (50-1)

Comment: I have to say that personally I'm not crazy about nuclear power but I am crazy about my air conditioning. And I am crazy about leaving my lights on all the time. And I live in a world of huge consumption of power. And because of that, we all need to have access to power. (51-1)

Comment: The bad news is we have an energy crisis in this country and in the world... The worse news is the projections indicate that the demand for energy in the United States will increase by about 50 percent in the next 15 years. And by 2040, the world's energy use will double. Under the current trends, this is very bad news indeed. (66-2)

Comment: Beyond that, the need for power, here in Illinois, make no mistake, we're not going to have brownouts here if we don't build another nuclear power plant. You guys are already generating way more electricity than, than Illinois uses. And you're exporting it to other states. And you've got the option to either build another nuclear plant here for, for this base load power need, that Exelon has identified. But that's just letting Exelon set the terms of the debate. (68-5)

Comment: I wanted to mention that there's a greater issue of energy demand. And you could argue about whether or not our need for electricity is justifiable or not. I believe that it is. There are, there are disadvantages, there are consequences to our large energy demand. But it has so significantly contributed to our quality of life, that I believe that this trade off is worth it. (78-1)

Comment: But if our population grows, if our economy grows, and most importantly, if we want to replace our gasoline powered vehicles with ones running on hydrogen, produced from non-hydro carbon sources, then we need another carbon free energy source to make up the difference. And the only such source that we know of today, is nuclear power, like the power that the new Clinton power plant would provide. (79-2)

Comment: Our position is that the building of the next generation of nuclear power plants is very important, to provide the electricity that will be needed in the year 2020. (84-1)

Comment: The draft Environmental Impact Statement (DEIS) for Exelon's Early Site Permit (ESP) application is incomplete for a variety of reasons. First, it avoids consideration of important siting factors. Specifically, the need for power in the central Illinois region was not examined. (87-1)

Comment: It will help people of all demographics gain electricity. The nation's energy needs are raising and the number of operation plants should rise with it to provide relief to the energy crisis. I don't want Illinois to become another California with rolling blackouts. (93-2)

Comment: We must have adequate, stable, and low cost power to provide for the jobs of the future in Illinois...We must now plan for the future as our jobs rely upon having good power supplies. (99-4)

Response: *According to NRC regulations at 10 CFR 52.18, the need for power does not have to be addressed in the ESP process because no decision involving the need for power is being made. The filing of an application for an ESP is a process that is separate from the filing of an application for a construction permit (CP) or a combined operating license (COL) for such a facility. The ESP application makes it possible to evaluate and resolve some safety and environmental issues related to siting before the applicant makes large commitments of resources. If the ESP is approved, the applicant can "bank" the site for up to 20 years for future reactor siting. The ESP does not authorize construction or operation of a nuclear power plant. At the point when the ESP holder believes that it wants to proceed with construction, it must obtain a CP or a COL, which will be a major Federal action that requires a separate EIS, which, among other things, addresses the need for power and cost of power. No change was made to the EIS as a result of these comments.*

E.2.29 Comments Concerning Operational Safety

Comment: Safety continues to be sacrificed in favor of higher profits by both the industry and the NRC. (09-7)

Comment: Nuclear plants are vulnerable to human error, equipment failure. (10-3)

Comment: The hazards of nuclear power plants and their demonstrated lack of compliance with minimum safety standards is not an acceptable risk. (12-4)

Comment: ...but we live with its threat to our communities. The track record of the current plant is not good. They need to demonstrate to us, on a consistent basis, that the facility is safe and cost-effective - and run by an organization who knows what they are doing. (22-3)

Comment: February 2005, the Commission briefing on fuel cladding and fuel performance indicates that as much as one third of the nation's reactors are now operating with failed fuel, where the cladding has been either split open or there are leaks where radioactive isotopes are now coming into the cooling. But more importantly, the first barrier in this so called Defense in Depth has been breached.

Exelon itself disclosed that it operated 11 of its 16 reactors with failed fuel. And this, what's interesting here is that the failed fuel is an indication of a nuclear waste gambit that only raises the, the threat with regard to unanalyzed condition and staff assumptions that follow for storage in the pool, transportation, dry cask and, ultimately, where ever this stuff will go. (67-3)

Response: *The issues raised in the comments are outside the scope of the environmental review and are not addressed in this EIS. That said, the following are examples of how NRC addresses operational safety issues. NRC maintains resident inspectors at each reactor site. These inspectors monitor the day-to-day operations of the plant and perform inspections to ensure compliance with NRC requirements. In addition, the NRC has an operational experience program that ensures that the safety issues that are found at one plant are properly addressed at the others, as appropriate. Finally, the design of any new reactors or storage facility will have already benefitted from lessons learned at existing reactors and incorporate new safety features that would be impracticable to backfit onto existing plants. The NRC will only issue a license or permit if it can conclude that there is reasonable assurance: (1) that the activities authorized by the license or permit can be conducted without endangering the health and safety of the public, and (2) that such activities will be conducted in compliance with the rules and regulations of the Commission. No change was made to the EIS as a result of these comments.*

Comment: The current operators of the Clinton station has proven the ability to operate the station reliable and safely. (01-3)

Comment: Amergen and its parent company, Exelon, have many years of nuclear expertise with many plants. That can be used to plan and build a more efficient, economical and safer plant than what is here today. This unit could be used as a prototype for a new generation of power plants that can eventually replace the aging fossil fuel plants. Once operating, a nuclear plant produces electricity more efficiently and cleaner than a coal-fired plant. (106-4)

Comment: I have seen a tremendous amount of redundancy in safety and operations systems. As mentioned earlier, Exelon brings a great deal of expertise and talent to the table. They are also not going to want to risk the financial well-being and reputation of their company be building an unsafe and unreliable nuclear facility. The excellent safety record of the company is a matter of public record. (106-6)

Comment: As a physician who has worked at the Clinton Power plant on and off for 3 yrs (91-93) I feel they have a very safe and well run plant there. I've had their training and safety courses and found their workers and staff very dedicated and safety minded. I would have no objection to another unit being built. Incidently I live only about 4 miles from the plant. (16-1)

Comment: I have raised my family in the shadow of the Clinton Nuclear Power Plant and consider it a safe place to live. Everyone has to accept risk with where they live and work. (27-2)

Comment: I've been involved with the security and some of the operation people out here in safety of the plant for most of my career in law enforcement. And I have the utmost respect for both the operational people and the security people in keeping our community safe. And I can honestly say that I don't feel that our community has been at risk at any point in my career. (43-2)

Comment: And I'd like to focus on safety here for a minute. The nuclear industry has had an amazing track record. Not only is it one of the safest industries in this country, but while maintaining this safety culture, they have been constantly improving their capacity factors along the way. (76-3)

Comment: Two or one is safe. (95-2)

Comment: I have raised my family in the shadow of the Clinton Nuclear Power Plant and consider it a safe place to live. Everyone has to accept risk with where they live and work. (97-2)

Response: These comments provide general information regarding safety issues at the currently operating Clinton Nuclear Plant, Unit 1, provided to support the Exelon ESP application. Because these comments do not relate to the environmental effects of the proposed action, they will not be assessed further. No change was made to the EIS as a result of these comments.

E.2.30 Comments Concerning Other Issues

Comment: Section 4.1.2, Page 4-3, Lines 16-19. "Exelon indicated that as a result of receiving an ESP, agreements would be made with the Regional Transmission Operator (RTO) and, if required, transmission lines would be upgraded in the event that the power demands and power production exceeded the line capabilities." ER-Section 4.1.2 Transmission Corridors and Off-Site Areas – "As described in Section 3.7, an RTO or the owner, both regulated by FERC, will bear the ultimate responsibility for defining the nature and extent of system improvements, as well as the design and routing of connecting transmission." The ER statement correctly places sole responsibility on the RTO or owner, whereas the DEIS suggests responsibility on both the RTO and Exelon. (141-54)

Response: If an ESP were granted, Exelon would initiate the process for obtaining access to the grid outlined in Section 3.3 of this EIS. Transmission lines would be upgraded by the Regional Transmission Operator (RTO), if needed. The full extent of potential land-use impacts in the transmission line rights-of-way can be estimated only after following the Federal Energy Regulatory Commission (FERC) process for connecting new large-generation facilities to the grid. To facilitate the analysis of impacts for the EIS, the staff utilized information from the applicant characterizing the likely transmission corridor upgrade that would be required, including a doubling of the width of the existing corridors. No change was made to the EIS as a result of the comment.

Comment: Station Operation Impacts, Section 5.8.3, Acute Effects of Electromagnetic Fields, pages 5-45, 5-46. Clarification of the inclusion of studies' results that have been published subsequent to NUREG-1437, need to be addressed. (172-27)

Comment: Station Operation Impacts, Section 5.8.4, Chronic Effects of Electromagnetic Fields, page 5-46. The Final EIS should include studies' results that have been published subsequent to the referenced reports. (172-28)

Response: The National Institute of Environmental Health Sciences directs research related to the effects of electromagnetic fields. It still has not reached a conclusion related to the chronic effects of electromagnetic fields. It would not serve the purposes of this EIS to present a detailed discussion of inconclusive research, particularly in this area, because it is highly unlikely that there would be significant differences in the chronic effects of electromagnetic fields among the proposed and alternative sites. Results of research on the chronic effects of electromagnetic fields through mid-2005 are summarized at http://www.mcw.edu/gcrc/cop/static-fields-cancer-FAQ/toc.html (accessed January 6, 2006). No change was made to the EIS as a result of these comments.

Comment: Section 2.4, Page 2-16, Line 34. Discussion of geology. As a point of clarification, while alluvium from stream deposits may be present over the glacially consolidated soils, much of the upper soil layer is dominated by loess, a wind blown silty to fine sand deposit. (141-16)

Response: The staff agrees and the text in Section 2.4 has been revised to mention the loess.

Comment: One major accident or attack anywhere in the world will derail nuclear for another two decades, if not longer. More importantly, public pressure to immediately shut down power plants would be very strong and possibly undeniable, resulting in a huge economic impact from the sudden loss of electricity. Diversification away from nuclear is critical, especially in Illinois where 50 percent is nuclear. (169-4)(179-4)

Comment: WHEN IT COMES TO AGGRESSIVELY PURSUING THESE APPROPRIATE RENEWABLE ENERGY TECHNOLOGIES (AND THIS GOAL MAKES A STATEWIDE REPS IMPERATIVE), AN ESP FOR CLINTON UNIT 2 WILL HELP TO CREATE A "POLITICALLY STIFLING CLIMATE" FURTHER NEGATING THIS IMPLEMENTATION. (25-10)

Comment: I also urge you to deny a license to any nuclear plant, for which the contractor is unwilling to assume full liability for serious accidents, as a contractor would for any other energy technology. (88-2)

Comment: It's been 50 years since the passage of the Price Anderson Act, which could originally be justified as a infant industry a situation where we didn't know much about the technology. But I see no reason to continue this protection. We usually allow the market and the private sector to assess the risks of different technologies. We do this by holding the firms liable for any hazards they might cause. And requiring them to get insurance, to make sure they can meet that liability. If this technology cannot find private insurers, then it should not go forward. If it can get insurance, it should pay for it. Now, if you had a house on a flood plain and couldn't get private insurance, what would that tell you? If the private sector is unwilling to assume the full liability for accidents, what is that telling you? (88-4)

Response: *These comments are noted. The comments did not provide new information relevant to the EIS and will not be evaluated further. No change was made to the EIS as a result of these comments.*

Comment: The General Assembly has already passed a state moratorium on building any more nuclear power plants until the issue of where to finally put the high-level radioactive waste is settled. Knowing Exelon's arrogance, the company's lobbyists will probably ask the state legislature for an exemption from this nuclear reactor construction moratorium...this moratorium is SOUND. THE USNRC SHOULD RESPECT THIS VALID "STATE'S RIGHTS ISSUE," AND DENY EXELON'S "EARLY SITE PERMIT." (25-5)

Comment: First off, there is a State moratorium on new nuclear reactor construction already passed by the General Assembly. The State moratorium calls for no new construction of nuclear plants until the issue of where to finally store a high level of radioactive waste is settled. Knowing Exelon, it will probably ask Springfield for an exemption, from this moratorium, but my view maintains that this moratorium is sound. The NRC should respect this State's rights issue, and deny Exelon's application for an early site permit. (71-6)

Comment: The Draft EIS does not address the impact of the Illinois nuclear moratorium law, 220 ILCS 5/8-406(c), which deems all potential sites in Illinois unacceptable for new nuclear power plants. (170-3)

Comment: Acknowledges that human health impacts from natural gas are "not expected . . . [to] be detectable," (DEIS at 8-13), and the NRC Staff have not claimed that natural gas presents the type of accident risks that nuclear power does. As with wind, it is arbitrary and capricious to suggest that an energy source that presents human health and accident risks is environmentally preferable to a clean energy alternative that does not. Certainly, those energy sources in combination, along with energy efficiency efforts, could not be considered to have greater environmental impacts than new nuclear power. Therefore, the NRC Staff must reconsider its rejection of clean energy alternatives, and engage in the rigorous and objective analysis of such alternatives that is required by NEPA but not found in the Draft EIS. The IEPA has not made any such finding. Nor could the IEPA legitimately do so because no license for the suggested Yucca Mountain facility has been applied for, much less "approved." In fact, the Department of Energy missed its plan to apply for such a license by the end of 2004, and recently delayed the planned filing even more. In addition, a federal court of appeals last year struck down the U.S. EPA's radiation safety guidelines for analyzing the Yucca Mountain proposal, and there have been recent allegations that various scientific studies used to justify the geologic suitability of the site were falsified. Plainly, there is little chance that a high-level waste repository will be approved, much less opened, in the near future. In addition, even if Yucca Mountain is approved, that site does not have the capacity to store all of the high-level wastes that will be created by existing nuclear power plants, much less a proposed new Clinton 2 plant. Given these facts, it is plain that this ESP proceeding is premature and that Exelon's ESP application should be denied until such time as Illinois lifts its moratorium. In essence, the moratorium answers with a resounding "no" the question presented in this ESP proceeding: Is the Clinton site (or any other site in Illinois) appropriate for a new nuclear power plant? Therefore, the NRC cannot approve the Clinton site and must deny the ESP at this time. Amazingly, despite the clear import of the Illinois nuclear moratorium, the Draft EIS does not even mention, much less analyze, the moratorium. This omission is especially glaring given that the Draft EIS includes an entire Appendix listing the "authorizations, permits, and certifications" that Exelon would have to obtain before construction the proposed Clinton 2 plant. Plainly, the NRC Staff must address the moratorium as part of the ESP process. (170-11)

Response: The State law does not apply to the process of reviewing and issuing an ESP. The Illinois moratorium law is applicable to the State agencies responsible for issuing permits required for construction and operation of a nuclear power plant.

The safety and environmental effects of long-term storage of spent fuel onsite have been assessed by the NRC, and, as set forth in the Waste Confidence Rule (10 CFR 51.23), the Commission generically determined that such storage could be accomplished without significant environmental impact. In the Waste Confidence Rule, the Commission determined that spent fuel can be stored onsite for at least 30 years beyond the license operating life, which may include the term of a renewed license. At or before the end of that period, the fuel would be removed to a permanent repository. In its Statement of Consideration for the 1990 update of the Waste Confidence Rule (55 FR 38472), the Commission addresses the impacts of both license renewal and potential new reactors. The current rule does cover new reactors and can be used in the staff's review of an ESP application. The rule was last reviewed by the Commission in 1999, when it reaffirmed the findings in the rule (64 FR 68005, dated December 6, 1999). No change was made to the EIS as a result of these comments.

E.2.31 Comments Concerning NRC's Administrative Process

Comment: I'm here tonight because of radiation knows no city limits. And we have been given an opportunity to publicly comment, because of the NRC isn't hold a, isn't holding public hearings in the other communities that will also be impacted by a new nuclear reactor. I think the NRC needs to hold more heavily publicized hearings. For this hearing, we had three different locations advertised that, you know, was very hectic trying to correct that. There needs to be better preparation going into these hearings. (85-1)

Response: The location of the meeting was changed in accordance with the NRC's policies and procedures to accommodate the expectation of a larger group of participants than originally expected. The staff ensured that information was communicated on the change of location. The comment did not provide new information relevant to this EIS and will not be evaluated further. No change was made to the EIS as a result of the comment.

Comment: Although I commend the Nuclear Regulatory Commission (NRC) for holding this meeting, I encourage the NRC to hold additional public meetings so the concerns of the region's citizens may be heard before granting an Early Site Permit to Exelon. (17-4)

Comment: For more than two months I have been preparing for this meeting, reading, studying and learning and discussing. And there is something rather disheartening about having three minutes to speak of what my heart is full of. (46-1)

Comment: This procedure reminds me of a comment Autwand Bismark made. He talked about how he accommodated dissent. He said I let them say anything they like and I do anything I like. Although now we're only allowed to say three minutes of that which we like. (55-1)

Comment: While we view our participation in these NRC hearings about Clinton as necessary to preserve our standing in this process, we must almost point out how utterly insufficient these hearings are, especially in the context of getting out of the box of NRC and the nuclear industry mind set, inadequate and illusory regulations and outright self fulfilling prophecies. These proceedings simply fail to deal openly and sufficiently with issues that the public, not just some distant NRC staffers view as important. (56-2)

Comment: For more than two months I have been preparing for this meeting, reading, studying and learning and discussing. And there is something rather disheartening about having three minutes to speak of what my heart is full of. (94-1)

Comment: Although I commend the Nuclear Regulatory Commission (NRC) for holding this meeting, I encourage the NRC to hold additional public meetings so the concerns of the region's citizens may be heard before granting an Early Site Permit to Exelon. (104-4)

Comment: There are many other problems with this ESP, but because of the way the laws are written to "support" nuclear energy, I am sure they will be glossed over, ignored, or considered "Low Risk". (157-9)

Response: The NRC holds public meetings both at the early site permit (ESP) and combined license (COL) stages, to gather and provide information from the public. Additionally, during the comment period after the EIS is published, any member of the public may submit written comments to the NRC regarding the contents of the EIS. These comments are considered by the NRC and are addressed and dispositioned. The comments did not provide new information relevant to this EIS and will not be evaluated further. No change was made to the EIS as a result of these comments.

Comment: Ignoring or persecuting whistleblowers and members at the NRC staff with different professional opinions on issues of safety and security such as resident inspectors at Illinois reactors in the 1980's and in Connecticut in the 1990's. And security experts shortly before and after the 9-11 attacks. Ignoring for nearly ten years, prior to the September 11th attacks, the constant warnings and pleas to improve reactor safety from the public NGO's, like the Nuclear Control Institute and the Committee to Bridge the Gap, whose warnings were amply validated on 9-11. Yet, almost up to that fateful date, the NRC was promulgating plans to permit the nuclear industry to defacto regulate itself on security issues in spite of an operational history of failure. Also, pretending to promote balance between the public's right to know and participation in decisions on the one hand and security concerns on the other. Yet for the first 30 days after September 11th, the NRC did absolutely nothing to restrict the flow of information on the NRC web site. Then shut down the whole site under the guise of security just before the critical votes in the Congress on nuclear issues took place, which required access to critical information on the NRC web site. Cherry picking the factual information provided on reactor safety and security issues and dismissing what does not fit or worse, what outright embarrasses the prevailing agency mind set, just as the U.S. Department of Energy has done and continues to do at Yucca Mountain. Violating its own questionably inadequate regulations by approving construction permits for radioactive waste canisters before approving the actual designs for those canisters. In two cases this resulted in accidents which members of the public warned against but which the NRC dismissed as, quote, unlikely. Continuing to insist that 9-11, like attacks on reactors and spent pools using commercial jet liners are unrealistic scenarios. While integrations of al Quaida operatives and other evidence from al Quaida have confirmed that reactors were and presumably are indeed considered targets for such attacks. And while professionals at the National Academy of Sciences state an attack would be certainly no more difficult than the September 11th attacks. We can go on with those examples. With this documented record, the NRC is in no position to make quality assurance statements about the validity or reliability in this or any other matter regarding nuclear power, waste or safety. The NRC may retain the legal authority to do so but has long ago forfeited its credibility. It can go through the motions of filing its regulatory mandate by conducting hearings like this one tonight. But this will not add one iota of legitimacy to either the process or the information promulgated. The actions belie any claim to legitimate authority. Because this legitimate authority, the authority to be legitimate must be based on the cherished principles of this country of informed consent and a democratic process. And the NRC's actions have eviscerated both largely to the benefit of the nuclear industry. So because we chose to participate in these hearings, while we know that many people at the NRC do their jobs with the highest standards of operation and integrity in mind, the overall agency mind set and agenda will thwart such attempts at excellence every time. (108-2)

Comment: The NRC could step up to its stated mission of protecting public health, safety and the environment instead of clouding with a nuclear industry whose motive is profit, not safety. The NRC could act for safety by closing down aging reactors, approving no new ones and taking leadership in handling responsibly nuclear waste we have already created over 50,000 tons of high level waste.

And the NRC could develop and implement guidelines for ethical management of radioactive materials as already proposed by Johanna Masey. If the NRC would do this, it would be incredible. (46-10)

Comment: Here's the core of my complaint. On the NRC's web site they proclaim their statement of purpose. They exist to safeguard the health, welfare and safety of U.S. citizens. In fact, the NRC was created to end the abuses of its predecessor, the AEC, which became a cheerleader for the civilian reactor rather than its watch dog. The spirit with which the NRC was created was a good one. It was supposed to put citizen interests first. Unfortunately, the NRC has a bad habit of forgetting why it was created. It's become a letter of the law commission. Current NRC regulations are written to favor the nuclear industry, not U.S. Citizens. (49-2)

Comment: These regulations [existing radiation standards] stymie the NRC's ability to fulfill its mandate. (49-4)

Comment: I demand that the NRC re-embrace the spirit of the laws that brought it into existence. (49-6)

Comment: If the NRC wants me to retract my characterization of this process as a sham, then I want some proof that it is taking criticism of its process seriously. The NRC must become a watch dog, not a lap dog of the nuclear industry. (49-10)

Comment: Ignoring or persecuting whistleblowers and members at the NRC staff with different professional opinions on issues of safety and security such as resident inspectors at Illinois reactors in the 1980's and in Connecticut in the 1990's. And security experts shortly before and after the 9-11 attacks. Ignoring for nearly ten years, prior to the September 11th attacks, the constant warnings and pleas to improve reactor safety from the public NGO's, like the Nuclear Control Institute and the Committee to Bridge the Gap, whose warnings were amply validated on 9-11. Yet, almost up to that fateful date, the NRC was promulgating plans to permit the nuclear industry to defacto regulate itself on security issues in spite of an operational history of failure. Also, pretending to promote balance between the public's right to know and participation in decisions on the one hand and security concerns on the other. Yet for the first 30 days after September 11th, the NRC did absolutely nothing to restrict the flow of information on the NRC web site. Then shut down the whole site under the guise of security just before the critical votes in the Congress on nuclear issues took place, which required access to critical information on the NRC web site. Cherry picking the factual information provided on reactor safety and security issues and dismissing what does not fit or worse, what outright embarrasses the prevailing agency mind set, just as the U.S. Department of Energy has done and continues to do at Yucca Mountain. Violating its own questionably inadequate regulations by approving construction permits for radioactive waste canisters before approving the actual designs for those canisters. In two cases this resulted in accidents which members of the public warned against but which the NRC dismissed as, quote, unlikely. Continuing to insist that 9-11, like attacks on reactors and spent pools using commercial jet liners are unrealistic scenarios. While integrations of al Quaida operatives and other evidence from al Quaida have confirmed that reactors were and presumably are indeed considered targets for such attacks. And while professionals at the National Academy of Sciences state an attack would be certainly no more difficult than the September 11th attacks. We can go on with those examples. With this documented record, the NRC is in no position to make quality assurance statements about the validity or reliability in this or any other matter regarding nuclear power, waste or safety. The NRC may retain the legal authority to do so but has long ago forfeited its credibility. It can go through the motions of filing its regulatory mandate by conducting hearings like this one tonight. But this will not add one iota of legitimacy to either the process or the information promulgated. The actions belie any claim to legitimate authority. Because this legitimate authority, the authority to be legitimate must be based on the cherished principles of this country of informed consent and a democratic process. And the NRC's actions have eviscerated both largely to the benefit of the nuclear industry. So because we chose to participate in these hearings, while we know that many people at the NRC do their jobs with the highest standards of operation and integrity in mind, the overall agency mind set and agenda will thwart such attempts at excellence every time. (56-4)

Comment: The NRC could step up to its stated mission of protecting public health, safety and the environment instead of clouding with a nuclear industry whose motive is profit, not safety. The NRC could act for safety by closing down aging reactors, approving no new ones and taking leadership in handling responsibly nuclear waste we have already created over 50,000 tons of high level waste.

And the NRC could develop and implement guidelines for ethical management of radioactive materials as already proposed by Johanna Masey. If the NRC would do this, it would be incredible. (94-10)

Response: The NRC takes seriously its responsibility under the Atomic Energy Act to protect the health and safety of the public and the environment in regulating the U.S. nuclear power industry. More information on NRC's roles and responsibilities is available on the NRC's website at http://www.nrc.gov/what-we-do.html. The NRC was created by Congress and designed so that it would not report to the same part of the government that was in charge of setting energy policy (any current Administration). The comments did not provide new information relevant to this EIS and will not be evaluated further. No change was made to the EIS as a result of these comments.

Comment: Quality assurance is not merely the presence of standards that are both necessary and sufficient to protect the public and the environment. Quality assurance also requires the active presence of credible regulators. Agents willing to regulate assertively in the public interest and on the public's behalf. In this sense, the well documented historic record of the NRC's catering to every conceivable whim of the nuclear industry leaves this process without a credible agent. And by extension, quality assurance deficient. The safety issues and quality of this process simply can't be assured given its lack of credibility. The NRC's documented history includes systematically re-writing its public participation process in ways that continuously weaken or make irrelevant public participation in events like this meeting tonight. (108-1)

Comment: Quality assurance is not merely the presence of standards that are both necessary and sufficient to protect the public and the environment. Quality assurance also requires the active presence of credible regulators. Agents willing to regulate assertively in the public interest and on the public's behalf. In this sense, the well documented historic record of the NRC's catering to every conceivable whim of the nuclear industry leaves this process without a credible agent. And by extension, quality assurance deficient. The safety issues and quality of this process simply can't be assured given its lack of credibility. The NRC's documented history includes systematically re-writing its public participation process in ways that continuously weaken or make irrelevant public participation in events like this meeting tonight. (56-3)

Response: *The NRC takes seriously its responsibility under the Atomic Energy Act to protect the health and safety of the public and the environment in regulating the U.S. nuclear power industry. More information on NRC's roles and responsibilities is available on the NRC's Internet website at http://www.nrc.gov/what-we-do.html. The public has been given the opportunity to participate in the rulemaking process that established the regulations that govern its review process. The comments did not provide new information relevant to this EIS and will not be evaluated further. No change was made to the EIS as a result of these comments.*

Comment: And I think it's really indicative that the NRC being basically an unjust organization and not a true clinical independent organization because an independent organization would look into the affects on the environment that is all encompassing; the physical, the economical, the political, the social aspects. (35-4)

Response: *This comment provides general information in opposition to the NRC's early site permit process and will not be assessed further. The staff has carefully reviewed the application and relevant information to assess the environmental impacts, including physical, economical, and social aspects of the action, according to applicable regulations. No change was made to the EIS as a result of the comment.*

E.2.32 General Comments in Support of Nuclear Power

Comment: Illinois Farm Bureau has long had a policy supportive of the use of nuclear power generators as a source of needed energy. (103-1)

Comment: First of all I believe that nuclear power should be a larger part of our energy supply. (105-5)

Comment: For the people that are concerned about the safety of nuclear power, I ask what alternatives do you want? Lives are lost every year in coal mine accidents, people are killed running into coal trains, natural gas explosions kill people, and the alternative of wind power sounds good until you really need the power on those hot, sultry, humid and windless days in August. (105-7)

Comment: We believe that nuclear energy is safe, clean, reliable and cost effective, and as such, it should continue to be an important part of a balanced energy mix. (111-2)

Comment: Illinois Farm Bureau has long had a policy supportive of the use of nuclear power generators as a source of needed energy. (14-1)

Comment: We believe it is time to move forward to guarantee the future of nuclear generation with the construction of advanced-design reactors. It is our understanding that the proposed second reactor at Clinton is one of the advanced-design models. (26-4)

Comment: First of all I believe that nuclear power should be a larger part of our energy supply. (29-5)

Comment: For the people that are concerned about the safety of nuclear power, I ask what alternatives do you want? Lives are lost every year in coal mine accidents, people are killed running into coal trains, natural gas explosions kill people, and the alternative of wind power sounds good until you really need the power on those hot, sultry, humid and windless days in August. (29-7)

Comment: First of all, I believe that nuclear power should be a larger part of our energy supply. (42-4)

Comment: For the people that are concerned about the safety of nuclear power I ask what alternatives do you want? Lives are lost every year in coal mine accidents. People are killed running into coal trains. Natural gas explosions kill people. And the alternative wind power sounds good until you really need the power on those hot, sultry, windless days in August. (42-6)

Comment: Risk is a part of life. For the concerned folks from Bloomington and Champaign, I would offer that having a nuclear plant in Clinton must not be a great concern for these cities as they would not be the boom towns they are for the downstate Illinois. If you were really threatened by the fact of living 25 miles from a nuclear plant, would you be living where you live? Or do you live there because of a good quality of life, good jobs and an adequate power supply? All of us took a much greater risk in driving to this meeting tonight than in living next to nuclear power plant. (42-8)

Comment: As nuclear technology relates to electricity generation, we want to tell everyone the success story that is nuclear power in our country. Nuclear energy is safe, clean, reliable and is an important part of our balance energy mix. (48-1)

Comment: That is nuclear power has perhaps the smallest impact on environment including water, land, habitat, species and air resources. (48-6)

Comment: I understand that there are risks. But there are risks to all kinds of power which we need I spent much of my life working with people with respiratory diseases. And I can tell you there are a lot of complications and many issues surrounding fossil fuels. So to assume that we're not living every day with some of our consumption needs would be naive. (51-2)

Comment: Previously you saw on the map, Dresden, Braidwood and LaSalle Plants. I lived in the middle of that triangle. So I obviously, you know, it may have affected me a little bit but I don't think it was too bad. (53-1)

Comment: I would also have you consider the following information in support of why I believe nuclear power is clean, safe and a reliable source of base load energy generation. (61-2)

Comment: But in light of this [energy] crisis we cannot eliminate an energy with one of the smallest environmental footprints. (66-5)

Comment: We support nuclear power. (74-1)

Comment: We support building a new plant, a new nuclear plant in Clinton. (74-2)

Comment: I believe, personally, that it's time to address the need for safe, reliable, local and environmentally sound energy resources, by developing new nuclear energy options for Illinois, as the best option for the citizens of Illinois, including myself and my family. (75-1)

Comment: I'd like to go back and say that I believe that nuclear energy is safe and reliable. (75-3)

Comment: And I think it's important that we consider all of the benefits of this technology, and not focus completely on the, the very small risk that in reality exists, as we, we try to move forward and determine the best path for our state. (75-9)

Comment: Because I believe nuclear energy is a clean, affordable, reliable and safe way to generate electricity. (76-2)

Comment: I came because I admire what is being done here in Clinton, and I only hope that this will encourage Wisconsin to look at nuclear, look into nuclear as an option for our energy needs. (77-1)

Comment: The only practical options, for generating large base load power, would be nuclear power or fossil fuels. (78-2)

Comment: I'm a nuclear engineering student because I'm in favor of nuclear power. (78-4)

Comment: I'm here today to voice my support for nuclear power. (86-1)

Comment: We need a diverse energy mix for our future. (86-8)

Comment: Nuclear Power is a clean, reliable, cheap, efficient source of energy (93-1)

Response: These comments express support for nuclear power in general. Because they did not provide new information, no change was made to the EIS as a result of these comments.

Comment: I have an allegory for you tonight related to nuclear power. Like many of you, gas prices in the economy have me thinking about the new generation of hybrid vehicles. Let's say as a college student you do your homework and you decide that a hybrid car is the way to go. So over spring break with the whole family at the dinner table, you announce that you've applied for a permit to buy a new hybrid vehicle.

Immediately your mother says, they're too dangerous. And again brings up that horrible accident in Pennsylvania. You remind her that it was 25 years ago and no one got hurt. The driver shut off the back up cooling, the engine over heated and was ruined. But the car's safety systems worked and the damage was confined to under the hood.

You tell her that there were 103 other hybrids in operation still today, 11 in your own neighborhood. They all have upgraded instrumentation and every driver is trained not to shut off the back up cooling system. This new generation of hybrids are safer than ever and they're so much better for the environment than regular cars. But she's still worried. She always be. She's your mother.

Now, your sister, the economics major, says that they're heavily subsidized and it's just too expensive, that you'd be better off with a regular car. But you know that it's worth paying more money for an environmentally friendly car with rising fuel costs and rumors of a carbon tax. It's only a matter of time before hybrids cost the same or less to buy and operate than a regular car. Your sister always focused on the moment and was never one to plan ahead.

Your brother, the environmentalist, he applauds your desire to reduce pollution. But he reminds you that hybrids use lead and cameroon batteries that you'll have to replace frequently. And you'd be generating waste that's deadly for a million years. He's cynical about the government's storage facility for used batteries. He says it's unsafe and it may never open. You tell him that even if you have to store the used batteries yourself, that driving a hybrid is still better for the environment. The lead and cameroon is in stable solid form. It's in thick sealed cases and it's not going anywhere until the government can eventually take your batteries.

They even recycle batteries in Britain, France and Japan. And we'll still have that option one day if we decide to use it. You point out to your brother that it's not deadly waste if it can be stored safely and 95 percent of it can be reused.

Finally, your grandmother says, we're just too wasteful of our resources these days, that we don't really need cars at all, fossil fuel or hybrid. She tells you that she went to work or school. She walked or rode her bike. These alternative forms of transportation were free and they had no impact on the environment at all.

You explain that our generation travels a lot more than her's did and that you would still walk or ride your bike for short trips. But you need an all weather, reliable form of transportation that you can use every day. You like to drive at night after the sun goes down. You like an air conditioned car on those sweltering hot summer days when the wind isn't blowing at all. You have nothing against walking or riding bikes, but you will need a car and the hybrid has the least environmental impact of cars available today.

As you finish dinner with your family you think about how unique they all are. They're each shaped by different life experiences and this affects the way they reacted to your announcement. While their reactions are heartfelt, you've done the research. You know the facts. And you know that it's the right thing to do for the future of our planet. Your children and your grandchildren will know that and they will thank you too. (64-2)

Response: *This comment expresses support for nuclear power in general and acknowledges the diversity of opinions. Because it did not provide new information, no change was made to the EIS as a result of the comment.*

E.2.33 General Comments in Opposition to Nuclear Power

Comment: Please do not approve another nuclear reactor in Illinois—or anywhere else for that matter. (04-4)

Comment: I am totally against your plans to build more radiation making factories in our state. (05-1)

Comment: The people of this country have worked long and hard to stop the tragedy of nuclear based energy. It is dangerous, costly, unhealthy and really rather insane. (09-2)

Comment: As a nation, we can't afford to start down the road of nuclear power again, after a 30-year hiatus. (09-4)

Comment: This news disturbs me deeply. It is indisputable that nuclear energy poses grave dangers to human life and livelihood. A chain reaction could have catastrophic consequences similar to an atomic bomb explosion. (10-1)

Comment: We must recognize that nuclear energy is a terrible mistake. (10-7)

Comment: It is serious enough having one there. I would prefer none. But the prospect of a second one is totally unacceptable to me (13-2)

Comment: I do not believe that nuclear power is safe. (148-1)

Comment: Absolutely not! Central Illinois does not want another nuclear power plant. (22-1)

Comment: Nuclear power is touted as safe, clean, cheap and inexhaustible. It is none of these. Do not believe these myths. (46-5)

Comment: This is a night of enormous opportunity because we are gathered here standing on a fine line between past and future. Are we going backwards toward nuclear power plant proliferation or will we right here in Clinton tonight have the courage and foresight to turn the tide of history by saying no more to nuclear power until its long term effects, long term effects on health can be fully understood and the nuclear waste riddle can be solved. (46-9)

Comment: I'm not a fan of nuclear power. (47-2)

Comment: We're very concerned about the health and safety, health affects and safety problems with nuclear energy. (56-1)

Comment: And one thing I can see is or I believe I can see is that the scientist are on our side, and I'm obviously opposed to nuclear power, the scientists on our side are the equal to the scientists on the other side, in all possible ways except the scientists on our side don't have the same, cannot get the same connection to the money in the power structure. So that's the difference. But the two sides have very different points of view. (60-1)

Comment: What does nuclear power do? It makes sacrificial zones for 24,000 years. That's 5,000 generations literally making people take care of the waste for 5,000 generations ahead. Now this is a very evil carcinogenic technology and every plant should be shut down now immediately at once. And we shouldn't even be considering building another plant. The waste problem isn't solved. (65-5)

Comment: The pro-nuclear cheerleaders are hyping up the so-called nuclear renaissance, what they consider to be a nuclear rebirth of what I consider to be a failed technology. I say so-called because this nonsense is more accurately described as a nuclear relapse. Like a reoccurring nightmare from a B science fiction movie. (71-4)

Comment: There are pluses and minuses to a whole lot of things in life and nuclear power has been one of them, over the course of my lifetime. As a young boy, it was the, it was going to be the savior of energy. We were going to have un-metered electricity. It was going to be as cheap as water. Of course, over time, it's not really worked out that way. I've come over the course of my adult life, to a rather unhappy conclusion on the subject. I conclude that nuclear power is on balance a failed experiment. (89-1)

Comment: Nuclear power is touted as safe, clean, cheap and inexhaustible. It is none of these. Do not believe these myths. (94-5)

Comment: This is a night of enormous opportunity because we are gathered here standing on a fine line between past and future. Are we going backwards toward nuclear power plant proliferation or will we right here in Clinton tonight have the courage and foresight to turn the tide of history by saying no more to nuclear power until its long term effects, long term effects on health can be fully understood and the nuclear waste riddle can be solved. (94-9)

Response: *These comments express opposition to nuclear power in general. Because they did not provide new information, no change was made to the EIS as a result of these comments.*

E.2.34 Comments that are Outside the Scope of Early Site Permitting

Comment: I wanted to observe that the Commission has streamed line its process of developing the environmental impact statements and has essentially declared that 69 of the 92 issues are the same for all plants with similar features. And these issues are classified as Category 1. And among these are human health. (52-1)

Response: *The commenter appears to be talking about the License Renewal Generic Environmental Impact Statement (GEIS) (NRC 1996), which is used during the staff's review of the license renewal of nuclear plants. The categorization of issues identified in the GEIS does not apply to the staff's review of an ESP. The GEIS was used as a reference to identify and possibly resolve the types of issues that should be considered during the staff's review of the ESP applications. No change was made to the EIS as a result of the comment.*

Comment: Due to safety concerns today and lack of effective long-term storage for waste tomorrow I would like to see the existing power plant at Clinton de-comissioned. (168-1)

Comment: These early site permits are costing taxpayers millions of dollars because the government has subsidized the process to encourage big energy companies to invest in nuclear power. We should be investing in renewable and energy-efficient technologies, not 20th century technologies that suffer from the same fatal flaws now as they have for the past 50 years. (191-8)

Comment: The second example of Exelon's arrogance is characterized by Exelon CEO John Rowe's pooh-poohing of renewable resources (such as wind), and Exelon's lobbyists in Springfield fervently opposed to a statewide "Renewable Energy Portfolio Standard (REPS)." Legislation (Senate Bill 2321) is currently pending in the General Assembly which would establish such a REPS. Being THE MOST IMPORTANT renewable energy legislation that could ever be passed HERE, this REPS would set realistic goals to "ramp up" our use of renewably-generated electricity--requiring 3% of our electricity by 2007 to come from renewable sources (solar, wind, and biomass), and then 10% by 2012. THIS "RENEWABLE ENERGY PORTFOLIO STANDARD" SHOULD BE PASSED IN SPRINGFIELD FIRST, BEFORE EXELON'S PROPOSED SECOND CLINTON REACTOR IS EVEN CONSIDERED. In fact, Exelon should be MANDATED to help meet our electricity requirements using these renewable resources. Again, granting Exelon an ESP will help to EFFECTIVELY NEGATE such a REPS on the state level. (Exelon CEO John Rowe and Exelon's lobbyists will probably argue something like this: "Well, we are already going to construct a new reactor near Clinton, so why do we need this 'Renewable Energy Portfolio Standard' too?" As the expression goes, this kind of backward thinking "puts the cart before the horse.") Continuing on, nuclear power is a major part of our energy problem itself (rather than a solution to it). As someone who has researched the history of nuclear power, I can say that Commonwealth Edison (now an Exelon company) has had an integral role in developing and promoting nuclear power in Illinois and nationwide. [With the current number of nuclear reactors (and the growing quantities of deadly radioactive wastes), the label "NUCLEAR ILLINOIS" is NOT exaggerated activist rhetoric.] This long-time (and big-time) "political stranglehold" by ComEd (and now Exelon) has stifled the fullest practical implementation possible of the best electricity-saving technologies and well-designed/properly installed renewable energy systems. (25-7)

Comment: I call for Exelon to pay for Clinton's school district shortfall caused by the devaluation of the first reactor as a demonstration of good faith to this community that Exelon promises to enrich with a second reactor. (49-8)

Comment: I cry out along with many other scientists and activists that the time has come for the reactors to release for independent scientific scrutiny, the radiation emission data that they have been gathering for over 40 years. And I insist that the NRC and Exelon fund independent, extensive, epidemiological studies of Illinois populations and that these studies be those critical, that they, in part, will be allowed feedback from those critical current radiation emission standards. (49-9)

Comment: I would like to identify Item 7 in Congressman Markey's letter asking that the commissioner to please provide copies of all documents related to any unanticipated releases of tritium and/or radioactive containments from the Exelon Corporation's Dresden, Braidwood and LaSalle stations since 1990. Please correct me if I am wrong, but my understanding is as of yesterday's date, these documents have not been submitted by the applicant. (52-3)

Response: The comments are noted. They do not provide information relevant to the EIS and are outside the scope of the EIS. Therefore, they will not be evaluated further. No change was made to the EIS as a result of these comments.

E.3 References

10 CFR Part 20. Code of Federal Regulations, Title 10, *Energy*, Part 20, "Standards for Protection Against Radiation."

10 CFR Part 51. Code of Federal Regulations, Title 10, *Energy*, Part 51, "Environmental Protection Regulations for Domestic Licensing and Related Regulatory Functions."

10 CFR Part 52. Code of Federal Regulations, Title 10, *Energy*, Part 52, "Early Site Permits, Standard Design Certifications, and Combined Licenses for Nuclear Power Plants."

10 CFR 71. Code of Federal Regulations, Title 10, *Energy*, Part 71, "Packaging and Transportation of Radioactive Material."

40 CFR Part 61. Code of Federal Regulations. Title 40, *Protection of Environment*, Part 61, "National Emission Standards for Hazardous Air Pollutants." U.S. Environmental Protection Agency.

40 CFR Part 190. Code of Federal Regulations, Title 40, *Protection of Environment*, Part 190, "Environmental Radiation Protection Standards for Nuclear Power Operation."

59 FR 7629. "Federal Actions to Address Environmental Justice in Minority Populations and Low-Income Populations." Executive Order 12898, *Federal Register.* February 16, 1994.

64 FR 68005. December 6, 1999. "Status Report on the Review of the Waste Confidence Decision." U.S. Nuclear Regulatory Commission, *Federal Register.*

42 UCS 1421, et seq. National Environmental Policy Act of 1969, as amended. (NEPA).

Clinton Power Station (CPS). 2002. *Clinton Power Station Updated Safety Analysis Report.* Revision 10.

Council on Environmental Quality (CEQ). 1997. *Environmental Justice: Guidance Under the National Environmental Policy Act.* Executive Office of the President, Washington, D.C.

Edinger, J.E., Associates. 1989. Probabilistic Hydrothermal Modeling of Clinton Lake. Document No. 89-15-R. Wayne, Pennsylvania.

Exelon Generation Company, LLC (Exelon). 2004. Letter dated July 23, 2004, from M.C. Kray to the NRC submitting additional information in response to an NRC request dated May 11, 2004. Exelon Nuclear, Kennett Square, Pennsylvania.

Exelon Generation Company, LLC (Exelon). 2006. *Exelon Generation Company, LLC, Early Site Permit Application: Environmental Report, Rev. 4.* Exelon Nuclear, Kennett Square, Pennsylvania.

Fisher, L.E., C.K. Chou, M.A. Gerhard, C.Y. Kimora, R.W. Martin, R.W. Mensing, M.E. Mount, and M.C. Witte. 1987. *Shipping Container Response to Severe Highway and Railway Accident Conditions.* NUREG/CR-4829, prepared Lawrence Livermore National Laboratory, Livermore, California, for the U.S. Nuclear Regulatory Commission, Washington, D.C.

Health Physics Society (HPS). 2004. "Radiation Risk in Perspective." Health Physics Society, McLean, Virginia.

Idaho National Engineering and Environmental Laboratory (INEEL). 2003. *Early Site Permit Environmental Report Sections and Supporting Documentation.* Office of Nuclear Energy, Science, and Technology, U.S. Department of Energy, Washington, D.C.

Illinois Environmental Protection Agency (IEPA). 2000. *National Pollutant Discharge Elimination System (NPDES). CPS Permit to Discharge from IEPA.* Permit No. IL0036919. April 24, 2000.

Illinois Environmental Protection Agency (IEPA). 2004. "Section 303(d) List."
IEPA/BOW/04-00, IEPA, Bureau of Water, Water Management Section, Planning Unit,
Springfield, Illinois.

Illinois Power Company (IPC). 1992. Clinton Power Station Environmental Monitoring Program
Water Quality Report, *January 1978 through December 1991*. Illinois Power Company.

International Atomic Energy Agency (IAEA). 1992. *Effects of Ionizing Radiation on Plants and
Animals at Levels Implied by Current Radiation Protection Standards*. Technical Report Series
No. 332. Vienna, Austria.

Moulder, J.E. 2004. "Electromagnetic Fields and Human Health: Power Lines and Cancer
FAQs." Available at http://www.mcw.edu/gcrc/cop/powerlines-cancer-faq/toc.html#16B
(accessed December 3, 2004).

National Council on Radiation Protection and Measurements (NCRP). 1987. *Ionizing Radiation
Exposure of the Population of the United States*. Report No. 93, NCRP, Bethesda, Maryland.

National Council on Radiation Protection and Measurements (NCRP). 1988. *Exposure of the
Population in the United States and Canada from Natural Background Radiation*. Report
No. 94, NCRP, Bethesda, Maryland.

National Energy Policy Development Group (NEPDG). 2001. *Reliable, Affordable, and
Environmentally Sound Energy for America's Future*. Accessed on the Internet June 21, 2005,
at http://whitehouse.gov/energy/National-Energy-Policy.pdf.

National Institute of Environmental Health Sciences (NIEHS). 1999. *NIEHS Report on Health
Effects from Exposure to Power Line Frequency and Electric and Magnetic Fields*. Publication
No. 99-4493, Research Triangle Park, North Carolina.

National Research Council. 2006. *Health Risks from Exposure to Low Levels of Ionizing
Radiation: BEIR VII – Phase 2*. Committee to Assess Health Risks from Exposure to Low
Levels of Ionizing Radiation, National Research Council, National Academies Press,
Washington, D.C.

Sprung, J.L., D.J. Ammerman, N.L. Breivik, R.J. Dukart, F.L. Kanipe, J.A. Koski, G.S. Mills,
K.S. Neuhauser, H.D. Radloff, R.F. Weiner, and H.R. Yoshimura. 2000. *Reexamination of
Spent Fuel Shipment Risk Estimates*. NUREG/CR-6672, U.S. Nuclear Regulatory Commission,
Washington, D.C.

U.S. Atomic Energy Commission (AEC). 1974. *Environmental Survey of the Uranium Fuel
Cycle*. WASH-1248, AEC, Washington, D.C.

Appendix E

U.S. Global Change Research Program (USGCRP). 2000. U.S. National Assessment of the Potential Consequences of Climate Variability and Change, Mega-Region: Midwest. Accessed on the Internet, March 1, 2006, at: http://www.usgrcrp.gov/usgcrp/nacc/midwest.htm.

U.S. Department of Energy (DOE). 2003. *Uranium Industry Annual 2002.* DOE/EIA-0478 (2002), U.S. Department of Energy, Washington D.C. Accessed March 2, 2006, at: http://tonto.eia.doe.gov/FTPROOT/nuclear/04782002.pdf.

U.S. Nuclear Regulatory Commission (NRC). 1976. *Environmental Survey of the Reprocessing and Waste Management Portions of the LWR Fuel Cycle.* NUREG-0116 (Supplement 1 to WASH-1248), NRC, Washington, D.C.

U.S. Nuclear Regulatory Commission (NRC). 1977a. *Calculation of Annual Doses to Man from Routine Releases of Reactor Effluents for the Purpose of Evaluating Compliance with 10 CFR Part 50, Appendix I.* Regulatory Guide 1.109, NRC, Washington, D.C.

U.S. Nuclear Regulatory Commission (NRC). 1977b. *Final Environmental Statement on Transportation of Radioactive Material by Air and Other Modes.* NUREG-0170, Vol. 1, NRC, Washington, D.C.

U.S. Nuclear Regulatory Commission (NRC). 1996. *Generic Environmental Impact Statement for License Renewal of Nuclear Plants.* NUREG-1437, Vols. 1 and 2, NRC, Washington, D.C. Available at http://www.nrc.gov/reading-rm/doc-collections/nuregs/staff/sr1437/.

U.S. Nuclear Regulatory Commission (NRC). 2000. *Environmental Standard Review Plan.* NUREG-1555, Vol. 1, NRC, Washington, D.C.

U.S. Nuclear Regulatory Commission (NRC). 2004. Office of Nuclear Reactor Regulation (NRR). "Procedural Guidance for Preparing Environmental Assessments and Considering Environmental Issues." NRR Office Instruction LIC-203, NRC, Washington, D.C.

U.S. Nuclear Regulatory Commission (NRC). 2006. *Clinton Early Site Permit Final Safety Evaluation Report (SER).* ADAMS Accession No. ML060380324, NRC, Washington, D.C.

Appendix F

**Exelon Generation Company, LLC's (Exelon's)
Key Early Site Permit Consultation Correspondence**

Appendix F

Exelon Generation Company, LLC's (Exelon's)
Key Early Site Permit Consultation Correspondence

Correspondence received during the evaluation process of the early site permit application for the Exelon site is identified in Table F-1. Copies of the correspondence are included at the end of this appendix.

Source	Recipient	Date of Letter
United States Nuclear Regulatory Commission (Pao-Tsin Kuo)	Advisory Council on Historic Preservation (Don Klima)	December 18, 2003
Delaware Nation (Phyllis Wahahrockah-Tasi)	United States Nuclear Regulatory Commission (Pao-Tsin Kuo)	December 22, 2003
United States Nuclear Regulatory Commission (Pao-Tsin Kuo)	Illinois Historic Preservation Agency (Maynard Crossland)	December 23, 2003
United States Nuclear Regulatory Commission (Pao-Tsin Kuo)	Kickapoo of Oklahoma Business Committee (Honorable Kendall Scott)	December 30, 2003
United States Nuclear Regulatory Commission (Pao-Tsin Kuo)	Kickapoo Traditional Tribe of Texas (Honorable Raul Garza, Jr.	December 30, 2003
United States Nuclear Regulatory Commission (Pao-Tsin Kuo)	Kickapoo of Kansas Tribal Council (Honorable Carol Anske)	December 30, 2003
United States Nuclear Regulatory Commission (Pao-Tsin Kuo)	Delaware Tribe of Western Oklahoma (Honorable Lawrence F. Snake)	December 30, 2003
United States Nuclear Regulatory Commission (Pao-Tsin Kuo)	Peoria Tribe of Indians of Oklahoma (Honorable John P. Froman)	December 30, 2003
United States Nuclear Regulatory Commission (Pao-Tsin Kuo)	Eastern Delaware Tribe (Honorable Dee Ketchum)	December 30, 2003
Peoria Tribe of Indians of Oklahoma (Honorable John P. Froman	United States Nuclear Regulatory Commission (Pao-Tsin Kuo)	January 13, 2004
Delaware Tribe of Indians of Oklahoma (Brice Obermeyer)	United States Nuclear Regulatory Commission (Pao-Tsin Kuo)	January 13, 2004
United States Nuclear Regulatory Commission (Pao-Tsin Kuo)	National Oceanic and Atmosphere Administration Fisheries (Patricia Kurkul)	March 17, 2004

Appendix F

Source	Recipient	Date of Letter
United States Nuclear Regulatory Commission (Pao-Tsin Kuo)	Chicago Ecological Field Service Office, United States Fish and Wildlife Service (John Rogner)	March 17, 2004
United States Nuclear Regulatory Commission (Pao-Tsin Kuo)	Rock Island Ecological Field Service Office, United States Fish and Wildlife Service (Richard Nelson)	March 17, 2004
Rock Island Ecological Field Service Office, United States Fish and Wildlife Service (Richard Nelson)	United States Nuclear Regulatory Commission (Pao-Tsin Kuo)	April 6, 2004
Chicago Ecological Field Service Office, United States Fish and Wildlife Service (John Rogner)	United States Nuclear Regulatory Commission (Pao-Tsin Kuo)	April 12, 2004
United States Nuclear Regulatory Commission (Pao-Tsin Kuo)	Rock Island Ecological Field Service Office, United States Fish and Wildlife Service (Richard Nelson)	April 7, 2005
Illinois Historic Preservation Agency (Anne Haaker)	United States Nuclear Regulatory Commission (Pao-Tsin Kuo)	April 11, 2005
United States Department of Interior (Michael Chezik)	United States Nuclear Regulation Commission (Michael Lesar)	May 17, 2005
United States Ecological Field Service Office, United States Fish and Wildlife Service (Richard Nelson)	United States Nuclear Regulatory Commission (Pao-Tsin Kuo)	May 19, 2005
United States Nuclear Regulatory Commission (Pao-Tsin Kuo)	United States Army Corps of Engineers (Kenneth Barr)	October 12, 2005
United States Army Corps of Engineers (Kenneth Barr) [The application referred to in the letter can be found in ADAMS under ML060530049.]	United States Nuclear Regulatory Commission (Thomas Kenyon)	February 7, 2006

December 18, 2003

Mr. Don Klima, Director
Office of Federal Agency Programs
Advisory Council on Historic Preservation
Old Post Office Building
1100 Pennsylvania Avenue, NW, Suite 809
Washington, DC 20004

SUBJECT: EARLY SITE PERMIT REVIEW FOR THE CLINTON SITE

Dear Mr. Klima:

The U.S. Nuclear Regulatory Commission (NRC) staff is reviewing an application for an early site permit (ESP) submitted by Exelon Generation Company, LLC (Exelon) on September 25, 2003. An ESP allows an applicant to set aside a site for potential future construction of one or more new nuclear power plants, and provides the opportunity to resolve site safety and environmental issues before construction begins. An ESP does not allow actual construction of a nuclear plant, which must be requested through another application. The ESP site proposed by Exelon is on property co-located with the existing Clinton Power Station site near the town of Clinton in DeWitt County, Illinois. The application was submitted by Exelon pursuant to NRC requirements at Title 10 of the *Code of Federal Regulations* Part 52 (10 CFR Part 52).

Exelon has also included a site redress plan in its application in accordance with 10 CFR 52.17(c) and 52.25. If a site redress plan is included in an ESP approved by the NRC, the applicant may carry out certain site preparation and limited construction activities. Exelon would still be required to obtain the appropriate local, State, and other Federal permits required for these activities prior to starting work.

As part of its review of the application, the NRC staff will prepare an environmental impact statement (EIS) pursuant to 10 CFR Part 51, the NRC regulations that implement the National Environmental Policy Act of 1969 (NEPA). In accordance with 36 CFR 800.8, the EIS will include analyses of potential impacts to historic and cultural resources. A draft EIS is scheduled for publication in December 2004, and will be provided to you for review and comment.

If you have any questions or require additional information, please contact the Environmental Project Manager for the Clinton ESP project, Mr. Thomas Kenyon at 301-415-1120 or TJK2@nrc.gov.

Sincerely,
/RA/
Pao-Tsin Kuo, Program Director
License Renewal and Environmental Impacts
Division of Regulatory Improvement Programs
Office of Nuclear Reactor Regulation

Docket No.: 52-007

cc: See next page

Delaware Nation NAGPRA Office

P.O. Box 825, Anadarko, OK 73005
Phone: (405) 247-2448
Fax: (405) 247-9898

11/25/03
68 FR 44130
(1)

03 DEC 29 PM 11:44

22 December 2003

ATTN: Chief, Rules and Directive Branch
Division of Administration Services
Office of Administration
Mailstop T-6 D59
U.S. Nuclear Regulatory Commission
Washington, DC 20555-001

Re: Proposed project- Early Site Permit Application at Clinton, Illinois Site

To Whom It May Concern:

Thank you for contacting the Delaware Nation regarding the above referenced project. The Delaware Nation is committed to protecting archaeological sites that are important to tribal heritage, culture, and religion. Furthermore, the tribe is particularly concerned with archaeological sites that may contain human burial remains and associated funerary objects.

Given the location of the proposed project, we request that you conduct a file search in conjunction with the State Office of Historic Preservation and the state's Archaeological Survey. These state agencies will advise you of the potential for archaeological resources, particularly sites of significant cultural interest or sites that contain human remains. Should either of these agencies determine that there are potentially significant archaeological sites in the area and that these sites are related to the tribe's heritage, the Delaware Nation requests that you contact our offices. Together with the SHPO and State Archaeologist, we will develop a plan to best protect these archaeological resources.

Should either of these agencies recommend an archaeological survey or test excavation of the proposed construction site, we ask that the Delaware Nation be informed of the results of the survey. The Delaware Nation also requests copies of any accompanying site forms or reports.

Also, any changes to the above referenced project should be resubmitted to the NAGPRA Director of the Delaware Nation for review.

Should this project inadvertently uncover an archaeological site and/or human remains, even after an archaeological survey, we request that you immediately contact the appropriate state agencies, as well as the Delaware Nation. Also, we ask that you halt all construction activities until the tribe and these state agencies are consulted.

We appreciate your cooperation in contacting the Delaware Nation. Should you have any questions, feel free to contact me.

Sincerely,

Phyllis Wahahrockah-Tasi M.H.R.
NAGPRA Director

Fields = ADM-03
add = T. Kenyon (SKa)

Template = ADM-013

December 23, 2003

Mr. Maynard Crossland
Director
Illinois Historic Preservation Agency
Preservation Services Division
One Old State Capitol Plaza
Springfield, IL 62701

SUBJECT: EARLY SITE PERMIT (ESP) REVIEW FOR THE CLINTON ESP SITE

Dear Mr. Crossland:

The U.S. Nuclear Regulatory Commission (NRC) staff is reviewing an application for an ESP to set aside a site for the potential future construction of one or more new nuclear power plants. The NRC staff is currently seeking information from consulting parties, and other individuals and organizations likely to have knowledge of, or concerns with, historic properties in the area, to identify issues relating to the proposed undertaking's potential effects on historic properties.

If built, the new unit(s) would be co-located with the existing Clinton Power Station (CPS) site near the town of Clinton in DeWitt County, Illinois. The application for an ESP was submitted by Exelon Generation Company, LLC (Exelon), on September 25, 2003, pursuant to NRC requirements at Title 10 of the *Code of Federal Regulations* Part 52 (10 CFR Part 52). The application is available through the web-based version of the NRC's Agencywide Documents Access and Management System (ADAMS) which can be found at http://www.nrc.gov/reading-rm/adams.html. The application is listed under accession number ML032721596.

As part of its review of the application, the NRC staff will prepare an environmental impact statement (EIS) under the provisions of 10 CFR Part 51, the NRC rules that implement the National Environmental Policy Act of 1969 (NEPA). In accordance with 36 CFR 800.8, the EIS will include analyses of potential impacts to historic properties, and will document the NRC staff's determination regarding the suitability of the proposed site for the construction and operation of one or more new nuclear plants.

If approved, the ESP would not authorize the applicant to begin construction of the unit(s). However, in its review the NRC staff will evaluate the environmental impacts of construction and operation and will also consider alternatives, including alternative sites.

M. Crossland   2

Exelon has also included a site redress plan in its application in accordance with
10 CFR 52.17(c) and 52.25. If a site redress plan is incorporated in an ESP approved by the
NRC, the applicant may carry out certain site preparation and limited construction activities.
Exelon would still be required to obtain the appropriate local, State, and other Federal permits
required for these activities prior to starting work.

In the context of the National Historic Preservation Act of 1966, as amended, the NRC staff has
determined that the area of potential effect (APE) for this ESP review is the area at the power
plant site and its immediate environs which may be impacted by land-disturbing activities
associated with the construction and operation of the new unit(s), and construction and
operation of new transmission lines that may follow parallel with some of the existing
transmission line systems now serving CPS. The new lines would run from the Clinton ESP site
to an interconnect point at the Brokaw substation near Bloomington, Illinois, about 23 miles
north of the site; and from the Clinton ESP site to the Oreana substation, about 8 miles south of
the site. The power plant site is located in DeWitt county, Illinois, and the transmission line
corridors traverse DeWitt and McLean counties.

In its application, Exelon refers to a comprehensive cultural resource and historic property
investigation that was performed prior to the construction of CPS, approximately 30 years ago.
Exelon further states that any issues that were raised were resolved through removal of these
historic and cultural resources. No historic standing structures have been identified within the
Exelon ESP power block footprint, cooling tower footprint, or within the immediate vicinity of
CPS. Exelon states that the location of the ESP facility power block footprint appears to have
been heavily disturbed by previous construction of CPS. If the power block or cooling tower
footprint area was expanded or moved, there is a potential for impact to historic properties. The
applicant has committed to perform further evaluation to determine if additional archaeological
review is required if additional area within the ESP site will be required for development.

We invite you and your staff to participate in the review of the Clinton ESP application. We will
also be contacting any Native American Tribes that may have a potential interest in the
proposed undertaking, affording them the opportunity to participate in this process and identify
issues of concern to them. These tribes are being identified by records research with the
Bureau of Indian Affairs, State and local governments, tribal organizations, and through other
historical documentation.

On December 18, 2003, the NRC will conduct a public environmental scoping meeting at the
Revere Ware Room in the Vespasian Warner Public Library, 310 N. Quincy Street, Clinton,
Illinois. You and your staff are invited to attend. Your office will receive a copy of the draft EIS
along with a request for comments after it is issued. This draft EIS will include identification of
historic properties, assessment of impacts, and our preliminary determination. The anticipated
publication date for the draft EIS is December 2004.

M. Crossland   3

If you have any questions or require additional information, please contact the Environmental
Project Manager for the Clinton ESP project, Mr. Thomas Kenyon at 301-415-1120 or
TJK2@nrc.gov.

                              Sincerely,
                              /RA/
                              Pao-Tsin Kuo, Program Director
                              License Renewal and Environmental Impacts
                              Division of Regulatory Improvement Programs
                              Office of Nuclear Reactor Regulation

Docket No.: 52-007

cc: See next page

Appendix F

December 30, 2003

The Honorable Kendall Scott, Chair
Kickapoo of Oklahoma Business Committee
Post Office Box 70
McCloud, OK 74851

SUBJECT: EARLY SITE PERMIT (ESP) REVIEW FOR THE CLINTON ESP SITE

Dear Chairman Scott:

The U.S. Nuclear Regulatory Commission (NRC) staff is reviewing an application for an ESP to set aside a site for the potential future construction of one or more new nuclear power plants. The application was submitted by Exelon Generation Company, LLC (Exelon), on September 25, 2003, pursuant to NRC requirements at Title 10 of the *Code of Federal Regulations* Part 52 (10 CFR Part 52). If built, the new unit(s) would be co-located with the existing Clinton Power Station (CPS) site near the town of Clinton in DeWitt County, Illinois. Exelon has also included a site redress plan in its application in accordance with 10 CFR 52.17(c) and 52.25. If a site redress plan is incorporated in an ESP approved by the NRC, the applicant may carry out certain site preparation and limited construction activities. Exelon would still be required to obtain the appropriate local, State, and other Federal permits required for these activities before starting work.

As part of its review of the application, the NRC staff will prepare an environmental impact statement (EIS) under the provisions of 10 CFR Part 51, the NRC rules that implement the National Environmental Policy Act of 1969 (NEPA). The NRC environmental review process includes an opportunity for public participation in the environmental review. The Clinton ESP site is located on land that may be of interest to the Kickapoo of Oklahoma Business Committee. We want to ensure that you are aware of our efforts and, pursuant to our regulations at 10 CFR 51.28(b), the NRC invites the Kickapoo of Oklahoma Business Committee to provide input to the scoping process relating to the NRC's environmental review of the application. The following is a description of the application and the environmental review process.

The EIS will document the NRC staff's determination regarding the suitability of the proposed site for the construction and operation of one or more new nuclear plants. In addition, the staff will also consider alternatives to the proposed action, including alternative sites. The EIS will contain the results of the review of the environmental impacts on the area surrounding the Clinton ESP site that are related to terrestrial ecology, aquatic ecology, hydrology, socioeconomic issues, and historic properties (among others), and will contain a recommendation regarding the environmental acceptability of granting an ESP. If approved, the ESP would not authorize the applicant to begin construction of the unit(s).

As part of this review, and in accordance with 36 CFR 800.8, the EIS will include analyses of potential impacts to historic properties. Accordingly, pursuant to 10 CFR 51.28 and 36 CFR 800.2(c)(2), the NRC wishes to ensure that Indian tribes that might have an interest in any potential historic properties in the area of potential effect (APE) are afforded the opportunity to

The Honorable K. Scott   -2-

identify their concerns, provide advice on the identification and evaluation of historic properties, including those of traditional religious and cultural importance, and, if necessary, participate in the resolution of any adverse effects to such properties.

In the context of the National Historic Preservation Act of 1966, as amended, the APE for this ESP review is the area at the power plant site and its immediate environs that may be impacted by land-disturbing activities associated with the construction and operation of the new unit(s), and construction and operation of new transmission lines that may follow parallel with some of the existing transmission line systems now serving CPS. The new lines would run from the Clinton ESP site to an interconnect point at the Brokaw substation near Bloomington, Illinois, about 23 miles north of the site; and from the Clinton ESP site to the Oreana substation, about 8 miles south of the site. The power plant site is located in DeWitt County, Illinois, and the transmission line corridors traverse DeWitt and McLean counties. The application is available through the web-based version of the NRC's Agencywide Documents Access and Management System (ADAMS) which can be found at http://www.nrc.gov/reading-rm/adams.html. The application is listed under accession number ML032721596.

In its application, Exelon refers to a comprehensive cultural resource and historic property investigation that was performed prior to the construction of CPS, approximately 30 years ago. Exelon further states that any issues that were raised were resolved through removal of these historic and cultural resources. No historic standing structures have been identified within the Exelon ESP power block footprint, cooling tower footprint, or within the immediate vicinity of CPS. Exelon states that the location of the ESP facility power block footprint appears to have been heavily disturbed by previous construction of CPS. If the power block or cooling tower footprint area was expanded or moved, there is a potential for impact to historic properties. The applicant has committed to perform further evaluation to determine if additional archaeological review is required if additional area within the ESP site will be required for development.

As discussed with Mr. Collier of your staff on December 12, 2003, the NRC conducted a public environmental scoping meeting on December 18, 2003 at the Revere Ware Room in the Vespasian Warner Public Library, 310 N. Quincy Street, Clinton, Illinois. Representatives of your tribe were invited to attend.

Please submit any written comments your tribe may have to offer on the scope of the environmental review by January 9, 2004. Comments should be submitted either by mail to the Chief, Rules and Directives Branch, Division of Administrative Services, Mail Stop T-6 D59, U.S. Nuclear Regulatory Commission, Washington, D.C. 20555-0001, or by Internet to *ClintonEIS@nrc.gov*.

The Honorable K. Scott   -3-

At the conclusion of the scoping process, the NRC staff will prepare a summary of the significant issues identified and the conclusions reached, and will send a copy to you. In addition, after it is issued, your office will receive a copy of the draft EIS along with a request for comments. The anticipated publication date for the draft EIS is December 2004. If you have any questions or require additional information, please contact the Environmental Project Manager for the Clinton ESP project, Mr. Thomas Kenyon, at 301-415-1120 or *ClintonEIS@nrc.gov.*

> Sincerely,
> /RA/
> Pao-Tsin Kuo, Program Director
> License Renewal and Environmental Impacts
> Division of Regulatory Improvement Programs
> Office of Nuclear Reactor Regulation

Docket No.: 52-007

cc: See next page

December 30, 2003

The Honorable Raul Garza Jr., Chair
Kickapoo Traditional Tribe of Texas
HC 1, Post office Box 9700
Eagle Pass, TX 78853
Miami, OK 74355

SUBJECT: EARLY SITE PERMIT (ESP) REVIEW FOR THE CLINTON ESP SITE

Dear Chairman Garza:

The U.S. Nuclear Regulatory Commission (NRC) staff is reviewing an application for an ESP to set aside a site for the potential future construction of one or more new nuclear power plants. The application was submitted by Exelon Generation Company, LLC (Exelon), on September 25, 2003, pursuant to NRC requirements at Title 10 of the *Code of Federal Regulations* Part 52 (10 CFR Part 52). If built, the new unit(s) would be co-located with the existing Clinton Power Station (CPS) site near the town of Clinton in DeWitt County, Illinois. Exelon has also included a site redress plan in its application in accordance with 10 CFR 52.17(c) and 52.25. If a site redress plan is incorporated in an ESP approved by the NRC, the applicant may carry out certain site preparation and limited construction activities. Exelon would still be required to obtain the appropriate local, State, and other Federal permits required for these activities before starting work.

As part of its review of the application, the NRC staff will prepare an environmental impact statement (EIS) under the provisions of 10 CFR Part 51, the NRC rules that implement the National Environmental Policy Act of 1969 (NEPA). The NRC environmental review process includes an opportunity for public participation in the environmental review. The Clinton ESP site is located on land that may be of interest to the Kickapoo Traditional Tribe of Texas. We want to ensure that you are aware of our efforts and, pursuant to our regulations at 10 CFR 51.28(b), the NRC invites the Kickapoo Traditional Tribe of Texas to provide input to the scoping process relating to the NRC's environmental review of the application. The following is a description of the application and the environmental review process.

The EIS will document the NRC staff's determination regarding the suitability of the proposed site for the construction and operation of one or more new nuclear plants. In addition, the staff will also consider alternatives to the proposed action, including alternative sites. The EIS will contain the results of the review of the environmental impacts on the area surrounding the Clinton ESP site that are related to terrestrial ecology, aquatic ecology, hydrology, socioeconomic issues, and historic properties (among others), and will contain a recommendation regarding the environmental acceptability of granting an ESP. If approved, the ESP would not authorize the applicant to begin construction of the unit(s).

As part of this review, and in accordance with 36 CFR 800.8, the EIS will include analyses of potential impacts to historic properties. Accordingly, pursuant to 10 CFR 51.28 and 36 CFR 800.2(c)(2), the NRC wishes to ensure that Indian tribes that might have an interest in any potential historic properties in the area of potential effect (APE) are afforded the opportunity to

The Honorable R. Garza Jr.   -2-

identify their concerns, provide advice on the identification and evaluation of historic properties, including those of traditional religious and cultural importance, and, if necessary, participate in the resolution of any adverse effects to such properties.

In the context of the National Historic Preservation Act of 1966, as amended, the APE for this ESP review is the area at the power plant site and its immediate environs that may be impacted by land-disturbing activities associated with the construction and operation of the new unit(s), and construction and operation of new transmission lines that may follow parallel with some of the existing transmission line systems now serving CPS. The new lines would run from the Clinton ESP site to an interconnect point at the Brokaw substation near Bloomington, Illinois, about 23 miles north of the site; and from the Clinton ESP site to the Oreana substation, about 8 miles south of the site. The power plant site is located in DeWitt County, Illinois, and the transmission line corridors traverse DeWitt and McLean counties. The application is available through the web-based version of the NRC's Agencywide Documents Access and Management System (ADAMS) which can be found at http://www.nrc.gov/reading-rm/adams.html. The application is listed under accession number ML032721596.

In its application, Exelon refers to a comprehensive cultural resource and historic property investigation that was performed prior to the construction of CPS, approximately 30 years ago. Exelon further states that any issues that were raised were resolved through removal of these historic and cultural resources. No historic standing structures have been identified within the Exelon ESP power block footprint, cooling tower footprint, or within the immediate vicinity of CPS. Exelon states that the location of the ESP facility power block footprint appears to have been heavily disturbed by previous construction of CPS. If the power block or cooling tower footprint area was expanded or moved, there is a potential for impact to historic properties. The applicant has committed to perform further evaluation to determine if additional archaeological review is required if additional area within the ESP site will be required for development.

As discussed with Ms. M. Salazar of your staff on December 5, 2003, the NRC conducted a public environmental scoping meeting on December 18, 2003 at the Revere Ware Room in the Vespasian Warner Public Library, 310 N. Quincy Street, Clinton, Illinois. Representatives of your tribe were invited to attend.

Please submit any written comments your tribe may have to offer on the scope of the environmental review by January 9, 2004. Comments should be submitted either by mail to the Chief, Rules and Directives Branch, Division of Administrative Services, Mail Stop T-6 D59, U.S. Nuclear Regulatory Commission, Washington, D.C. 20555-0001, or by Internet to ClintonEIS@nrc.gov.

The Honorable R. Garza Jr.   -3-

At the conclusion of the scoping process, the NRC staff will prepare a summary of the significant issues identified and the conclusions reached, and will send a copy to you. In addition, after it is issued, your office will receive a copy of the draft EIS along with a request for comments. The anticipated publication date for the draft EIS is December 2004. If you have any questions or require additional information, please contact the Environmental Project Manager for the Clinton ESP project, Mr. Thomas Kenyon, at 301-415-1120 or *ClintonEIS@nrc.gov.*

> Sincerely,
> /RA/
> Pao-Tsin Kuo, Program Director
> License Renewal and Environmental Impacts
> Division of Regulatory Improvement Programs
> Office of Nuclear Reactor Regulation

Docket No.: 52-007

cc: See next page

Appendix F

December 30, 2003

The Honorable Carol Anske, Chair
Kickapoo of Kansas Tribal Council
Route 1, Box 157
Horton, KS 66439

SUBJECT: EARLY SITE PERMIT (ESP) REVIEW FOR THE CLINTON ESP SITE

Dear Chairwoman Anske:

The U.S. Nuclear Regulatory Commission (NRC) staff is reviewing an application for an ESP to set aside a site for the potential future construction of one or more new nuclear power plants. The application was submitted by Exelon Generation Company, LLC (Exelon), on September 25, 2003, pursuant to NRC requirements at Title 10 of the *Code of Federal Regulations* Part 52 (10 CFR Part 52). If built, the new unit(s) would be co-located with the existing Clinton Power Station (CPS) site near the town of Clinton in DeWitt County, Illinois. Exelon has also included a site redress plan in its application in accordance with 10 CFR 52.17(c) and 52.25. If a site redress plan is incorporated in an ESP approved by the NRC, the applicant may carry out certain site preparation and limited construction activities. Exelon would still be required to obtain the appropriate local, State, and other Federal permits required for these activities before starting work.

As part of its review of the application, the NRC staff will prepare an environmental impact statement (EIS) under the provisions of 10 CFR Part 51, the NRC rules that implement the National Environmental Policy Act of 1969 (NEPA). The NRC environmental review process includes an opportunity for public participation in the environmental review. The Clinton ESP site is located on land that may be of interest to the Kickapoo of Kansas Tribal Council. We want to ensure that you are aware of our efforts and, pursuant to our regulations at 10 CFR 51.28(b), the NRC invites the Kickapoo of Kansas Tribal Council to provide input to the scoping process relating to the NRC's environmental review of the application. The following is a description of the application and the environmental review process.

The EIS will document the NRC staff's determination regarding the suitability of the proposed site for the construction and operation of one or more new nuclear plants. In addition, the staff will also consider alternatives to the proposed action, including alternative sites. The EIS will contain the results of the review of the environmental impacts on the area surrounding the Clinton ESP site that are related to terrestrial ecology, aquatic ecology, hydrology, socioeconomic issues, and historic properties (among others), and will contain a recommendation regarding the environmental acceptability of granting an ESP. If approved, the ESP would not authorize the applicant to begin construction of the unit(s).

As part of this review, and in accordance with 36 CFR 800.8, the EIS will include analyses of potential impacts to historic properties. Accordingly, pursuant to 10 CFR 51.28 and 36 CFR 800.2(c)(2), the NRC wishes to ensure that Indian tribes that might have an interest in any potential historic properties in the area of potential effect (APE) are afforded the opportunity to

The Honorable C. Anske   -2-

identify their concerns, provide advice on the identification and evaluation of historic properties, including those of traditional religious and cultural importance, and, if necessary, participate in the resolution of any adverse effects to such properties.

In the context of the National Historic Preservation Act of 1966, as amended, the APE for this ESP review is the area at the power plant site and its immediate environs that may be impacted by land-disturbing activities associated with the construction and operation of the new unit(s), and construction and operation of new transmission lines that may follow parallel with some of the existing transmission line systems now serving CPS. The new lines would run from the Clinton ESP site to an interconnect point at the Brokaw substation near Bloomington, Illinois, about 23 miles north of the site; and from the Clinton ESP site to the Oreana substation, about 8 miles south of the site. The power plant site is located in DeWitt County, Illinois, and the transmission line corridors traverse DeWitt and McLean counties. The application is available through the web-based version of the NRC's Agencywide Documents Access and Management System (ADAMS) which can be found at http://www.nrc.gov/reading-rm/adams.html. The application is listed under accession number ML032721596.

In its application, Exelon refers to a comprehensive cultural resource and historic property investigation that was performed prior to the construction of CPS, approximately 30 years ago. Exelon further states that any issues that were raised were resolved through removal of these historic and cultural resources. No historic standing structures have been identified within the Exelon ESP power block footprint, cooling tower footprint, or within the immediate vicinity of CPS. Exelon states that the location of the ESP facility power block footprint appears to have been heavily disturbed by previous construction of CPS. If the power block or cooling tower footprint area was expanded or moved, there is a potential for impact to historic properties. The applicant has committed to perform further evaluation to determine if additional archaeological review is required if additional area within the ESP site will be required for development.

As communicated to Mr. Curtis Simon of your staff by voice mail on December 10, 2003, the NRC conducted a public environmental scoping meeting on December 18, 2003 at the Revere Ware Room in the Vespasian Warner Public Library, 310 N. Quincy Street, Clinton, Illinois. Representatives of your tribe were invited to attend.

Please submit any written comments your tribe may have to offer on the scope of the environmental review by January 9, 2004. Comments should be submitted either by mail to the Chief, Rules and Directives Branch, Division of Administrative Services, Mail Stop T-6 D59, U.S. Nuclear Regulatory Commission, Washington, D.C. 20555-0001, or by Internet to *ClintonEIS@nrc.gov*.

The Honorable C. Anske   -3-

At the conclusion of the scoping process, the NRC staff will prepare a summary of the significant issues identified and the conclusions reached, and will send a copy to you. In addition, after it is issued, your office will receive a copy of the draft EIS along with a request for comments. The anticipated publication date for the draft EIS is December 2004. If you have any questions or require additional information, please contact the Environmental Project Manager for the Clinton ESP project, Mr. Thomas Kenyon, at 301-415-1120 or *ClintonEIS@nrc.gov.*

                            Sincerely,
                            /RA/
                            Pao-Tsin Kuo, Program Director
                            License Renewal and Environmental Impacts
                            Division of Regulatory Improvement Programs
                            Office of Nuclear Reactor Regulation

Docket No.: 52-007

cc: See next page

December 30, 2003

The Honorable Lawrence F. Snake, President
Delaware Tribe of Western Oklahoma
Post Office Box 825
Anadardo, OK 73005

SUBJECT: EARLY SITE PERMIT (ESP) REVIEW FOR THE CLINTON ESP SITE

Dear President Snake:

The U.S. Nuclear Regulatory Commission (NRC) staff is reviewing an application for an ESP to set aside a site for the potential future construction of one or more new nuclear power plants. The application was submitted by Exelon Generation Company, LLC (Exelon), on September 25, 2003, pursuant to NRC requirements at Title 10 of the *Code of Federal Regulations* Part 52 (10 CFR Part 52). If built, the new unit(s) would be co-located with the existing Clinton Power Station (CPS) site near the town of Clinton in DeWitt County, Illinois. Exelon has also included a site redress plan in its application in accordance with 10 CFR 52.17(c) and 52.25. If a site redress plan is incorporated in an ESP approved by the NRC, the applicant may carry out certain site preparation and limited construction activities. Exelon would still be required to obtain the appropriate local, State, and other Federal permits required for these activities before starting work.

As part of its review of the application, the NRC staff will prepare an environmental impact statement (EIS) under the provisions of 10 CFR Part 51, the NRC rules that implement the National Environmental Policy Act of 1969 (NEPA). The NRC environmental review process includes an opportunity for public participation in the environmental review. The Clinton ESP site is located on land that may be of interest to the Delaware Tribe of Western Oklahoma. We want to ensure that you are aware of our efforts and, pursuant to our regulations at 10 CFR 51.28(b), the NRC invites the Delaware Tribe of Western Oklahoma to provide input to the scoping process relating to the NRC's environmental review of the application. The following is a description of the application and the environmental review process.

The EIS will document the NRC staff's determination regarding the suitability of the proposed site for the construction and operation of one or more new nuclear plants. In addition, the staff will also consider alternatives to the proposed action, including alternative sites. The EIS will contain the results of the review of the environmental impacts on the area surrounding the Clinton ESP site that are related to terrestrial ecology, aquatic ecology, hydrology, socioeconomic issues, and historic properties (among others), and will contain a recommendation regarding the environmental acceptability of granting an ESP. If approved, the ESP would not authorize the applicant to begin construction of the unit(s).

As part of this review, and in accordance with 36 CFR 800.8, the EIS will include analyses of potential impacts to historic properties. Accordingly, pursuant to 10 CFR 51.28 and 36 CFR 800.2(c)(2), the NRC wishes to ensure that Indian tribes that might have an interest in any potential historic properties in the area of potential effect (APE) are afforded the opportunity to identify their concerns, provide advice on the identification and evaluation of historic properties,

The Honorable L. Snake   -2-

including those of traditional religious and cultural importance, and, if necessary, participate in the resolution of any adverse effects to such properties.

In the context of the National Historic Preservation Act of 1966, as amended, the APE for this ESP review is the area at the power plant site and its immediate environs that may be impacted by land-disturbing activities associated with the construction and operation of the new unit(s), and construction and operation of new transmission lines that may follow parallel with some of the existing transmission line systems now serving CPS. The new lines would run from the Clinton ESP site to an interconnect point at the Brokaw substation near Bloomington, Illinois, about 23 miles north of the site; and from the Clinton ESP site to the Oreana substation, about 8 miles south of the site. The power plant site is located in DeWitt County, Illinois, and the transmission line corridors traverse DeWitt and McLean counties. The application is available through the web-based version of the NRC's Agencywide Documents Access and Management System (ADAMS) which can be found at http://www.nrc.gov/reading-rm/adams.html. The application is listed under accession number ML032721596.

In its application, Exelon refers to a comprehensive cultural resource and historic property investigation that was performed prior to the construction of CPS, approximately 30 years ago. Exelon further states that any issues that were raised were resolved through removal of these historic and cultural resources. No historic standing structures have been identified within the Exelon ESP power block footprint, cooling tower footprint, or within the immediate vicinity of CPS. Exelon states that the location of the ESP facility power block footprint appears to have been heavily disturbed by previous construction of CPS. If the power block or cooling tower footprint area was expanded or moved, there is a potential for impact to historic properties. The applicant has committed to perform further evaluation to determine if additional archaeological review is required if additional area within the ESP site will be required for development.

As discussed with Ms. P. Wahahrockahtasi of your staff on December 5, 2003, the NRC conducted a public environmental scoping meeting on December 18, 2003 at the Revere Ware Room in the Vespasian Warner Public Library, 310 N. Quincy Street, Clinton, Illinois. Representatives of your tribe were invited to attend.

Please submit any written comments your tribe may have to offer on the scope of the environmental review by January 9, 2004. Comments should be submitted either by mail to the Chief, Rules and Directives Branch, Division of Administrative Services, Mail Stop T-6 D59, U.S. Nuclear Regulatory Commission, Washington, D.C. 20555-0001, or by Internet to *ClintonEIS@nrc.gov*.

The Honorable L. Snake   -3-

At the conclusion of the scoping process, the NRC staff will prepare a summary of the significant issues identified and the conclusions reached, and will send a copy to you. In addition, after it is issued, your office will receive a copy of the draft EIS along with a request for comments. The anticipated publication date for the draft EIS is December 2004. If you have any questions or require additional information, please contact the Environmental Project Manager for the Clinton ESP project, Mr. Thomas Kenyon, at 301-415-1120 or *ClintonEIS@nrc.gov.*

       Sincerely,
       /RA?
       Pao-Tsin Kuo, Program Director
       License Renewal and Environmental Impacts
       Division of Regulatory Improvement Programs
       Office of Nuclear Reactor Regulation

Docket No.: 52-007

cc: See next page

Appendix F

December 30, 2003

The Honorable John P. Froman, Chief
The Peoria Tribe of Indians of Oklahoma
118 S. Eight Tribes Trail
P.O. Box 1527
Miami, OK 74355

SUBJECT: EARLY SITE PERMIT (ESP) REVIEW FOR THE CLINTON ESP SITE

Dear Chief Froman:

The U.S. Nuclear Regulatory Commission (NRC) staff is reviewing an application for an ESP to set aside a site for the potential future construction of one or more new nuclear power plants. The application was submitted by Exelon Generation Company, LLC (Exelon), on September 25, 2003, pursuant to NRC requirements at Title 10 of the *Code of Federal Regulations* Part 52 (10 CFR Part 52). If built, the new unit(s) would be co-located with the existing Clinton Power Station (CPS) site near the town of Clinton in DeWitt County, Illinois. Exelon has also included a site redress plan in its application in accordance with 10 CFR 52.17(c) and 52.25. If a site redress plan is incorporated in an ESP approved by the NRC, the applicant may carry out certain site preparation and limited construction activities. Exelon would still be required to obtain the appropriate local, State, and other Federal permits required for these activities before starting work.

As part of its review of the application, the NRC staff will prepare an environmental impact statement (EIS) under the provisions of 10 CFR Part 51, the NRC rules that implement the National Environmental Policy Act of 1969 (NEPA). The NRC environmental review process includes an opportunity for public participation in the environmental review. The Clinton ESP site is located on land that may be of interest to the Peoria Tribe of Indians of Oklahoma. We want to ensure that you are aware of our efforts and, pursuant to our regulations at 10 CFR 51.28(b), the NRC invites the Peoria Tribe of Indians of Oklahoma to provide input to the scoping process relating to the NRC's environmental review of the application. The following is a description of the application and the environmental review process.

The EIS will document the NRC staff's determination regarding the suitability of the proposed site for the construction and operation of one or more new nuclear plants. In addition, the staff will also consider alternatives to the proposed action, including alternative sites. The EIS will contain the results of the review of the environmental impacts on the area surrounding the Clinton ESP site that are related to terrestrial ecology, aquatic ecology, hydrology, socioeconomic issues, and historic properties (among others), and will contain a recommendation regarding the environmental acceptability of granting an ESP. If approved, the ESP would not authorize the applicant to begin construction of the unit(s).

As part of this review, and in accordance with 36 CFR 800.8, the EIS will include analyses of potential impacts to historic properties. Accordingly, pursuant to 10 CFR 51.28 and 36 CFR 800.2(c)(2), the NRC wishes to ensure that Indian tribes that might have an interest in any potential historic properties in the area of potential effect (APE) are afforded the opportunity to

The Honorable J. Froman   -2-

identify their concerns, provide advice on the identification and evaluation of historic properties, including those of traditional religious and cultural importance, and, if necessary, participate in the resolution of any adverse effects to such properties.

In the context of the National Historic Preservation Act of 1966, as amended, the APE for this ESP review is the area at the power plant site and its immediate environs that may be impacted by land-disturbing activities associated with the construction and operation of the new unit(s), and construction and operation of new transmission lines that may follow parallel with some of the existing transmission line systems now serving CPS. The new lines would run from the Clinton ESP site to an interconnect point at the Brokaw substation near Bloomington, Illinois, about 23 miles north of the site; and from the Clinton ESP site to the Oreana substation, about 8 miles south of the site. The power plant site is located in DeWitt County, Illinois, and the transmission line corridors traverse DeWitt and McLean counties. The application is available through the web-based version of the NRC's Agencywide Documents Access and Management System (ADAMS) which can be found at http://www.nrc.gov/reading-rm/adams.html. The application is listed under accession number ML032721596.

In its application, Exelon refers to a comprehensive cultural resource and historic property investigation that was performed prior to the construction of CPS, approximately 30 years ago. Exelon further states that any issues that were raised were resolved through removal of these historic and cultural resources. No historic standing structures have been identified within the Exelon ESP power block footprint, cooling tower footprint, or within the immediate vicinity of CPS. Exelon states that the location of the ESP facility power block footprint appears to have been heavily disturbed by previous construction of CPS. If the power block or cooling tower footprint area was expanded or moved, there is a potential for impact to historic properties. The applicant has committed to perform further evaluation to determine if additional archaeological review is required if additional area within the ESP site will be required for development.

As Mr. Kenyon discussed with you on December 5, 2003, the NRC conducted a public environmental scoping meeting on December 18, 2003 at the Revere Ware Room in the Vespasian Warner Public Library, 310 N. Quincy Street, Clinton, Illinois. Representatives of your tribe were invited to attend.

Please submit any written comments your tribe may have to offer on the scope of the environmental review by January 9, 2004. Comments should be submitted either by mail to the Chief, Rules and Directives Branch, Division of Administrative Services, Mail Stop T-6 D59, U.S. Nuclear Regulatory Commission, Washington, D.C. 20555-0001, or by Internet to *ClintonEIS@nrc.gov*.

The Honorable J. Froman   -3-

At the conclusion of the scoping process, the NRC staff will prepare a summary of the significant issues identified and the conclusions reached, and will send a copy to you. In addition, after it is issued, your office will receive a copy of the draft EIS along with a request for comments. The anticipated publication date for the draft EIS is December 2004. If you have any questions or require additional information, please contact the Environmental Project Manager for the Clinton ESP project, Mr. Thomas Kenyon, at 301-415-1120 or *ClintonEIS@nrc.gov.*

> Sincerely,
> **/RA/**
> Pao-Tsin Kuo, Program Director
> License Renewal and Environmental Impacts
> Division of Regulatory Improvement Programs
> Office of Nuclear Reactor Regulation

Docket No.: 52-007

cc: See next page

December 30, 2003

The Honorable Dee Ketchum, Chief
Eastern Delaware Tribe
220 Northwest Virginia
Bartlesville, OK 74003

SUBJECT: EARLY SITE PERMIT (ESP) REVIEW FOR THE CLINTON ESP SITE

Dear Chief Ketchum:

The U.S. Nuclear Regulatory Commission (NRC) staff is reviewing an application for an ESP to set aside a site for the potential future construction of one or more new nuclear power plants. The application was submitted by Exelon Generation Company, LLC (Exelon), on September 25, 2003, pursuant to NRC requirements at Title 10 of the *Code of Federal Regulations* Part 52 (10 CFR Part 52). If built, the new unit(s) would be co-located with the existing Clinton Power Station (CPS) site near the town of Clinton in DeWitt County, Illinois. Exelon has also included a site redress plan in its application in accordance with 10 CFR 52.17(c) and 52.25. If a site redress plan is incorporated in an ESP approved by the NRC, the applicant may carry out certain site preparation and limited construction activities. Exelon would still be required to obtain the appropriate local, State, and other Federal permits required for these activities before starting work.

As part of its review of the application, the NRC staff will prepare an environmental impact statement (EIS) under the provisions of 10 CFR Part 51, the NRC rules that implement the National Environmental Policy Act of 1969 (NEPA). The NRC environmental review process includes an opportunity for public participation in the environmental review. The Clinton ESP site is located on land that may be of interest to the Eastern Delaware Tribe. We want to ensure that you are aware of our efforts and, pursuant to our regulations at 10 CFR 51.28(b), the NRC invites the Eastern Delaware Tribe to provide input to the scoping process relating to the NRC's environmental review of the application. The following is a description of the application and the environmental review process.

The EIS will document the NRC staff's determination regarding the suitability of the proposed site for the construction and operation of one or more new nuclear plants. In addition, the staff will also consider alternatives to the proposed action, including alternative sites. The EIS will contain the results of the review of the environmental impacts on the area surrounding the Clinton ESP site that are related to terrestrial ecology, aquatic ecology, hydrology, socioeconomic issues, and historic properties (among others), and will contain a recommendation regarding the environmental acceptability of granting an ESP. If approved, the ESP would not authorize the applicant to begin construction of the unit(s).

As part of this review, and in accordance with 36 CFR 800.8, the EIS will include analyses of potential impacts to historic properties. Accordingly, pursuant to 10 CFR 51.28 and 36 CFR 800.2(c)(2), the NRC wishes to ensure that Indian tribes that might have an interest in any potential historic properties in the area of potential effect (APE) are afforded the opportunity to identify their concerns, provide advice on the identification and evaluation of historic properties,

The Honorable D. Ketchum   -2-

including those of traditional religious and cultural importance, and, if necessary, participate in the resolution of any adverse effects to such properties.

In the context of the National Historic Preservation Act of 1966, as amended, the APE for this ESP review is the area at the power plant site and its immediate environs that may be impacted by land-disturbing activities associated with the construction and operation of the new unit(s), and construction and operation of new transmission lines that may follow parallel with some of the existing transmission line systems now serving CPS. The new lines would run from the Clinton ESP site to an interconnect point at the Brokaw substation near Bloomington, Illinois, about 23 miles north of the site; and from the Clinton ESP site to the Oreana substation, about 8 miles south of the site. The power plant site is located in DeWitt County, Illinois, and the transmission line corridors traverse DeWitt and McLean counties. The application is available through the web-based version of the NRC's Agencywide Documents Access and Management System (ADAMS) which can be found at http://www.nrc.gov/reading-rm/adams.html. The application is listed under accession number ML032721596.

In its application, Exelon refers to a comprehensive cultural resource and historic property investigation that was performed prior to the construction of CPS, approximately 30 years ago. Exelon further states that any issues that were raised were resolved through removal of these historic and cultural resources. No historic standing structures have been identified within the Exelon ESP power block footprint, cooling tower footprint, or within the immediate vicinity of CPS. Exelon states that the location of the ESP facility power block footprint appears to have been heavily disturbed by previous construction of CPS. If the power block or cooling tower footprint area was expanded or moved, there is a potential for impact to historic properties. The applicant has committed to perform further evaluation to determine if additional archaeological review is required if additional area within the ESP site will be required for development.

As discussed with Mr. Obemeyer of your staff on December 11, 2003, the NRC conducted a public environmental scoping meeting on December 18, 2003 at the Revere Ware Room in the Vespasian Warner Public Library, 310 N. Quincy Street, Clinton, Illinois. Representatives of your tribe were invited to attend.

Please submit any written comments your tribe may have to offer on the scope of the environmental review by January 9, 2004. Comments should be submitted either by mail to the Chief, Rules and Directives Branch, Division of Administrative Services, Mail Stop T-6 D59, U.S. Nuclear Regulatory Commission, Washington, D.C. 20555-0001, or by Internet to ClintonEIS@nrc.gov.

The Honorable D. Ketchum   -3-

At the conclusion of the scoping process, the NRC staff will prepare a summary of the significant issues identified and the conclusions reached, and will send a copy to you. In addition, after it is issued, your office will receive a copy of the draft EIS along with a request for comments. The anticipated publication date for the draft EIS is December 2004. If you have any questions or require additional information, please contact the Environmental Project Manager for the Clinton ESP project, Mr. Thomas Kenyon, at 301-415-1120 or *ClintonEIS@nrc.gov.*

                              Sincerely,
                              /RA/
                              Pao-Tsin Kuo, Program Director
                              License Renewal and Environmental Impacts
                              Division of Regulatory Improvement Programs
                              Office of Nuclear Reactor Regulation

Docket No.: 52-007

cc: See next page

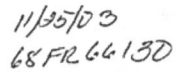

11/25/03
68 FR 66130
20

DELAWARE TRIBE OF INDIANS

220 N.W. VIRGINIA • BARTLESVILLE, OKLAHOMA 74003
TELEPHONE: (918) 336-5272 • FAX: (918) 336-5513

January 13, 2004

Chief
Attn: Rules and Directives Branch
Division of Administrative Services
Mail Stop T-6 D59
U.S. Nuclear Regulatory Commission
Washington, D.C. 20555-0001

Re: Early Site Permit (ESP) Review For the Clinton ESP Site

Dear Chief,

Our review indicates that this project is located in an area that was not
inhabited by the Delaware Tribe. As such, there is little potential for
impacting unknown archaeological sites culturally affiliated with the
Delaware Tribe and we have no particular objection to the proposal.

If you have any questions, feel free to contact this office by phone at
(918) 336-5272.

Sincerely,

Brice Obermeyer

Brice Obermeyer
NAGPRA Director
Delaware Tribe of Indians

E-RTDS = ADU 03
adu = T. Kenyon (TJK0)

Template = ADU-03

March 17, 2004

Ms. Patricia A. Kurkul, Regional Administrator
NOAA Fisheries
Northeast Regional Office
One Blackburn Drive
Gloucester, MA 01930-2298

SUBJECT: APPLICATION FOR AN EARLY SITE PERMIT FOR THE CLINTON ESP SITE

Dear Ms. Kurkul:

The U.S. Nuclear Regulatory Commission (NRC) staff is reviewing an application submitted by
Exelon Generation Company, LLC (Exelon) for an early site permit (ESP) for the potential future
construction of one or more new nuclear power plants. As part of the review of this application
the NRC is preparing an environmental impact statement (EIS). The impact analyses in the EIS
will include the potential impacts of the construction and operation of one or more new nuclear
power plants at the preferred site or at one of six alternate sites, including the potential impacts
to fish and wildlife and threatened and endangered species.

The preferred location of the new unit(s), if built, would be the existing Clinton Power Station
(CPS) site on Lake Clinton near the town of Clinton in DeWitt County, Illinois, approximately
6 miles east of the City of Clinton along the shore of Clinton Lake. Exelon also has a list of six
alternate sites that will be evaluated in the EIS. The six alternate sites are existing generating
stations owned and operated by Exelon: Braidwood Nuclear Station, in Will County, Illinois (on
Braidwood Lake); Zion Nuclear Power Station, in Lake County, Illinois (on the western shore of
Lake Michigan); Byron Generating Station, in Ogle County, Illinois (on the Rock River); Dresden
Nuclear Power Station, in Grundy County, Illinois (on the south shoreline of the Illinois River at
the confluence of the Des Plaines and Kankakee Rivers); LaSalle County Generating Station, in
LaSalle County, Illinois (on LaSalle Lake); and Quad Cities Nuclear Power Station, in Rock
Island County, Illinois (on the east bank of Pool 14 of the Mississippi River).

The application for an ESP was submitted by Exelon on September 25, 2003, pursuant to NRC
requirements at Title 10 of the *Code of Federal Regulations* Part 52 (10 CFR 52). If approved,
the ESP will document the NRC staff's determination regarding the suitability of the proposed
site for the construction and operation of one or more new nuclear plants. The ESP would not
authorize the applicant to begin construction of the new unit(s). However, in its review the NRC
staff will evaluate the environmental impacts of construction and operation and will also
consider alternatives, including the above alternative sites.

P. Kurkul   2

To support the EIS preparation process and to ensure compliance with Section 7 of the Endangered Species Act of 1973, the NRC requests a list of endangered, threatened, candidate, and proposed species, and designated and proposed critical habitat under the jurisdiction of NOAA Fisheries, that may be in the vicinity of the CPS site and its transmission line corridors (McLean and DeWitt counties only), and the six alternate sites listed above. In addition, please provide any information you consider appropriate under the provisions of the Fish and Wildlife Coordination Act of 1934.

If you have any questions concerning the ESP application or other aspects of this project, please contact Mr. Thomas Kenyon, Senior Environmental Senior Project Manager, at (301) 415-1120 or by e-mail at TJK2@nrc.gov.

Sincerely,
/RA/
Pao-Tsin Kuo, Program Director
License Renewal and Environmental Impacts
Division of Regulatory Improvement Programs
Office of Nuclear Reactor Regulation

Docket No.: 52-007

cc: See next page

March 17, 2004

Mr. John Rogner, Field Supervisor
U.S. Fish and Wildlife Service
1250 S. Grove Avenue, Suite 103
Barrington, IL 60010

SUBJECT: APPLICATION FOR AN EARLY SITE PERMIT FOR THE CLINTON ESP SITE

Dear Mr. Rogner:

The U.S. Nuclear Regulatory Commission (NRC) staff is reviewing an application submitted by Exelon Generation Company, LLC (Exelon) for an early site permit (ESP) for the potential future construction of one or more new nuclear power plants. As part of the review of this application the NRC is preparing an environmental impact statement (EIS). The impact analysis in the EIS includes the potential impacts of the construction and operation of one or more new nuclear power plants at the preferred or at one of six alternate sites, including the potential impacts to fish and wildlife and threatened and endangered species.

The preferred location of the new unit(s), if built, would be the existing Clinton Power Station (CPS) site near the town of Clinton in DeWitt County, Illinois, approximately 6 miles east of the City of Clinton along the shore of Clinton Lake. Two of the six alternate sites that will be evaluated in the EIS are located within the area serviced by your office. The two alternate sites are existing generating stations owned and operated by Exelon; Braidwood Nuclear Station, in Will County, Illinois and Zion Nuclear Power Station, in Lake County, Illinois. Please note that the NRC will submit separate correspondence to the Rock Island Ecological Field Services office regarding the other four alternate sites; Byron Generating Station, in Ogle County, Illinois; Dresden Nuclear Power Station, in Grundy County, Illinois; LaSalle County Generating Station, in LaSalle County, Illinois; and Quad Cities Nuclear Power Station, in Rock Island County, Illinois.

The application for an ESP was submitted by Exelon on September 25, 2003, pursuant to NRC requirements at Title 10 of the *Code of Federal Regulations* Part 52 (10 CFR 52). If approved, the ESP will document the NRC staff's determination regarding the suitability of the proposed site for the construction and operation of one or more new nuclear plants. The ESP would not authorize the applicant to begin construction of the new unit(s). However, in its review the NRC staff will evaluate the environmental impacts of construction and operation and will also consider alternatives, including the above alternative sites.

J. Rogner   2

To support the EIS preparation process and to ensure compliance with Section 7 of the Endangered Species Act of 1973, the NRC requests a list of endangered, threatened, candidate, and proposed species, and designated and proposed critical habitat, that may be in the vicinity of the two alternate sites. In addition, please provide any information you consider appropriate under the provisions of the Fish and Wildlife Coordination Act of 1934.

If you have any questions concerning the ESP application or other aspects of this project, please contact Mr. Thomas Kenyon, Environmental Project Manager, at (301) 415-1120 or by e-mail at TJK2@nrc.gov.

> Sincerely,
> /RA/
> Pao-Tsin Kuo, Program Director
> License Renewal and Environmental Impacts
> Division of Regulatory Improvement Programs
> Office of Nuclear Reactor Regulation

Docket No.: 52-007

cc: See next page

March 17, 2004

Mr. Richard Nelson, Field Supervisor
U.S. Fish and Wildlife Service
4469 48th Avenue Court
Rock Island, IL 61201

SUBJECT: APPLICATION FOR AN EARLY SITE PERMIT FOR THE CLINTON ESP SITE

Dear Mr. Nelson:

The U.S. Nuclear Regulatory Commission (NRC) staff is reviewing an application submitted by
Exelon Generation Company, LLC (Exelon) for an early site permit (ESP) for the potential future
construction of one or more new nuclear power plants. As part of the review of this application
the NRC is preparing an environmental impact statement (EIS). The impact analysis in the EIS
includes the potential impacts of the construction and operation of one or more new nuclear
power plants at the preferred or at one of six alternate sites, including the potential impacts to
fish and wildlife and threatened and endangered species.

The preferred location of the new unit(s), if built, would be the existing Clinton Power Station
(CPS) site near the town of Clinton in DeWitt County, Illinois, approximately 6 miles east of the
City of Clinton along the shore of Clinton Lake. Four of the six alternate sites that will be
evaluated in the EIS are located within the area serviced by your office. The four alternate sites
are existing generating stations owned and operated by Exelon: Byron Generating Station, in
Ogle County, Illinois; Dresden Nuclear Power Station, in Grundy County, Illinois; LaSalle County
Generating Station, in LaSalle County, Illinois; and Quad Cities Nuclear Power Station, in Rock
Island County, Illinois. Please note that the NRC will submit separate correspondence to the
Chicago Ecological Field Services office regarding the other two alternate sites; Braidwood
Nuclear Station, in Will County, Illinois and Zion Nuclear Power Station, in Lake County, Illinois.

The application for an ESP was submitted by Exelon on September 25, 2003, pursuant to NRC
requirements at Title 10 of the *Code of Federal Regulations* Part 52 (10 CFR 52). If approved,
the ESP will document the NRC staff's determination regarding the suitability of the proposed
site for the construction and operation of one or more new nuclear plants. The ESP would not
authorize the applicant to begin construction of the new unit(s). However, in its review the NRC
staff will evaluate the environmental impacts of construction and operation and will also
consider alternatives, including the above alternative sites.

R. Nelson   2

To support the EIS preparation process and to ensure compliance with Section 7 of the Endangered Species Act of 1973, the NRC requests a list of endangered, threatened, candidate, and proposed species, and designated and proposed critical habitat, that may be in the vicinity of the CPS site and its transmission line corridors (McLean and DeWitt counties only), and the four alternate sites. In addition, please provide any information you consider appropriate under the provisions of the Fish and Wildlife Coordination Act of 1934.

If you have any questions concerning the ESP application or other aspects of this project, please contact Mr. Thomas Kenyon, Environmental Project Manager, at (301) 415-1120 or by e-mail at TJK2@nrc.gov.

Sincerely,
/RA/
Pao-Tsin Kuo, Program Director
License Renewal and Environmental Impacts
Division of Regulatory Improvement Programs
Office of Nuclear Reactor Regulation

Docket No.: 52-007

cc: See next page

United States Department of the Interior

FISH AND WILDLIFE SERVICE
Rock Island Field Office
4469 48th Avenue Court
Rock Island, Illinois 61201
Phone: (309) 793-5800 Fax: (309) 793-5804

IN REPLY REFER
TO.
FWS/RIFO

52-007

April 6, 2004

Mr. Pao-Tsin Kuo, Program Director
License Renewal and Environmental Impacts
U.S. Nuclear Regulatory Commission
Washington, DC 20555-0001

Dear Mr. Kuo:

This responds to your letter of March 17, 2004, requesting our comments on the application
for an early site permit for the Clinton ESP Site submitted by Exelon Generation Company,
LLC. The preferred location of the new unit(s), if built, would be the existing Clinton Power
Station (CPS) site near the town of Clinton in DeWitt County, Illinois. In this letter we will
provide information regarding the presence of threatened and/or endangered species for
DeWitt County, as well as Ogle, Grundy, LaSalle and Rock Island Counties.

To facilitate compliance with Section 7(c) of the Endangered Species Act of 1973, as amended,
Federal agencies are required to obtain from the Fish and Wildlife Service information
concerning any species, listed or proposed to be listed, which may be present in the area of a
proposed action. Therefore, we are furnishing you the following list of species which may be
present in the concerned area:

Classification	Common Name (Scientific Name)	Habitat
Protected	Bald eagle (*Haliaeetus leucocephalus*)	Breeding, wintering
Endangered	Indiana bat (*Myotis sodalis*)	Caves, mines (hibernacula); small stream corridors with well developed riparian woods; upland forests (foraging)
Endangered	Karner blue butterfly (*Lycaeides Melissa samuelis*)	Pine barrens and oak savannas on sandy soils and containing wild lupines

DO73

Mr. Pao-Tsin Kuo   2

		(*Lupinus perennis*), the only known food plant of the larvae
Endangered	Higgins' eye pearly mussel (*Lampsilis higginsi*)	Mississippi River; Rock River to Steel Dam
Threatened	Prairie bush clover (*Lespedeza leptostachya*)	Dry to mesic prairies with gravelly soil
Threatened	Eastern prairie fringed orchid (*Platanthaera leucophaea*)	Mesic to wet prairies

The threatened bald eagle (*Haliaeetus leucocephalus*) is listed as breeding Ogle County. It is also listed as wintering along large rivers, lakes and reservoirs in DeWitt, Grundy, LaSalle, Ogle, and *Rock Island Counties in Illinois (* counties that contain night roosts).

During the winter, this species feeds on fish in the open water areas created by dam tailwaters, the warm water effluents of power plants and municipal and industrial discharges, or in power plant cooling ponds. The more severe the winter, the greater the ice coverage and the more concentrated the eagles become. They roost at night in groups in large trees adjacent to the river in areas that are protected from the harsh winter elements. They perch in large shoreline trees to rest or feed on fish. There is no critical habitat designated for this species. The eagle may not be harassed, harmed, or disturbed when present nor may nest trees be cleared.

The endangered Indiana bat (*Myotis sodalis*) is known to occur in LaSalle County, Illinois. **Potential habitat for this species occurs statewide, therefore, Indiana bats are considered to potentially occur in any area with forested habitat.**

Indiana bats migrate seasonally between winter hibernacula and summer roosting habitats. Winter hibernacula include caves and abandoned mines. Females emerge from hibernation in late March or early April to migrate to summer roosts. Females form nursery colonies under the loose bark of trees (dead or alive) and/or cavities, where each female gives birth to a single young in June or early July. A maternity colony may include from one or more individuals. A single colony may utilize a number of roost trees during the summer, typically a primary roost tree and several alternates. Some males remain in the area near the winter hibernacula during the summer months, but others disperse throughout the range of the species and roost individually or in small numbers in the same types of trees as females. The species or size of tree does not appear to influence whether Indiana bats utilize a tree for roosting provided the appropriate bark structure is present. However, the use of a particular tree does appear to be influenced by weather conditions, such as temperature and precipitation.
During the summer, the Indiana bat frequents the corridors of small streams with riparian woods as well as mature upland forests. It forages for insects along stream corridors, within the canopy of floodplain and upland forests, over clearings with early successional vegetation

:

Mr. Pao-Tsin Kuo   3

(old fields), along the borders of croplands, along wooded fencerows, over farm ponds and in pastures.

Suitable summer habitat in Illinois is considered to have the following characteristics within a ½ mile radius of a project site:

1)   forest cover of 15% or greater;
2)   permanent water;
3)   one or more of the following tree species: shagbark and shellbark hickory that may be dead or alive, and dead bitternut hickory, American elm, slippery elm, eastern cottonwood, silver maple, white oak, red oak, post oak, and shingle oak with slabs or plates of loose bark;
4)   potential roost trees with 10% or more peeling or loose bark

If the project site contains any habitat that fits the above description, it may be necessary to conduct a survey to determine whether the bat is present. If Indiana bats are known to be present, they must not be harmed, harassed or disturbed when present. Large-scale habitat alterations within known or potential Indiana bat habitat should not be permitted without a bat survey and/or consultation with this office as indicated below.

If the project site contains any habitat that fits the above description, it may be necessary to conduct a survey to determine whether the bat is present. If Indiana bats are known to be present, they must not be harmed, harassed or disturbed when present. Large-scale habitat alterations within known or potential Indiana bat habitat should not be conducted without a bat survey and consultation with this office. "Mist Netting Guidelines" can be obtained from our office.

Minor alterations of Indiana bat habitat (i.e., timber stand improvement or clearing of small stands) should be limited to non-maternity periods between the dates of September 16 and April 14.

The endangered Karner blue butterfly (*Lycaeides melissa samuelis*) is currently known to occur only in Lake County, Illinois. However, potential habitat may be found in Ogle County, based on the historic distribution of the wild lupine plant *Lupinus perennis*. This plant is the only known food source for the larval stage of this species.

The endangered Higgins' eye pearly mussel (*Lampsilis higginsi*) is listed for the Mississippi River north of Lock and Dam 20 which includes Rock Island County, Illinois. This species prefers sand/gravel substrates with a swift current and is most often found in the main channel border or an open, flowing side channel.

While there is no designated critical habitat, the Higgins' eye Recovery Team has designated habitats essential to the recovery of the species. These areas include Cordova, Rock Island

Mr. Pao-Tsin Kuo   4

County, Illinois (river mile 503-505.4L); and Sylvan Slough, Rock Island, Illinois (river mile 485.4-486L).

The State of Illinois has also designated certain mussel refuge areas that contain this species. Their regulations would affect the commercial harvest of mussels on these refuges. If project is located near a known Higgins' eye mussel bed, it may be necessary to conduct a survey to determine the presence of the species.

The prairie bush clover (*Lespedeza leptostachya*) is listed as threatened in Ogle County, Illinois. It occupies dry to mesic prairies with gravelly soil. There is no critical habitat designated for this species. Federal regulations prohibit any commercial activity involving this species or the destruction, malicious damage or removal of this species from Federal land or any other lands in knowing violation of State law or regulation, including State criminal trespass law. This species should be searched for whenever prairie remnants are encountered.

The eastern prairie fringed orchid (*Platanthera leucophaea*) is listed as threatened for Grundy County, Illinois. It may potentially occur in Ogle County, Illinois. It occupies wet grassland habitats. There is no critical habitat designated for this species. Federal regulations prohibit any commercial activity involving this species or the destruction, malicious damage or removal of this species from Federal land or any other lands in knowing violation of State law or regulation, including State criminal trespass law. This species should be searched for whenever wet prairie remnants are encountered.

These comments provide technical assistance only and do not fulfill the requirements under Section 7 of the Endangered Species Act of 1973, as amended, unless you have been designated, in writing, to the Regional Director of the U.S. Fish and Wildlife Service, Region 3, by the appropriate Federal agency, as a non-Federal representative for the purposes of conducting informal consultation on the subject Federal action, pursuant to 50 CFR 402.08. This letter provides comments under the authority of and in accordance with provisions of the Fish and Wildlife Coordination Act (48 Stat. 401, as amended; 16 U.S.C. 661 et seq.). If you have questions, please contact Heidi Woeber of my staff at (309)793-5800, ext. 209.

Sincerely,

Richard C. Nelson
Supervisor

S:\Office Users\Heidi\tanuke.doc

United States Department of the Interior

FISH AND WILDLIFE SERVICE
Chicago Ecological Services Field Office
1250 South Grove Avenue, Suite 103
Barrington, Illinois 60010
Phone: (847) 381-2253   Fax: (847) 381-2285

52-007

IN REPLY REFER TO:
FWS/AES-CIFO/4-1200

April 12, 2004

Mr. / Ms. Pao-Tsin Kuo
United States Nuclear Regulatory Commission
Washington, D.C. 20555-0001

Dear Sir or Madame:

This responds to your letter dated March 17, 2004 requesting information on endangered and threatened species on or near two proposed alternate sites for one or more nuclear power plants by Exelon Generation Company, LLC. Both alternate sites are existing generating stations owned and operated by Exelon; Braidwood Nuclear Station in Will County, Illinois and Zion Nuclear Power Station in Lake County, Illinois.

The Zion Nuclear Power Station in Lake County, Illinois appears to lie directly adjacent to Illinois Beach State Park where we have three federal species of concern and one species with critical habitat designation.

The federally threatened eastern prairie white fringed orchid (*Plantanthera leucophaea*) is located at Illinois Beach State Park. Possible habitat of the Eastern prairie white fringed orchid includes but is not restricted to mesic prairie, sedge meadows, marsh edges and bogs. Soils of these habitats include glacial soils, lake plain deposits, muck, and peat. We request that a search for these types of habitat be conducted. If any of these aforementioned habitat remnants are found within any of the project areas, we request that searches for this species be conducted between June 28 and July 11, as this is when the orchid typically flowers and is most identifiable. If any eastern prairie white fringed orchids are found, this office should be notified immediately.

The federally endangered Karner blue butterfly *(Lycaeides Melissa samuelis)* is also believed to occur at Illinois Beach State Park. This butterfly was historically associated with native barrens and savanna ecosystems, but it is now associated with remnant barrens and savannas, highway and powerline right-of-ways, gaps within forest stands, young forest stands, forest roads and trails, airports, and military camps. These areas all have soils that are suitable for lupine growth, an open canopy, and management that causes soil disturbance or suppression of perennial shrub and herbaceous vegetation (such as by mowing, brush-hogging, logging, chemical control, or prescribed fire). These habitats can be very diverse vegetationally, and support herbaceous species that co-occur with lupine in the native remnant barrens and savanna habitats. Almost all

DOB

Mr. / Ms. Pao-Tsin Kuo   2

of these contemporary habitats can be described as having a broken or scattered tree canopy that varies within habitats from 0 to between 50 and 80 percent canopy cover. The habitats have lupine, the sole larval food source, nectar plants for feeding adults, critical microhabitats, and attendant ants. Illinois Beach State Park provides each of these criteria.

The federally threatened Pitcher's thistle *(Cirsium pitcheri)* also occurs at Illinois Beach State Park. Pitcher's thistle is part of a dynamic dune ecosystem. It is found most frequently in the near-shore plant communities. Potential *Cirsium pitcheri* habitat includes beach, foredune, interdunal trough, and secondary dune areas. Pitcher's thistle colonizes patches of open, windblown areas of the landscape, and gradually declines locally as the density of vegetation and ground litter increases through plant succession. It is dependent on continually colonizing the mosaic of open habitats within the Great Lakes dunes. The species is patchily distributed with varying population sizes in all open zones of the dunes vegetation. Plant populations decline in stabilized, late successional secondary dune sites and in areas heavily used by people. We strongly caution you to avoid impacts to any of these ecosystems.

In addition, portions of Illinois Beach State Park are designated as critical habitat for the Great Lakes breeding population of the piping plover *(Charadrius melodus)*. Piping plovers are listed as endangered under the Endangered Species Act of 1973, as amended. Critical habitat is a specific geographic area that is essential for the conservation of a threatened or endangered species and that may require special management and protection. Critical habitat may include an area that is not currently occupied by the species but that will be needed for its recovery. Please ensure that your proposed project actions will not destroy or adversely modify critical habitat. Exact boundaries of proposed construction showing no adverse effect to beach areas should be considered.

At this time there are no known federal occurrences of listed species in or near the Braidwood Nuclear Station, in Will County, Illinois.

This letter only addresses federally listed species; the Illinois Department of Natural Resources should be contacted for information on State-listed species. Any impacts to wetlands or waters of the United States may require a permit from the U.S. Army Corps of Engineers. This letter does not preclude separate evaluation and comment by the U.S. Fish and Wildlife Service on wetland impacts proposed for section 404, Clean Water Act authorization.

If you have any questions, please contact Ms. Cathy Pollack at 847/381-2253 ext. 239, or Ms. Karla Kramer at 847/381-2253 ext. 230.

Sincerely,

John D. Rogner
Field Supervisor

April 7, 2005

Mr. Richard Nelson, Field Supervisor
Rock Island Field Office
U.S. Fish and Wildlife Service
4469 48th Avenue Court
Rock Island, IL 61201

SUBJECT: BIOLOGICAL ASSESSMENT FOR AN EARLY SITE PERMIT (ESP) FOR THE
EXELON ESP SITE AND A REQUEST FOR INFORMAL CONSULTATION

Dear Mr. Nelson:

The U.S. Nuclear Regulatory Commission (NRC) has prepared the enclosed biological
assessment (BA) to evaluate whether the proposed action of issuing an ESP for the Exelon
ESP site would have adverse effects on listed species. The Exelon ESP site is located within
the Clinton Power Station (CPS) site along the shore of Clinton Lake, near the town of Clinton
in DeWitt County, Illinois. The proposed Federal action is the issuance, under provisions of
Title 10 of the *Code of Federal Regulations,* Part 52 (10 CFR Part 52), of an ESP for a new
nuclear unit at the Exelon ESP site, which would authorize Exelon to conduct site preparation
and limited construction activities. The site preparation and limited construction activities
allowed by 10 CFR 52.25 include clearing, grading, and constructing non-safety-related
facilities. The proposed action does not include approval to construct and operate a new
nuclear unit. Therefore, the BA does not analyze the environmental impacts that could result
from such construction and operation. Impacts associated with construction and operation of a
new nuclear unit will be assessed during the NRC staff's review of an application for a
combined license or construction permit, should an applicant choose to go forward with the
project.

Exelon has indicated that the addition of four new 345-kilovolt (-kV) transmission lines likely
would be required to accommodate the proposed new nuclear unit, and that these would likely
be sited within the existing utility rights-of-way. Because the site preparation and limited
construction activities noted above may include the addition of new transmission lines to and
expansion of the existing CPS transmission corridor, these impacts are analyzed in the
enclosed BA.

By letter dated March 17, 2004, the NRC requested that the U.S. Fish and Wildlife Service
(FWS) provide a list of Federally threatened or endangered species that may occur in the
vicinity of the Exelon ESP site and transmission line corridors. In a letter dated April 6, 2004,
the FWS provided a list of Federally listed species. The FWS identified one threatened
species, the bald eagle (*Haliaeetus leucocephalus*); and one endangered species, the Indiana
bat *(Myotis sodalis)* as being present. For documentation purposes, the NRC has addressed
the potential impacts of site preparation and limited construction activities on these two species
in its BA.

R. Nelson   -2-

The NRC has determined that site preparation and limited construction activities are not likely to adversely affect the bald eagle because bald eagles are not known to nest or roost in DeWitt County, and no concentrations of foraging eagles have been reported on or in the vicinity of the Exelon ESP site. The staff has also determined that site preparation and limited construction activities are not likely to adversely affect the Indiana bat. Indiana bats potentially occur anywhere in Illinois where there is forest habitat. Indiana bats likely would not be adversely affected if habitat suitability and occupancy were determined prior to construction and appropriate actions were taken to avoid disturbing the species.

We are placing this BA in our project files and are requesting your concurrence with our determination. In reaching our conclusion, the NRC staff relied on information provided by the applicant, on the independent review performed by NRC staff, and on information provided by the FWS in its correspondence to NRC.

If you have any questions regarding this BA or the staff's request, please contact Mr. Thomas Kenyon, Environmental Project Manager, at 301-415-1120, or by e-mail at tjk2@nrc.gov.

Sincerely,

/RA/
Pao-Tsin Kuo, Program Director
License Renewal and Environmental Impacts Program
Division of Regulatory Improvement Programs
Office of Nuclear Reactor Regulation

Docket No.: 52-007

Enclosure: As stated

cc w/encl.: See next page

R. Nelson   -2-

The NRC has determined that site preparation and limited construction activities are **not likely to adversely affect the bald eagle** because bald eagles are not known to nest or roost in DeWitt County, and no concentrations of foraging eagles have been reported on or in the vicinity of the Exelon ESP site. The staff has also determined that site preparation and limited construction activities are not likely to adversely affect the Indiana bat. Indiana bats potentially occur anywhere in Illinois where there is forest habitat. Indiana bats likely would not be adversely affected if habitat suitability and occupancy were determined prior to construction and appropriate actions were taken to avoid disturbing the species.

We are placing this BA in our project files and are requesting your concurrence with our determination. In reaching our conclusion, the NRC staff relied on information provided by the applicant, on research performed by NRC staff, and information provided by FWS in its correspondence to NRC.

If you have any questions regarding this BA or the staff's request, please contact Mr. Thomas Kenyon, Environmental Project Manager, at 301-415-1120, or by e-mail at tjk2@nrc.gov.

> Sincerely,
> /RA/
> Pao-Tsin Kuo, Program Director
> License Renewal and Environmental Impacts Program
> Division of Regulatory Improvement Programs
> Office of Nuclear Reactor Regulation

Docket No.: 52-007

Enclosure: As stated

cc w/encl.: See next page

<u>DISTRIBUTION</u>: See next page

Adams accession no.: **ML050980127**

E:\Filenet\ML050980127.wpd

OFFICE	GS:RLEP	LA:RLEP	PM:RLEP	SC:RLEP	OGC (NLO w/comments)	PD:RLEP
NAME	JDavis	MJenkins	TKenyon	AKugler	AHodgdon	PTKuo
DATE	03/25/05	03/09/05	04/1/05	04/6/05	03/31/05	04/7/05

OFFICIAL RECORD COPY

Biological Assessment

Exelon Generation Company, LLC

Early Site Permit

Exelon Early Site Permit Site

DeWitt County, Illinois

April 2005

Docket Number 52-007

U.S. Nuclear Regulatory Commission
Rockville, Maryland

Biological Assessment of the Potential Effects on Threatened or Endangered Species from the Proposed Exelon Generation Company Early Site Permit (ESP) for the Exelon ESP Site

1.0 Introduction

The U.S. Nuclear Regulatory Commission (NRC) is reviewing an application submitted by Exelon Generation Corporation, LLC (Exelon) for an early site permit (ESP) for the potential future construction and operation of a new nuclear power unit. As part of the review of this application pursuant to Title 10 of the *Code of Federal Regulations* Part 51 (10 CFR Part 51), the NRC has issued a draft environmental impact statement (EIS) (NRC 2005). The impact analysis in the EIS includes an assessment of the potential environmental impacts of the construction and operation of a new nuclear power unit at the proposed site, including potential impacts to threatened or endangered species. The proposed location of the new unit, if built, would be on the existing Clinton Power Station (CPS) site near the town of Clinton in DeWitt County, Illinois, located approximately 10 kilometers (km) (6 miles [mi]) east of the City of Clinton along the shore of Clinton Lake (Figure 1). If approved, the ESP would not authorize the applicant to begin construction of the new unit; however, it would authorize limited site-preparation activities. Thus, only site-preparation activities are considered in this biological assessment (BA). The site under consideration is hereafter referred to as the Exelon ESP site.

Exelon submitted its ESP application on September 25, 2003, pursuant to NRC requirements in 10 CFR Part 52. By letter dated March 17, 2004 (NRC 2004), the NRC requested that the U.S. Fish and Wildlife Service (FWS) Rock Island, Illinois, Field Office provide information regarding Federally listed species at the proposed Exelon ESP site. The Rock Island Field Office responded by letter dated April 6, 2004 (FWS 2004).

2.0 Proposed Action

If approved, an ESP would authorize the permit holder to perform, at its discretion, the limited preconstruction site-preparation activities described in the site redress plan (Exelon 2003a). The site redress plan describes the measures that may be necessary to restore (redress) the site to a condition suitable for other appropriate use in the event the project does not proceed to construction or the site is abandoned.

Prerequisites to preconstruction activities that must be fulfilled include, but are not limited to, documentation of existing site conditions within the Exelon ESP site and acquisition of the necessary permits (e.g., local building permits, Illinois Environmental Protection Agency [IEPA] National Pollutant Discharge Elimination System [NPDES] permit, IEPA Clean Water Act permit, IEPA General Stormwater Permit, etc.).

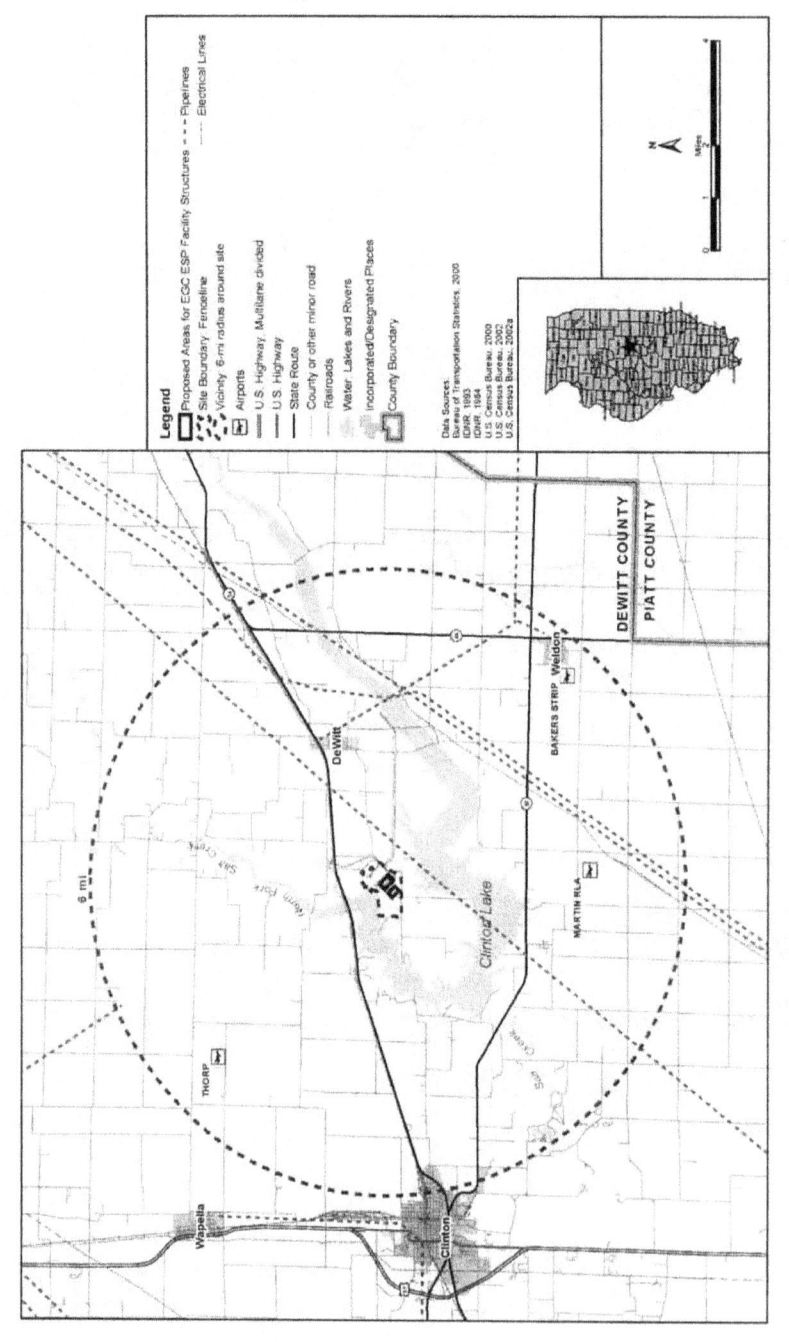

Figure 1. Location of the proposed Exelon ESP site

Once these prerequisites have been achieved, site-preparation activities may proceed and may include none, some, or all of the following activities (Exelon 2003a):

- preparation of the site for construction of a new nuclear unit (including clearing, grading, and construction of temporary access roads and borrow areas)

- installation of temporary construction-support facilities

- evacuation plans for facility/structures

- construction of service facilities

- drilling sample/monitoring wells

- construction of plant cooling towers that are not safety related

- construction of plant intake structures that are not safety related

- installation of non-safety-related fire detection and protection equipment

- expansion of the existing CPS switchyard

- expansion of the existing CPS transmission system

- modification of the existing CPS discharge flume

- construction of any other additional structures, systems, and components that do not prevent or mitigate the consequences of postulated accidents that could cause undue risk to the health and safety of the public.

Site redress activities are specific to the effects of preconstruction site-preparation activities. Redress activities also reflect specific land use and zoning requirements of local municipal, county, and state jurisdictions, in addition to more broadly applicable Federal requirements and industry standards. Redress activities take into account both pre-existing site conditions and a range of potential future-use scenarios, including habitat replacement, recontouring, revegetating, and replanting cleared areas. If necessary (e.g., because the new nuclear unit is not constructed or the site is abandoned), Exelon would restore the site to pre-existing conditions or to the specifications of the future owners in accordance with applicable regulations (contained in 10 CFR Part 50 and 10 CFR Part 52). The protection of critical ecological elements would be maintained in compliance with applicable regulations (Exelon 2003a).

3.0 Potential Environmental Impacts of Preconstruction Site-Preparation Activities

3.1 Potential Habitat Destruction on the Proposed Exelon ESP Site

A total of 187 hectares (ha) (461 acres [ac]) are located within the Exelon ESP site boundary. Most of the footprint of the proposed new nuclear unit consists of areas that would be occupied by the power block structures, normal heat sink cooling towers, switchyard expansion, new intake structures, and safety-related cooling towers. Figure 2 depicts the anticipated footprint of these structures. Existing access roads and infrastructure would primarily be used for construction of the proposed new nuclear unit (Figure 2). The site preparation activities described in Section 2.0 could disturb up to approximately 39 ha (96 ac), most of which would occur about 213 meters (m) (700 feet [ft]) south of the CPS (Figure 2) (Exelon 2003b).

Preconstruction site-preparation activities for the new nuclear unit would occur primarily in previously-disturbed areas that currently support virtually no biota (e.g., impervious surfaces, crushed stone, existing structures, etc.) and in open fields (e.g., previously used as equipment lay-down areas during construction of the CPS, etc.) (Exelon 2003b). However, it would be necessary to clear two small forest stands of about 1.25 ha (3 ac) and about 0.2 ha (0.5 ac) in the northern corner of the power block footprint and within the new intake footprint (Figure 2), respectively.

Site-preparation activities for the new nuclear unit would not be anticipated to adversely affect wetlands (and hence any associated forests) onsite. The four minor wetlands (less than 0.4 ha [1 ac]) listed in the National Wetlands Inventory database that are located on the CPS site do not occur within the power block, cooling tower, switchyard expansion, or new intake footprint areas, and, therefore, would not be impacted by preconstruction site-preparation activities for these structures (Exelon 2003b). However, there is currently no proposed route for the conduit from the new intake to the new power block. Thus, ground clearing for this conduit could impact some wetland and forest habitat, depending on its ultimate location.

3.2 Potential Habitat Destruction along the Transmission Line Corridor Supporting the Proposed New Nuclear Unit

The actual need for and nature of any transmission-system improvements would be determined definitively by the transmission and distribution system owner and operator (currently Illinois Power Company) under Federal Energy Regulatory Commission (FERC) Order 2003 (18 CFR Part 35), *Standardization of Generator Interconnection Agreements and Procedures* (FERC 2003). This order mandates performance of feasibility, system impact, and facilities studies when there is a proposed load increase on the existing transmission system of at least 20 megawatts electric (MWe). Any transmission system improvement studies required by FERC Order 2003 would be carried out prior to construction of the transmission line improvements. The location, nature, and magnitude of environmental impacts associated with the construction of any transmission system improvements would be definitively established by the transmission and distribution system owner and operator at that time. However, to support the

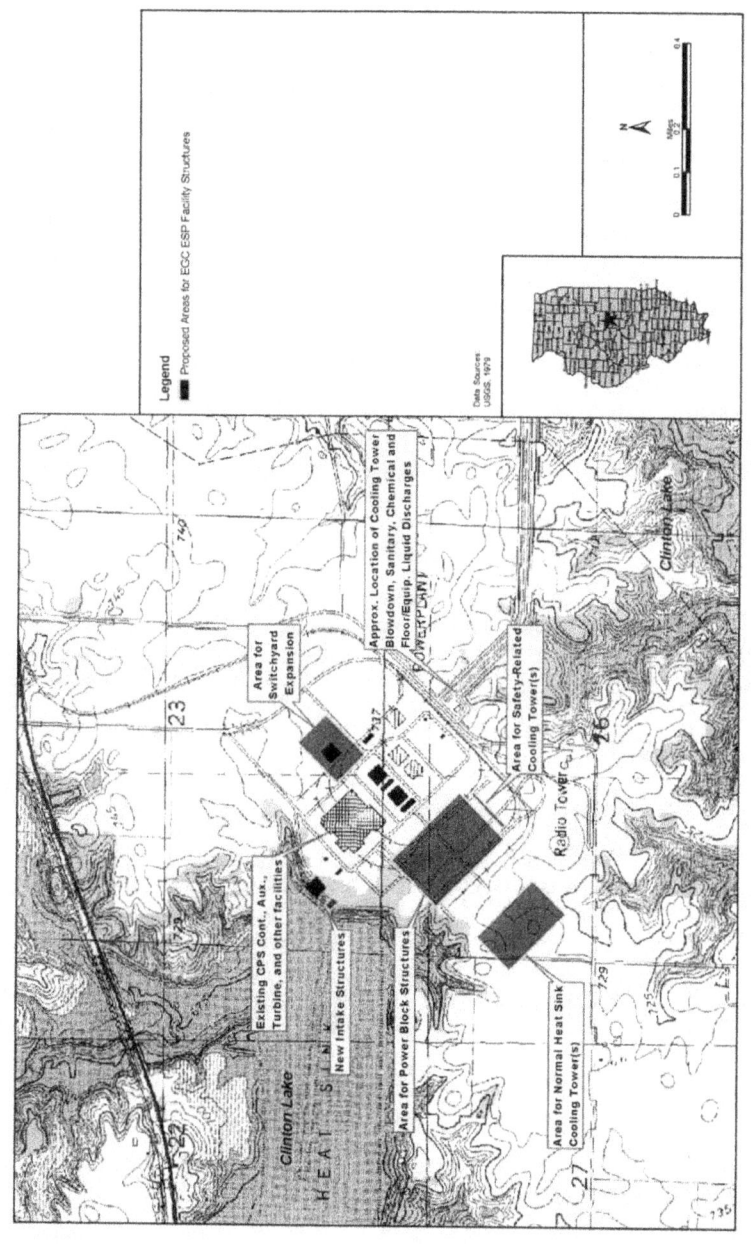

Figure 2. Footprint of the proposed new nuclear unit

April 2005   - 5 -   Exelon ESP Biological Assessment

environmental review of the ESP application, Exelon has stated that it is likely that the addition of four new 345-kilovolt (-kV) transmission lines (two parallel, double-circuit lines running north to the Brokaw Substation near Bloomington and two running south to the Oreana Substation [Figure 3]) would be required to accommodate the proposed new nuclear unit (Exelon 2003b).

The actual amount of disturbance associated with any preconstruction site-preparation activities for anticipated transmission system improvements would be contingent on the techniques used. It is anticipated that any site-preparation activities for transmission system modifications would be located within or immediately adjacent to the existing CPS substation (requiring the switchyard expansion) and within or along the existing transmission line corridor (Figure 3). Transmission line improvements, such as the addition of two new lines and support structures, such as towers, would be sited within the existing CPS transmission corridor to the greatest extent possible. However, it is anticipated that widening the existing corridor about 40 m (130 ft) to 76 m (250 ft) would be required (Exelon 2003b).

Based on the assumption that a maximum of 12 percent of the total length (69 km [43 mi]) of the existing transmission line corridor crosses forest habitat, and the width of the potential expansion is 36 m [120 ft], a loss of no more than about 30 ha (74 ac) of forest habitat is expected by Exelon. It should be noted that forest cutting could be tapered to minimize disturbance and eliminate the need to clear-cut the entire width of the corridor expansion area (Exelon 2003b).

3.3 Traffic-Related Wildlife Mortality

Daily traffic on IL Route 54 and IL Route 10 near the Exelon ESP site currently consists of 2,750 and 2,000 vehicles (cars and trucks), respectively. During construction, daily traffic on Routes 54 and 10 would be expected to increase by an additional 1650 cars and trucks on each highway (an increase of about 60 and 83 percent for Routes 54 and 10, respectively) (Exelon 2003b). The increase in daily traffic for preconstruction site-preparation activities would be substantially less than during construction. Thus, traffic-related wildlife mortalities would be expected to increase marginally.

3.4 Avian Collisions with Transmission Lines

Exelon currently anticipates adding two new transmission lines to the existing CPS transmission corridor to support a new nuclear unit (Exelon 2003b), as discussed in Section 3.2. This could be done as part of site preparation activities (see Section 2.0). These two new lines present additional opportunities for avian collisions beyond those of the existing transmission lines.

4.0 Description of the Project Area

4.1 Exelon ESP Site and Vicinity

The proposed Exelon ESP site consists of 187 ha (461 ac) located on Clinton Lake, which is 10 km (6 mi) east of the City of Clinton, Illinois (Figure 1). The 1983-ha (4900-ac) Clinton Lake was designed and built to be a source of cooling water for the CPS, by constructing an earthen

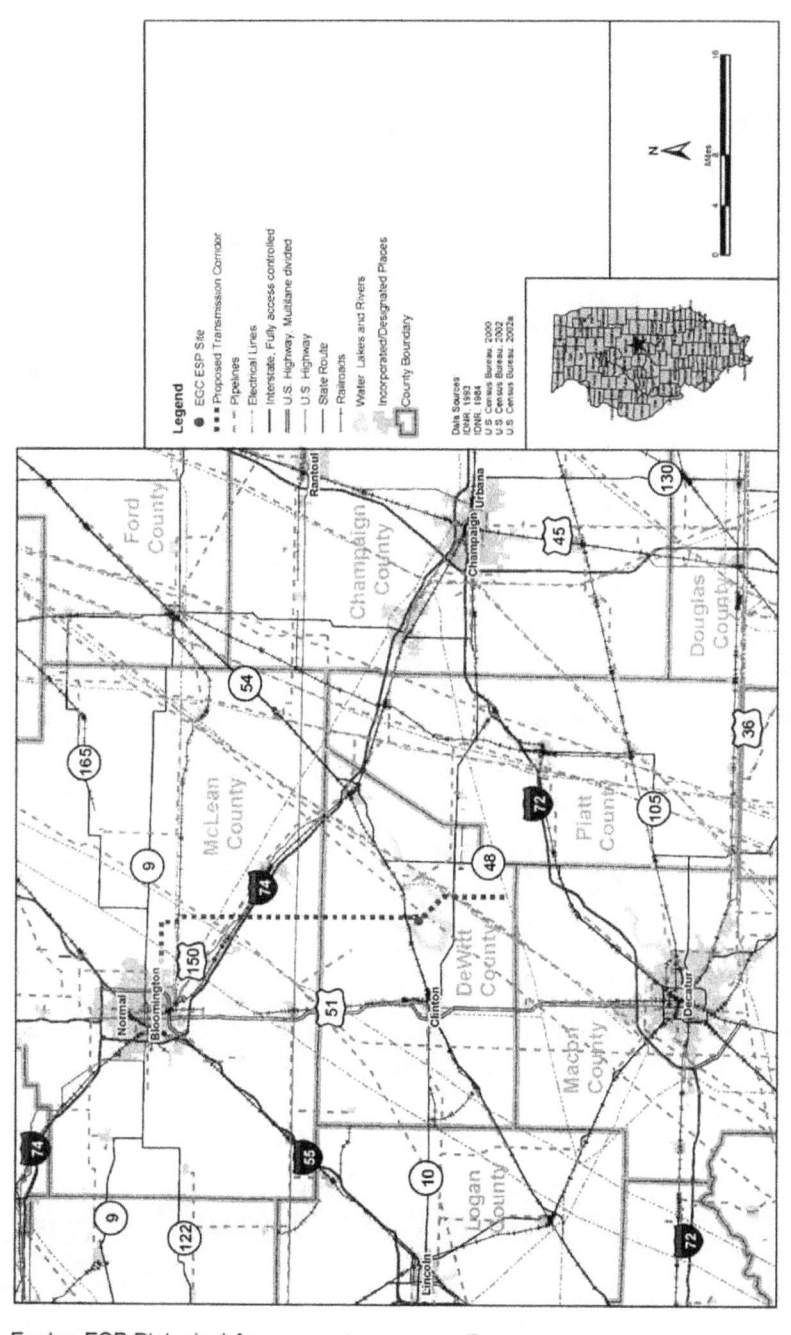

Figure 3. Location of the transmission line corridor for the proposed new nuclear unit

Exelon ESP Biological Assessment — 7 — April 2005

dam 366 m (1200 ft) below the confluence of North Fork Salt Creek and Salt Creek (Figure 1), about 90 km (56 mi) east of the confluence of Salt Creek and the Sangamon River (Exelon 2003b).

Clinton Lake is the main attraction for the Clinton Lake State Recreation Area, a 3764-ha (9300-ac) facility. The park land is owned by AmerGen Energy Company, LLC (AmerGen), the owner and operator of the CPS. Since 1978, the Illinois Department of Natural Resources (IDNR) has operated the recreation area through a lease agreement with AmerGen (IDNR 2004a).

Clinton Lake and its environs are situated within the Central Cornbelt Plains ecoregion (Omernik 1987). This ecoregion consists of glaciated plains that were once dominated by extensive prairie communities intermixed with oak-hickory forests. Farms are now extensive over the ecoregion where little native prairie remains.

Topography at and in the vicinity of the proposed Exelon ESP site is generally flat, except along Salt Creek and North Fork of Salt Creek, where it is gently rolling to steeply sloped. A variety of vegetation communities in various stages of ecological succession occur on the site and in the vicinity. Agriculture (including hay, row crops, and small grains) is the predominant land use (82 percent) within 10 km (6 mi) of the site. Open lands that are not used for active agricultural purposes are commonly used as pasture. Open field habitats dominate the landscape on and adjacent to the site. Upland forest communities in the vicinity harbor overstory and herbaceous species that are common and typical of the region (Exelon 2003b).

Besides agriculture, other land use on and in the vicinity of the proposed Exelon ESP site is recreational (about 17 percent), which is accounted for by the Clinton Lake State Recreation Area, and industrial (about 1 percent).

Important terrestrial and aquatic habitats in the vicinity of the proposed Exelon ESP site include the Clinton Lake State Recreation Area, portions of Tenmile Creek and Salt Creek, Weldon Springs State Recreation Area, and several small wetland areas. Major habitat types of the Clinton Lake State Recreation Area include forest (38 percent of the area), grassland (32 percent), shrubs (21 percent), cropland (6 percent), and wetlands (3 percent). In addition, there are several habitats that are important for a variety of birds, including wet meadows, pine forest, and a marsh (Exelon 2003b).

Illinois designates some environmentally-sensitive areas as natural areas under the jurisdiction of the Illinois Nature Preserves Commission. Two environmentally-sensitive areas are near the proposed Exelon ESP site. The first includes a portion of Tenmile Creek west of the city of Clinton and approximately 8 km (5 mi) from the site. It is designated as an important medium-gradient creek by the IDNR and as a unique aquatic resource by the Illinois Environmental Protection Agency. The second environmentally-sensitive area is along Salt Creek, approximately 5 km (3 mi) from the site (Exelon 2003b).

Several wetland and 100-year-floodplain areas are located within 10 km (6 mi) of the proposed Exelon ESP site and transmission line corridor, and contain forest, emergent, and scrub-shrub

Exelon ESP Biological Assessment   - 8 -   April 2005

communities. These generally are associated with small tributaries of Salt Creek and North Fork of Salt Creek (Exelon 2003b).

Important terrestrial and aquatic habitats within the Exelon ESP site boundary include four minor wetlands (less than 0.4 ha [1 ac]) documented in the National Wetland Inventory database (http://wetland.fws.gov/). These generally consist of open water in association with constructed sediment basins and have palustrine unconsolidated bottom (Exelon 2003b). Some of these are used by the IDNR as fish-rearing ponds.

4.2 Transmission Line Corridor Supporting the New Nuclear Unit

The anticipated transmission line corridor for the proposed new nuclear unit is an existing corridor used to transmit power generated by the CPS. The transmission line corridor is divided into two sections. The northern section is approximately 37 km (23 mi) long and the southern section is approximately 30 km (20 mi) long; both sections are about 40 m (130 ft) wide. The northern section runs north of the proposed Exelon ESP site, and then turns west and terminates at the Brokaw Substation, just west of Bloomington. The southern section runs southeast of the site past Clinton Lake, and then turns south and terminates at the Oreana Substation, just north of Decatur. Figure 3 depicts this transmission line corridor.

Land use within the existing transmission line corridor is predominantly agricultural (about 88 percent), with a small portion (about 1 percent) being industrial, consisting primarily of CPS structures and rail and highway crossings. A portion of the southern section of the transmission line corridor crosses Clinton Lake (Figure 3). About 11 percent of the land use is recreational.

Exelon currently anticipates that four new 345-kV transmission lines (two parallel, double-circuit lines running north to the Brokaw Substation near Bloomington and two running south to the Oreana Substation) would be required to accommodate the bounding case of an output of 2200 MWe from the new nuclear unit at the Exelon ESP site. The actual amount of disturbance associated with any transmission system improvements would be contingent on, among other factors, the construction techniques used. Transmission-line improvements, such as the addition of two new lines and support structures, would be sited within the existing CPS rights-of-way to the greatest extent possible. However, it is anticipated that widening the existing rights-of-way from 40 m (130 ft) to 76m (250 ft) would be required (Exelon 2003b).

5.0 Species Potentially Affected

The FWS identified one threatened species, the bald eagle (*Haliaeetus leucocephalus*) which is currently proposed for delisting (64 FR 36453-36464 [1999]); and one endangered species, the Indiana bat *(Myotis sodalis)* as potentially occurring at the Exelon ESP site and along the transmission line corridor (FWS 2004). No critical habitat has been designated for the bald eagle. Although critical habitat has been designated for the Indiana bat (FWS 2004), the only critical habitat in Illinois is the Blackball Mine in LaSalle County (41 FR 41914 [1976]) located approximately 160 km (100 mi) from CPS.

April 2005   - 9 -   Exelon ESP Biological Assessrnent

5.1 Bald eagle (*Haliaeetus leucocephalus*), threatened (60 FR 35999-36010 [1995])

The bald eagle is a bird often associated with aquatic ecosystems. It frequents estuaries, large lakes, reservoirs, major rivers, and some coastal habitats. Fish is the major component of its diet, but the bald eagle also eats waterfowl, seagulls, and carrion. The species may also use prairies if adequate food is available. Bald eagles usually nest in trees near water but are known to nest on cliffs and (rarely) on the ground. Nest sites are usually in large trees along shorelines in relatively remote areas that are free of disturbance. In winter, bald eagles often congregate at specific wintering sites that are generally close to open water and offer good perch trees and night roosts (64 FR 36453 [1999]).

The bald eagle is known to winter along large rivers, lakes, and reservoirs in DeWitt County (FWS 2004), and has been observed in the vicinity of the Exelon ESP site (Exelon 2003b). During the winter, this species feeds on fish in open-water areas created by dam tailwaters, warm water effluents of power plants and municipal and industrial discharges, or power plant cooling ponds. The more severe the winter, the greater the ice coverage and the more concentrated the eagles become. Bald eagles generally roost at night in groups in large trees adjacent to such associated bodies of water in areas that are protected from harsh winter weather. During the day, they perch in large shoreline trees to rest or feed on fish (FWS 2004).

5.2 Indiana Bat (*Myotis sodalis*), endangered (32 FR 4001 [1967])

The Indiana bat is known to occur in LaSalle County. Because potential habitat for this species occurs statewide, however, Indiana bats are considered to potentially occur in any area with forested habitat (FWS 2004). Consequently, this species could occur at and in the vicinity of the Exelon ESP site, although there are no records of its occurrence within 16 km (10 mi) (IDNR 2004b).

During the summer, the Indiana bat frequents the corridors of small streams with well-developed riparian woods as well as mature upland forests. It forages for insects along stream corridors, within the canopy of floodplain and upland forests, over clearings with early successional vegetation (old fields), along the borders of croplands, along wooded fence rows, over farm ponds, and in pastures. The foraging range (up to 33 ha [81 ac]) for the species varies by season, age, and sex. It roosts and rears its young beneath the loose bark of large dead or dying trees, and the species tends to be philopatric, i.e., returning to the same roosting area year after year. Indiana bats winter in caves and abandoned mines (FWS 2004).

Suitable summer habitat in Illinois is considered to have the following four characteristics within a 0.8-km (0.5-mi) radius of any project site:

- forest cover of 15 percent or greater

- permanent water

- one or more of the following tree species: shagbark and shellbark hickory (*Carya ovata* and *C. laciniosa*, respectively) that may be dead or alive, and dead bitternut hickory (*C. cordiformis*), American elm (*Ulmus americana*), slippery elm

Exelon ESP Biological Assessment   - 10 -   April 2005

(*U. rubra*), eastern cottonwood (*Populus deltoides*), silver maple (*Acer saccharinum*), white oak (*Quercus alba*), red oak (*Q. rubra*), post oak (*Q. stellata*), and shingle oak (*Q. imbricaria*) with slabs or plates of loose bark

- at least one potential roost tree per one ha (2.5 ac), where potential roost trees have greater than ten percent coverage of loose bark (FWS 2004).

6.0 Evaluation of Potential Impacts

6.1 Bald eagle (*Haliaeetus leucocephalus*)

Although the bald eagle is known to winter along large rivers, lakes, and reservoirs in DeWitt County, there are no known nests or night roosts in DeWitt County (FWS 2004). Therefore, bald eagles would not be impacted by removal of potential nest/roost trees along Clinton Lake during preconstruction site-preparation activities [see Section 2.0].

No concentrations of foraging eagles have been reported on or in the vicinity of the Exelon ESP site (IDNR 2004b; Exelon 2003b; FWS 2004). Thus, it is likely that only low numbers of eagles utilize food resources in the vicinity of the Exelon ESP site, and then only on an infrequent basis. These would be expected to be minimally affected, if at all, by disturbance (e.g., noise, human presence, etc.) associated with preconstruction site-preparation activities on the Exelon ESP site and along the associated transmission line corridor.

Few bald eagles frequent the Exelon ESP site, and eagles only occasionally feed on carrion. Hence, any eagle mortalities associated with an increase in traffic during preconstruction site-preparation activities would be expected to be negligible.

Bird collisions with transmission lines were evaluated previously in the *Generic Environmental Impact Statement for License Renewal of Nuclear Plants* (NRC 1996) and were found not to be a problem at operating nuclear power plants, which have variable numbers of transmission corridors and variable numbers of lines within corridors. Thus, incremental eagle mortalities, if any, due to collisions with the additional two transmission lines that would support the new nuclear unit would be expected to be negligible.

In summary, the impacts of preconstruction site-preparation activities on bald eagles are expected to be negligible.

6.2 Indiana bat (*Myotis sodalis*)

Suitable Indiana bat summer habitat has the specific characteristics previously described in Section 5.2. At least some tree species, such as shagbark hickory, silver maple, several species of oak and elm (Exelon 2003b), and upland white oak forest (Illinois Power 1982), listed previously as being important for roosts and rearing of young are known to occur in the vicinity of the Exelon ESP site. If suitable summer habitat occurs in any area of the Exelon ESP site or associated transmission line corridor where preconstruction site-preparation activities would occur and the species is present, then impacts could occur if the forests are cleared. Large-scale habitat alterations within known or potential Indiana bat summer habitat should not

April 2005 - 11 - Exelon ESP Biological Assessment

be undertaken without a bat survey and/or consultation with the FWS Rock Island, Illinois, Field Office. Minor alterations of summer habitat (e.g., clearing of small timber stands such as those described in Section 3.1) should be limited to non-maternity periods between September 16 and April 14. However, before initiating forest-clearing activities, all potentially suitable habitat should be surveyed to determine if it is suitable and, if so, whether it is occupied by Indiana bats, following the steps recommended by the FWS (FWS 2004) and outlined below.

The staff expects that Exelon would determine, prior to initiation of site preparation activities, whether the small stands of forest within the footprint of the new power block and intake structure and potential expansion area for the transmission line corridor are potentially suitable for Indiana bats, i.e., whether they potentially satisfy all four criteria listed in Section 5.2 or any other contemporaneous criteria established by the FWS. If these forest stands are not potentially suitable, they may be cleared without any timing restrictions for Indiana bats. (Note that compliance with other timing restrictions imposed for other species that are not subject of this BA, e.g., migratory birds, would still be necessary). However, if the forest stands are determined to be potentially suitable, they would be surveyed by Exelon to determine their suitability (e.g., to determine percent forest canopy cover, tree species present, density of potential roost trees, percentage of loose bark of potential roost trees, etc.). If these forest stands are found to be unsuitable, they may be cleared without any timing restrictions. However, if they are found to be suitable, it would be determined whether or not they are occupied. If the forest habitat is suitable and unoccupied, clearing could be undertaken during the non-maternity period between September 16 and April 14. However, if the habitat is suitable and occupied, the staff expects that Exelon would first consult with the FWS Rock Island, Illinois, Field Office before undertaking any forest clearing activities.

Indiana bats winter in caves and abandoned mines (FWS 2004), but such habitat features are not known to occur on the Exelon ESP site or along its transmission line corridor.

In summary, if the above guidelines regarding the determination of presence and occupancy of summer habitat are followed, adverse impacts to the Indiana bat from site preparation activities are expected to be negligible.

Exelon ESP Biological Assessment   - 12 -   April 2005

7.0 Conclusions

7.1 Bald eagle (*Haliaeetus leucocephalus*)

Bald eagles are not known to nest or roost in DeWitt County (FWS 2004) and no concentrations of foraging eagles have been reported on or in the vicinity of the Exelon ESP site (IDNR 2004b; Exelon 2003b; FWS 2004). Thus, it is likely that only a low number of eagles utilize food resources in the vicinity of the Exelon ESP site, and then only on an infrequent basis. These would be expected to be minimally affected, if at all, by disturbance associated with preconstruction site-preparation activities on the Exelon ESP site (including mortality due to increased traffic) and along the associated transmission line corridor (including mortality due to collisions with new power lines).

The staff concludes that preconstruction site-preparation activities for a new nuclear unit are not likely to adversely affect the bald eagle.

7.2 Indiana bat Indiana Bat (*Myotis sodalis*)

Indiana bats are considered to occur potentially anywhere in Illinois in any area with forested habitat (FWS 2004). Consequently, this species could occur on and in the vicinity of the Exelon ESP site; however, there are no records of its occurrence within 16 km (10 mi) (IDNR 2004b).

The staff concludes that preconstruction site preparation activities for an Exelon ESP new nuclear unit are not likely to adversely affect the Indiana bat. However, this determination is contingent on Exelon's implementation of the measures recommended by FWS and described in Section 6.2 above for determination of the presence and occupancy of summer habitat.

8.0 References

32 FR 4001. 1967. Endangered Species List - 1967. *Federal Register*, U.S. Fish and Wildlife Service (FWS).

41 FR 41914. 1976. "Determination of Critical Habitat for American Crocodile, California Condor, Indiana Bat, and Florida Manatee." *Federal Register*, U.S. Fish and Wildlife Service (FWS).

60 FR 35999. 1995. Final Rule to Reclassify the Bald Eagle From Endangered to Threatened in All of the Lower 48 States. *Federal Register*, U.S. Fish and Wildlife Service (FWS).

64 FR 36454. 1999. "Endangered and Threatened Wildlife and Plants; Proposed Rule to Remove the Bald Eagle in the Lower 48 States from the List of Endangered and Threatened Wildlife." *Federal Register*, U.S. Fish and Wildlife Service (FWS).

Exelon Generation Company (Exelon). 2003a. *Site Redress Plan for the EGC Early Site Permit*. DEL-096-REV0. September.

Appendix F

Exelon Generation Company (Exelon). 2003b. *Environmental Report for the Exelon Generation Company Early Site Permit.* September.

Illinois Department of Natural Resources (IDNR). 2004a. Accessed on the internet June 2, 2004 at: http://dnr.state.il.us/lands/Landmgt/PARKS/R3/Clinton.htm.

Illinois Department of Natural Resources (IDNR). 2004b. "Provision of electronic data regarding the locations of federal and state listed threatened and endangered species within 2 mi and 10 mi of the Exelon ESP site." E-mail from IDNR (Springfield, Illinois) to Pacific Northwest National Laboratory (Richland, Washington). February 19.

Illinois Power. 1982. *Clinton Power Station Environmental Report – Operating License Stage, Supplement 3.* April.

Omernik, J. 1987. "Ecoregions of the conterminous United States. Map (Scale 1:7500000)." *Annals of the Association of American Geographers* 77(1):118-125.

U.S. Federal Energy Regulatory Commission (FERC). 2003. *Standardization of Generator Interconnection Agreements and Procedures.* Order No. 2003. 18 CFR Part 35. Accessed on the Internet May 7, 2004, at: http://www.ferc.gov/whats-new/comm-meet/072303/E-1.pdf.

U.S. Fish and Wildlife Service (FWS). 2004. "Provision of information regarding federal listed threatened or endangered species, that may occur in the vicinity of the Exelon ESP site." Letter from FWS (Rock Island, Illinois, Field Office) to the U.S. Nuclear Regulatory Commission (Washington, DC), April 6. Accession No. ML041180181.

U.S. Nuclear Regulatory Commission (NRC). 1996. *Generic Environmental Impact Statement for License Renewal of Nuclear Plants.* NUREG-1437, NRC, Washington, DC.

U.S. Nuclear Regulatory Commission (NRC). 2004. "Request for information regarding federal listed species that may occur in the vicinity of the Exelon ESP site." Letter from NRC, (Washington, DC) to U.S. Fish and Wildlife Service (Rock Island, Illinois, Field Office), March 17. Accession No. ML040770896.

U.S. Nuclear Regulatory Commission (NRC). 2005. *Environmental Impact Statement for an Early Site Permit (ESP) at the Exelon ESP Site.* Draft Report for Comment. NUREG-1815, NRC, Washington, DC.

RD00 received 5/16/05

3/10/05 (18)

70 FR 12022

Illinois Historic Preservation Agency

Voice (217) 782-4836

1 Old State Capitol Plaza • Springfield, Illinois 62701-1507 • Teletypewriter Only (217) 524-7128

DeWitt County
Clinton
Clinton Lake

PLEASE REFER TO: IHPA LOG #009030705

NRC
Draft EIS, Exelon ESP Site

April 11, 2005

Pao-Tsin Kuo
United States Nuclear Regulatory Commission
License Renewal and Environmental Impacts Programs
Division of Regulatory Improvement Programs
Office of Nuclear Reactor Regulation
Washington, DC 20555-0001

Dear Mr. Kuo:

Thank you for requesting comments from our office concerning the possible effects of the project referenced above on cultural resources. Our comments are required by Section 106 of the National Historic Preservation Act of 1966 (16 USC 470), as amended, and its implementing regulations, 36 CFR 800: "Protection of Historic Properties".

The project area has not been surveyed and may contain prehistoric/historic archaeological resources. Accordingly, a Phase I archaeological reconnaissance survey to locate, identify, and record all archaeological resources within the project area will be required. This decision is based upon our understanding that there has not been any large scale disturbance of the ground surface (excluding agricultural activities) such as major construction activity within the project area which would have destroyed existing cultural resources prior to your project. If the area has been heavily disturbed prior to your project, please contact our office with the appropriate written and/or photographic evidence.

The area(s) that need(s) to be surveyed include(s) all area(s) that will be developed as a result of the issuance of the federal agency permit(s) or the granting of the federal grants, funds, or loan guarantees that have prompted this review.

Enclosed you will find an attachment briefly describing Phase I surveys and a list of archaeological contracting services. THE IHPA LOG NUMBER OR A COPY OF THIS LETTER SHOULD BE PROVIDED TO THE SELECTED PROFESSIONAL ARCHAEOLOGICAL CONTRACTOR TO ENSURE THAT THE SURVEY RESULTS ARE CONNECTED TO YOUR PROJECT PAPERWORK.

If you have questions, please contact David J. Halpin, Staff Archaeologist, at 217-785-4998.

Sincerely,

Anne E. Haaker

Anne E. Haaker
Deputy State Historic
     Preservation Officer

AEH:DJH

Enclosure: Archaeological Contractors List

E-EIDS = ADM-03

All = T. Kenyon (TJK2)

SISP Review Complete

Template = ADM-013

Illinois Historic
Preservation Agency

Voice (217) 782-4836
1 Old State Capitol Plaza • Springfield, Illinois 62701-1507 • Teletypewriter Only (217) 524-7128

PROTECTING ILLINOIS' CULTURAL RESOURCES
An Introduction to Archaeological Surveys

Prepared by
ILLINOIS STATE HISTORIC PRESERVATION OFFICE

When you read the accompanying letter, you were notified that your Federal or State permitted, funded, or licensed project will require an archaeological survey. We also review projects that use public land. The purpose of this survey will be to determine if prehistoric or historic resources are present within the project area. If you are the average applicant you have had little or no experience with such surveys – this short introduction is designed to help you fulfill the Federal/State requirements and complete the process.

WHY PROTECT HISTORIC RESOURCES? Historic preservation legislation grew out of the public concern for the rapid loss of our prehistoric and historic heritage in the wake of increasingly large-scale Federal/State and private development. The legislation is an attempt to protect our heritage while at the same time allowing economic development to go forward.

WHAT IS THE LEGAL BASIS? The basis for all subsequent historic preservation legislation lies within the national Historic Preservation Act of 1966 (NHPA). Section 106 of NHPA requires all Federal Agencies "undertakings" to "take into account" their effect on historic properties. As of January 1, 1990, the State Agency Historic Resources Preservation Act (Public Act 86-707) requires the same for all private or public undertakings involving state agencies. An "undertaking" is defined to cover a wide range of Federal or State permitting, funding, and licensing activities. It is the responsibility of Federal/State Agencies to ensure the protection of historic resources and the State Historic Preservation Office (SHPO) regulates this effort. In Illinois the SHPO is part of the Illinois Historic Preservation Agency (IHPA).

WHAT IS AN ARCHAEOLOGICAL SURVEY? An archaeological survey includes both (1) an examination of the written records, such as county plat books, published and unpublished archaeological reports, state site files, and (2) a field investigation of the project area to determine if prehistoric or historic resources are present. This process of resource identification is called a Phase I survey.

WHAT DOES A PHASE I SURVEY REQUIRE? Archaeological evidence is normally buried beneath the surface of the ground. To determine if an archaeological site is present it is necessary to get below this surface. The most efficient way is by plowing. If the project area is or can be plowed then the artifactual evidence will be brought to the surface and systematic pedestrian surveys (walkovers) will determine if a site is present. These walkovers are best done when the vegetation is low in the fall or spring. If the project area is covered with vegetation then small shovel probes (1' sq.) are excavated on a systematic grid pattern (usually 50' intervals) to sample the subsurface deposits. Where deeply buried sites may be present, such as in floodplains, deep coring or machine trenching may be required.

WHO DOES ARCHAEOLOGICAL SURVEYS? Professional archaeologists who meet the Federal standards set forth in the Secretary of the Interior's <u>Professional Qualifications Standards</u> (48 FR 44738-9) may conduct Federal surveys, while those meeting the State standards set forth in the Archaeological and Paleontological Resources Protection Act (20 ILCS 3435) may conduct surveys on public land in the State (see the other side of this sheet for information on obtaining the services of a contract archaeologist). The applicant is responsible for obtaining and paying for such services.

AFTER THE SURVEY – WHAT NEXT? When the field investigations are completed the archaeologist will submit a report of their findings and recommendations to the applicant. **IT IS THE RESPONSIBILITY OF THE APPLICANT TO FORWARD TWO (2) COPIES TO THE SHPO FOR EVALUATION AND FINDINGS.** If no sites were found or the sites found are not eligible for the National Register the project may proceed. Occasionally, a significant archaeological site may be encountered. In such a case the SHPO and the Federal or State Agency will work with the applicant to protect both the cultural resources and to facilitate the completion of your project.

NEED FURTHER ASSISTANCE? The IHPA is here to assist you and the Federal/State agencies in complying with the mandates of the historic preservation legislation. If you have questions or need assistance with archaeological resources protection or Federal/State compliance, please contact the Archaeology Section, Preservation Services Division, Illinois Historic Preservation Agency, 500 East Madison Street, Springfield, Illinois 62701 (217/785-4512).

OVER

02/09/05

Illinois Historic Preservation Agency

Voice (217) 782-4836

1, Old State Capitol Plaza • Springfield, Illinois 62701-1507 • Teletypewriter Only (217) 524-7128

Illinois Historic Preservation Agency – Archaeology Section
Information for Developers and Agencies about general procedures for Phase 2 archaeology projects

Anyone notified of an archaeological site subject to Phase 2 testing in their project area, has several options:

1   Preserve the site by planning your project to avoid or greenspace the site, a deed covenant maybe necessary
        depending on the land ownership and the law the project is being reviewed under.
2.   Hire an archaeological firm to conduct a Phase 2 project on the site.
3.   Choose a different location for the project (generally means starting review process over from scratch, but
        there will be rare occasions when this is actually the fastest and cheapest option). This is something you may
        wish to consider if there are burials in the project area, or an extremely large or dense site in the project area.

Phase 2 archaeological projects consist of fieldwork, analysis, and report by the archaeological firm, and then review of
the report by the IHPA and sometimes also by the funding or permitting agency, with additional work required part of
time depending on the significance of the site(s). However, if a project has no significant sites after a Phase 2 project
has been completed and reviewed, then the archaeology is completed as soon as IHPA accepts the report. If a project
area has more than 1 site, each one is reviewed independently, in other words, one could be determined not significant
and while another one is determined significant or potentially significant.

Phase 2 field work generally consists of obtaining good artifact type and location data from the site surface by methods
such as grid collections, piece plotting, etc., this is followed by a small scale excavation. In some cases the fieldwork
(commonly called test units) can be done with assistance of machines like backhoes or occasionally even large
equipment like belly scrapers (plowed or partially disturbed sites), but sometimes it is necessary to dig by hand
(mounds, unplowed sites, or inaccessible locations). The test units are excavated to the base of the plowzone or
topsoil, and then the base of the unit is checked for presence of archaeological features (foundations, pits, hearths,
burials, middens, etc.) If features are present, a small number (generally not more than 5-10) of them are excavated to
provide information about the site's age, function, integrity, etc. Samples of soil from each feature for botanical and
zoological analysis are usually taken. Also on floodplains of large rivers, several additional "deep" trenches are
usually necessary to check for buried sites. The amount of time required for fieldwork is highly dependent on the size
of a site, on whether machines can be used, and on the density of features, as well as the weather.

Analysis at Phase 2 consists of identifying and inventorying all of the artifacts recovered and preparing data recorded
in the field for a report. The length of time needed is again highly variable based on the factors listed above. The
report describes the field and lab information, provides a preliminary interpretation of the site, and makes
recommendations concerning the significance of the site.

The archaeology staff at the State Historic Preservation Office (IHPA in Illinois) and sometimes the archaeologists at
the lead funding or permitting agency review the report. Based on the report and their knowledge of regional
archaeological, they determine (following criteria outlined in the appropriate law and regulations for each project) if
the work done was acceptable, and whether the site(s) are not significant and need no further investigation or are
significant. If a site is significant (meets the eligibility criteria for the National Register of Historic Places), the choices
are mitigation (generally by complete excavation) or preservation.

Joseph S. Phillippe, Chief Archaeologist (1-1-2005)

02/09/05

ILLINOIS-BASED CONSULTING SERVICES WITH PROFESSIONAL ARCHAEOLOGISTS (by zip code order, 2/15/2005 update)
In order to assist agencies, engineering firms, and others who require professional archaeological services the Illinois Historic Preservation Agency (IHPA) has listed below Illinois-based firms with professional archaeologists currently performing contract archaeological compliance work. Based on documentation supplied by them these individuals appear to meet current Federal qualifications. This list is provided for your assistance, however, you may use any archaeologist who meets the minimum qualifications as set forth in Secretary of the Interior's Professional Qualifications Standards (36 CFR 61). Federal and state regulations require a completed graduate degree with an emphasis in archaeology and 16 months of professional archaeological experience (**BOLD** names below). If you have any questions please contact IHPA at 217/785-4512. THE INCLUSION OF INDIVIDUALS OR ORGANIZATIONS ON THIS LIST DOES NOT CONSTITUTE ANY RECOMMENDATION OR ENDORSEMENT OF THEIR PROFESSIONAL EXPERTISE OR PERFORMANCE RECORD BY THE IHPA.

CHICAGO METRO REGION

Contact Dr. Kevin P. McGowan
Public Service Archaeology Prgm
Chicagoland Office (UI-UC)
Post Office Box 7085
Grayslake, Illinois 60030
847-548-7961 (fax same)

Ms. Lynn M. Gierek
ENSR International
27755 Diehl Road
Warrenville, Illinois 60555-3998
630-839-5332 / 836-1711 (fax)
lgierek@ensr.com

Mr. Stephen Parrish
Archaeological Research, Inc.
1005 Greta Avenue
Woodstock, Illinois 60098
815-334-8077 / 0530 (fax)

Contact Dr. Mark W. Mehrer
Northern Illinois University
Contract Archaeology Program
Department of Anthropology
102 Stevens Building
DeKalb, Illinois 60115
815-753-7544 / 7027 (fax)
mmehrer@niu.edu

Contact Dr. Rochelle Lurie
Dr. M. Catherine Bird
Midwestern Archaeological
  Research Services, Inc.
505 North State Street
Marengo, Illinois 60152
815-568-0680 / 0681 (fax)

Dr. Cynthia L. Balek
Archaeology & Geomorphology Services
2220 Mayfair Avenue
Westchester, Illinois 60154
708-531-1445 / 562-7314 (fax)
cbalek@msn.com

Mr. Douglas Kullen
Allied Archeology
239 South Calumet Avenue
Aurora, Illinois 60506
630-896-9375 / 897-9682 (fax)
archon2001@hotmail.com

02/17/05

CHICAGO METRO REGION CON'T

Contact Mr. Jeff Schuh
Patrick Engineering, Inc.
4970 Varsity Drive
Lisle, Illinois 60532
630/795-7200 / 434-8400 (fax)

Contact Mr. David Keene
Archaeological Research, Inc.
1735 North Paulina Street, Suite 113
Chicago, Illinois 60622
773-384-8134 / 8286 (fax)

NORTHERN REGION

Contact Dr. Thomas E. Berres
OurHeritage Archaeological Srvs, Inc.
983 Quail Run
DeKalb, Illinois 60115-6117
815-754-9611

Contact Ms. K. Shane Vanderford
ITARP Northern Illinois Survey Division
6810 Forest Hills Road
Loves Park, Illinois 61111
815-282-0762 / 0754 (fax)

CENTRAL REGION

Mr. Keith L. Barr
Archaeological & Architectural Surveys
Old Inn Farm
Rural Route 1
Fairview, Illinois 61432
309-778-2536

Mr. Lawrence A. Conrad
Western Illinois University
Archaeology Lab
201 Tillman Hall
Macomb, Illinois 61455
309-298-1188

CENTRAL REGION CON'T

Dr. Michael D. Wiant
Dickson Mounds Museum
10956 North Dickson Mounds Road
Lewistown, Illinois 61542
309-547-3721

Dr. Brian Adams
Public Service Archaeology Program
716 North Ashland Avenue
West Peoria, Illinois 61604
309-671-0876 (fax same)

Dr. Charles L. Rohrbaugh
Archaeological Consultants
320 Robert Drive
Normal, Illinois 61761
309-454-6590

Contact Dr. Paul P. Kreisa
University of Illinois
Anthropology Department
Public Service Archaeology Program
109 Davenport Hall
607 South Matthews Avenue
Urbana, Illinois 61801
217-333-1636 / 217-244-1911 (fax)

Contact Mr. Dale McElrath
University of Illinois Champaign-Urbana
UIUC-ITARP Statewide Office
23 East Stadium Drive
209 Nuclear Physics Lab (MC 571)
Champaign, Illinois 61820
217-333-0667 / 244-7458 (fax)

Dr. Fred A. Finney
Upper Midwest Archaeology
Post Office Box 106
St. Joseph, Illinois 61873-0106
217-469-0106 (voice/fax same)
cell 217-778-0348
FAFinney@aol.com

Dr. William E. Whittaker
118 Paddock Drive
Savoy, Illinois 61874
217-840-6245
wewhittaker@hotmail.com

More Central Listings – Over

CENTRAL REGION CON'T

Ms. Gail Anderson
Center for American Archeology
(Kampsville Archeological Center)
Post Office Box 22
Kampsville, Illinois 62053
618-653-4316 / 4232 (fax)

Contact Mr. David J. Nolan
ITARP Western Illinois Survey Division
604 East Vandalia
Jacksonville, Illinois 62650
217-243-9491 / 7991 (fax)
Macomb Lab
309-833-3097
Springfield Lab
217-522-4295 / 4395 (fax)

Contact Dr. Michael Sheehay
Illinois State Museum Society
1011 East Ash Street
Springfield, Illinois 62703
217-785-0037 / 2857 (fax)

Mr. Floyd Mansberger
Fever River Research
Post Office Box 5234
Springfield, Illinois 62705
217-525-9002 / 6093 (fax)

Mr. Joseph Craig
Environmental Compliance Consultants
Post Office Box 5603
Springfield, Illinois 62705-5603
217-544-4881 / 4988 (fax)
jcraig@eccinc.org

METRO EAST REGION

Contact Mr. Joseph L. Harl
Archaeological Research Center
 Of St. Louis (Illinois Office)
313 Filmore Street
Post Office Box 444
Worden, Illinois 62097-0444
314-426-2577 / 2599 (fax)
or 800-340-4CRM

Contact Dr. Charles W. Markman
Markman & Associates, Inc.
4618 North Illinois, Suite 178
Fairview Heights, Illinois 62208
314-862-6117 / 6712 (fax)
charlie@markmaninc.com

02/17/05

METRO EAST REGION CON'T

Contact Dr. Steve Dasovich
SCI Engineering, Inc.
15 Executive Drive
Fairview Heights, Illinois 62208
636-949-8200 / 8269 (fax)

Contact Mr. Brad Koldehoff
UIUC – ITARP
American Bottom Survey Division
6608 West Main Street
Belleville, Illinois 62223
618-397-5096 / 5097 (fax)

Dr. John Kelly
Central Mississippi Valley
 Archaeological Research Institute
Post Office Box 413
Columbia, Illinois 62236
618-281-8205

SOUTHERN REGION

Contact Mr. Michael J. McNerney
Mr. Steve Titus
American Resources Group, Ltd.
127 North Washington Street
Carbondale, Illinois 62901
618-529-2741 / 457-5070 (fax)

Contact Dr. Brian M. Butler
Southern Illinois University
Center for Archaeological Investigations
Mail Code 4527
Carbondale, Illinois 62901
618-453-5031 / 8467 (fax)

Appendix F

United States Department of the Interior

OFFICE OF THE SECRETARY
Office of Environmental Policy and Compliance
Custom House, Room 244
200 Chestnut Street
Philadelphia, Pennsylvania 19106-2904

2005 MAY 24 AM 9: 24

TAKE PRIDE
IN AMERICA

IN REPLY REFER TO:

May 17, 2005

3/10/05
70 FR 12022
(30)

ER 05/245

Mr. Michael T. Lesar
Chief, Rules and Directives Branch
U.S. Nuclear Regulatory Commission
Mail Stop T6-D59
Washington, D.C. 20555-0001

Received RDB
2005 MAY 24 AM 9: 24

Mr. Michael T. Lesar:

The U.S. Department of the Interior (Department) has reviewed the February 2005 Draft Environmental Impact Statement (DEIS), NUREG-1815, for an Early Site Permit at the Exelon Generating Company, LLC, DeWitt County, Illinois. Exelon has applied to the U.S. Nuclear Regulatory Commission (NRC) for approval of a site within the existing Clinton Power Station boundaries as suitable for the construction and operation of a new nuclear power generating facility. The NRC staff has also provided a Biological Assessment to the U. S. Fish and Wildlife Service (FWS) with a determination of the effects of the proposed action on federally listed threatened or endangered species. The FWS will provide a response to the NRC under separate cover.

The DEIS adequately discusses potential impacts of the project alternatives on fish and wildlife resources, as well as species protected by the Endangered Species Act. The greatest potential for impacts is associated with the possible need for modifications to transmission line rights-of-way, with a maximum loss of no more than 74 acres of forested habitat expected. These potential impacts will be addressed further in the construction permit application stage. Exelon has also agreed to contact the FWS before beginning any construction activities to ascertain whether previous determinations regarding threatened and endangered species remain valid or whether further evaluation would be needed. The Department appreciates this commitment.

For continued consultation and coordination on fish and wildlife matters and threatened and endangered species, please contact the Field Supervisor, U.S. Fish and Wildlife Service, 4469 48th Avenue Court, Rock Island, Illinois 61201; Telephone: (309) 793-5800.

We appreciate the opportunity to provide these comments.

Sincerely,

Michael T. Chezik
Regional Environmental Officer

E-REDS = ADM-03
Adm = T. Kenyon (TJK2)

SFS/ Review Complete

Template = ADM-013

United States Department of the Interior

FISH AND WILDLIFE SERVICE
Rock Island Field Office
4469 48ᵗʰ Avenue Court
Rock Island, Illinois 61201
Phone: (309) 793-5800 Fax: (309) 793-5804

IN REPLY REFER
TO:
FWS/RIFO

52-007

May 19, 2005

Pao-Tsin Kuo, Program Director
License Renewal and Environmental Impacts Program
Division of Regulatory Improvement Programs
Office of Nuclear Reactor Regulation
U.S. Nuclear Regulatory Commission
Washington, DC 20555-0001

Dear Sir or Madam:

We have reviewed your April 2005, Biological Assessment for an early site permit (ESP) for
the Exelon ESP site, located within the Clinton Power Station site along the shore of Clinton
Lake, near the town of Clinton in DeWitt County, Illinois, and your request for informal
consultation. We have the following comments.

We concur with your findings that, with appropriate avoidance measures, the proposed project
is not likely to adversely affect federally listed endangered species. Should the project be
modified or new information indicate endangered species may be affected, consultation should
be initiated.

Since the formation of the Avian Power Line Interaction Committee (APLIC) in 1989, the
electric utility industry and the U.S. Fish and Wildlife Service (USFWS) have worked together
to reduce avian electrocution and collision mortality. This has resulted in the cooperative
development of the recently released voluntary guidelines designed to help electrical utilities
protect and conserve migratory birds. The guidelines for Avian Protection Plans (APP), when
voluntarily implemented by a utility, will greatly reduce avian risk, as well as its own risk of
enforcement under the Migratory Bird Treaty Act (MBTA). An APP is utility-specific and is
designed to reduce avian and operational risks that result from avian interactions with utility
facilities by using the latest technology and science to tailor a voluntary APP that meets
specific utility needs at its facilities. The guidance document is available at
http://migratorybirds.fws.gov/, and references the latest industry standards for preventing
avian power line interactions.

Pao-Tsin Kuo, Program Director   2

Thank you for the opportunity to provide comments. If you have any additional questions or concerns, please contact Heidi Woeber of my staff at extension 209.

Sincerely,

Richard C. Nelson
Field Supervisor

cc: FWS RIFO (Lundh)

S:\Office Users\Heidi\concurexelon.doc

October 12, 2005

Mr. Kenneth Barr, Branch Chief
U.S. Army Corps of Engineers, Rock Island District
Clock Tower Building
Rodman Avenue
Rock Island, IL 61201

SUBJECT: CLINTON EARLY SITE PERMIT REVIEW (TAC NO. MC1125)

Dear Mr. Barr:

The U.S. Nuclear Regulatory Commission (NRC) staff is reviewing an application submitted by Exelon Generation Company, LLC (Exelon) for an early site permit (ESP). The proposed action requested in Exelon's application is for the NRC to: (1) approve a site within the existing Clinton Power Station (CPS) boundaries as suitable for the construction and operation of a new nuclear power generating facility; and (2) issue an ESP for the proposed site located adjacent to the CPS. An ESP does not authorize construction or operation of a nuclear power plant. Rather, the ESP application and review process makes it possible to evaluate and resolve certain safety and environmental issues related to siting before the applicant makes large commitments of resources. If the ESP is approved, the applicant can "bank" the site for up to 20 years for future reactor siting. To construct or operate a nuclear power plant, an ESP holder must obtain a construction permit and an operating license, or a combined license.

As part of its environmental review of Exelon's ESP application, the NRC prepared a draft environmental impact statement (DEIS) in accordance with 10 CFR 52.18. The DEIS includes the NRC staff's analysis of the environmental impacts of constructing and operating a nuclear unit at the Exelon ESP site, or at alternative sites. It also includes the staff's preliminary recommendation to the Commission regarding the proposed action. In addition, as described in the DEIS, if the ESP includes a site redress plan, the ESP holder can conduct certain site preparation and preliminary construction activities allowed by Title 10 of the *Code of Federal Regulations* Section 50.10(e)(1) (10 CFR 50.10 (e)(1)), provided the final EIS concludes that such activities will not result in any significant environmental impact that cannot be redressed. Exelon has included a site redress plan in its application. If the ESP is approved, Exelon will be allowed to conduct site preparation and preliminary construction activities pursuant to 10 CFR 52.25 and 10 CFR 50.10(e)(1), subject to receipt of any other necessary Federal, State, and/or local approvals. Exelon has stated that it does not plan to conduct such activities at this time. However, these activities, if performed, could include dredging and other activities potentially subject to Clean Water Act requirements. The environmental impacts of these activities are discussed in the DEIS.

K. Barr   -2-

Pursuant to the "Memorandum of Understanding Between the Corps of Engineers, United States Army, and the United States Nuclear Regulatory Commission for Regulation of Nuclear Power Plants" (40 FR 37110 (dated August 25, 1975)), we request that the Army Corps of Engineers review and provide to the NRC any comments on the DEIS.

Enclosed is a copy of NUREG-1815 "The Draft Environmental Impact Statement for an Early Site Permit (ESP) at the Exelon ESP Site." We request your comments no later than December 16, 2005. Enclosed to aid in your review is a CD containing Exelon's application for an ESP. If you have any questions concerning the ESP application or other aspects of this project, please contact Mr. Thomas Kenyon, Senior Environmental Project Manager, at 301-415-1120 or by e-mail at TJK2@nrc.gov.

                         Sincerely,
                         /RA/
                         Pao-Tsin Kuo, Program Director
                         License Renewal and Environmental Impacts Program
                         Division of Regulatory Improvement Programs
                         Office of Nuclear Reactor Regulation

Docket No.: 52-007

Enclosure: As stated

cc wo/encl.: See next page

DEPARTMENT OF THE ARMY
ROCK ISLAND DISTRICT. CORPS OF ENGINEERS
CLOCK TOWER BUILDING - P.O. BOX 2004
ROCK ISLAND, ILLINOIS 61204-2004

February 7, 2006

52-007

REPLY TO
ATTENTION OF:

Planning, Programs, and
    Project Management Division

Mr. Thomas Kenyon
Senior Environmental Project Manager
OWFN 11 F-1
United States Nuclear Regulatory Commission
Washington, D.C. 20555-001

Dear Mr. Kenyon:

I received your letter dated October 12, 2005, requesting comments regarding the
Environmental Impact Statement for an Early Site Permit (ESP) at the Exelon ESP Site (EIS).
Rock Island District staff reviewed the information you provided and have the following
comments:

a. Your proposal does not involve Rock Island District Corps of Engineers (Corps)
administered land; therefore, no further Rock Island District Corps real estate coordination is
necessary.

b. Any proposed placement of fill or dredged material into waters of the United States
(including wetlands) requires Department of the Army authorization under Section 404 of
the Clean Water Act. A Section 404 permit will be required for this project. Prior to submission
of the permit application to the Corps, we suggest that project proponents contact this District for
pre-application discussions on content of the application, timing, etc. When detailed information
is available, please complete and submit the enclosed application packet to the Rock Island
District for processing (enclosure). The application should include determinations of wetlands
and other waters of the United States, size estimations of impacts to those areas, and wetland
types and relative functions.

c. The Responsible Federal Agency should coordinate with Ms. Anne Haaker, the Illinois
State Historic Preservation Officer, 1 Old State Capitol Plaza, Springfield, Illinois 62704 to
determine impacts to historic properties.

d. The Rock Island Field Office of the U.S. Fish and Wildlife Service should be contacted
to determine if any federally listed endangered species are being impacted and, if so, how to
avoid or minimize impacts. The Rock Island Field Office address is: 4469 - 48th Avenue Court,
Rock Island, Illinois 61201. Mr. Rick Nelson is the Field Supervisor. You can reach him by
calling 309/793-5800.

DO73

--

e. The Illinois Emergency Management Agency should be contacted to determine if the proposed project may impact areas designated as floodway. Mr. Ron Davis is the Illinois State Hazard Mitigation Officer. His address is: 1035 Outer Park, 2nd Floor, Springfield, Illinois 62704. You can reach him by calling 217/782-8719.

f. The EIS does not provide any information regarding the lack of effects from the project. The Corps suggests close coordination with U.S. EPA and Illinois EPA to ensure water quality requirements are met.

No other concerns surfaced during our review. Thank you for the opportunity to comment on your proposal. If you need more information, please call Mr. Matt Campbell of our Economic and Environmental Analysis Branch, telephone 309/794-5827.

You may find additional information about the Corps' Rock Island District on our web site at **http://www.mvr.usace.army.mil**. To find out about other Districts within the Corps, you may visit web site: **http://www.usace.army.mil/divdistmap.html**.

Sincerely,

Kenneth A. Barr
Chief, Economic and
    Environmental Analysis Branch

Enclosure

Appendix G

Environmental Impacts of Transportation

Appendix G

Environmental Impacts of Transportation

This appendix discusses the potential environmental impacts of transporting reactor fuel and radioactive waste to and from potential early site permit (ESP) sites including North Anna Power Station, Clinton Nuclear Power Station, Grand Gulf Nuclear Station, and their associated alternative sites. Section G.1 briefly discusses the effects of transporting unirradiated fuel to ESP sites, and Section G.2 discusses the effects of transporting spent fuel from ESP sites to a spent fuel disposal facility. Section G.3 discusses the environmental effects of radioactive waste shipments.

G.1 Unirradiated Fuel Shipping

This section addresses the number and characteristics of shipments of unirradiated fuel to ESP sites relative to the conditions in Title 10 of the Code of Federal Regulations (CFR) Part 51.52. Comparisons are also made against Table S–4 in 10 CFR 51.52(c) and WASH-1238 (AEC 1972), which provided the data that supports Table S–4. Section G.1.1 presents the basic unirradiated fuel shipping requirements for each advanced reactor design. These data were extracted from Idaho National Engineering and Environmental Laboratory (INEEL) (2003). Section G.1.2 presents the comparisons to 10 CFR 51.52 conditions.

G.1.1 Advanced Reactor Unirradiated Fuel Shipping Data

In WASH-1238 (AEC 1972), a reference boiling water reactor (BWR) and pressurized water reactor (PWR) were used to formulate the basic numbers of unirradiated fuel shipments required for initial core loading and refueling. Both reference reactor types had a net electrical output of 1100 MW(e). The reference BWR assumed an initial core loading of 150 metric tons of uranium (MTU), and the reference PWR assumed a 100 MTU initial loading. Both reactor types resulted in 18 truck shipments of unirradiated fuel per reactor for initial core loading. Annual reload quantities were assumed to be 30 MTU/yr for both reactor types, which resulted in an additional six truck shipments per year per reactor. In total, about 252 truck shipments of unirradiated fuel would be required over a 40-year reactor life, including the initial core and 39 years of reloads, for both reactor types.

The initial fuel loading and annual reload quantities for the Advanced Boiling Water Reactor (ABWR), a 1500-MW(e) reactor, and the Economic Simplified Boiling Water Reactor (ESBWR) are approximately the same: 156.96 MTU per reactor initial core loading and 32.76 MTU/yr per reactor reload quantities (INEEL 2003). This equates to about 872 unirradiated fuel assemblies in the initial core and 213 assemblies per year for refueling. Truck shipment capacities were stated in INEEL (2003) to be 28 to 30 unirradiated fuel assemblies per truck shipment.

Assuming 30 fuel assemblies per truck shipment, approximately 30 shipments of unirradiated fuel would be required to load the initial core and 6.1 truck shipments per year would be needed for refueling. If 28 fuel assemblies per truck shipment are used, the initial core load would require about 32 shipments of unirradiated fuel and annual refueling would require about 6.5 truck shipments per year.

The surrogate AP1000 is an 1150-MW(e) advanced PWR. The initial core load was estimated to be 84.5 MTU per reactor, and the annual reload requirement was estimated to be 24.4 MTU/yr per reactor. The data in INEEL (2003) also indicated that the average uranium mass in an unirradiated surrogate AP1000 fuel assembly would be 0.583 MTU and that 12 fuel assemblies per truck shipment would be transported. Therefore, about 14 truck shipments would be needed to supply the initial core and about 3.8 truck shipments per year would be needed to support refueling. For a site with two reactors, these estimates would be doubled.

The ACR-700 is an Advanced CANDU (CANada Deuterium Uranium) Reactor assumed to generate 731 MW(e). It was stated in INEEL (2003) that the initial core load for the ACR-700 is 61.3 MTU per reactor, and the annual refueling requirement is 33.1 MTU/yr per reactor. Each fuel assembly contains 18 kg of uranium (INEEL 2003). This corresponds to 3406 fuel assemblies in the initial core loading and 1839 fuel assemblies per year for refueling. The range of truck shipment capacities given by INEEL (2003) was 180 to 240 fuel assemblies per truck shipment. This equates to 15 to 19 truck shipments needed to supply the initial core load and from 7.7 to 10.2 annual refueling shipments. For a site with two reactors, these estimates would be doubled.

The International Reactor Innovative and Secure (IRIS) design is a 335-MW(e) advanced PWR. It requires an initial core load of 48.67 MTU or 89 fuel assemblies per unit (546.9 kg of uranium per fuel assembly) (INEEL 2003). For refueling, the IRIS reactor was assumed to require an additional 6.26 MTU/yr of unirradiated fuel per reactor or about 40 unirradiated fuel assemblies every 3.5 years. INEEL (2003) indicates that a "typical" site may contain three reactors. Assuming each truck shipment carries eight fuel assemblies, the initial core load would require 34 truck shipments per three-reactor site, and annual refueling would require an additional 4.3 truck shipments per year per three-reactor site.

The Gas Turbine–Modular Helium Reactor (GT-MHR) is a gas-cooled reactor that uses a substantially different fuel design than current and advanced LWRs. The reactor's thermal power level is rated at 600 MW(t) per reactor, and the electric generation capacity is rated at 285 MW(e) per reactor. A standard GT-MHR site is assumed to be composed of four reactors. INEEL (2003) states that the initial core load for a single reactor would be about 1020 fuel assemblies. Annual average reload requirements would be 510 fuel assemblies per reactor. INEEL (2003) also indicates that each truck shipment could carry 80 fuel assemblies, so for all four reactors, about 51 truck shipments would be required to transport the initial core load and about 20 truck shipments per year would be required for the annual reload requirements.

The Pebble Bed Modular Reactor (PBMR) is a gas-cooled reactor that is rated at 400 MW(t) (165 MW(e)) per reactor. A typical PBMR site is assumed to consist of eight reactors. The PBMR uses a substantially different fuel design than a typical LWR. INEEL (2003) states that each reactor requires 260,000 fuel spheres for its initial core load; 120,000 fuel spheres per reactor are required for annual average reloads. A total of 48,000 fuel spheres is assumed to be transported in a typical truck shipment. As a result, it would take about 44 shipments of fuel spheres to transport the initial core load for all eight reactors and about 20 shipments per year to transport the annual reload quantity for all eight reactors.

To make comparisons to Table S–4, the environmental impacts were normalized to a reference reactor year. The reference reactor is an 1100 MW(e) reactor that has an 80 percent capacity factor, for a total electrical output of 880 MW(e) per year. The environmental impacts can be adjusted to calculate impacts per site by multiplying the normalized impacts by the ratio of the total electrical output for the advanced reactor sites to the electrical output of the reference reactor.

G.1.2 Analysis of the Environmental Impacts of Unirradiated Fuel Shipments

As required by 10 CFR 51.52, applicants for a construction permit are required to submit a statement that the reactor and the transportation of fuel and waste to and from the reactor meet all the conditions specified in 10 CFR 51.52(a) or 10 CFR 51.52(b). An ESP is a partial construction permit (10 CFR 52.21). The conditions specified in 10 CFR 51.52(a) that apply to unirradiated fuel include the following:

(1) The reactor core has a thermal loading less than 3800 MW. [51.52(a)(1)]

(2) The reactor fuel is in the form of sintered UO_2 pellets not exceeding 4 percent uranium-235 by weight, and the pellets are encapsulated in zircaloy rods. [51.52(a)(2)]

(3) Unirradiated fuel is shipped to the reactor by truck. [51.52(a)(5)]

(4) The environmental impacts of transportation of fuel and waste are as set forth in Summary Table S–4 in 10 CFR 51.52(c). [51.52(a)(6)]

If these conditions are not met, 10 CFR 51.52(b) requires the applicant to provide a full description and detailed analysis of the environmental impacts of transporting fuel and waste to and from the reactor, including values for the environmental impact under normal conditions of transport and the environmental risk from accidents in transport.

Unirradiated fuel shipment information for the advanced reactors is discussed below for each of these criteria.

Appendix G

G.1.2.1 Reactor Core Thermal Loading

The thermal output ratings of the seven advanced reactor types, as given in INEEL (2003), are as follows:

- ABWR – 4300 MW(t) (single reactor)
- ESBWR – 4000 MW(t) (single reactor)
- Surrogate AP1000 – 3400 MW(t) per reactor x two reactors per site = 6800 MW(t) per site
- ACR-700 – 1982 MW(t) per reactor x two reactors per site = 3964 MW(t) per site
- IRIS – 1000 MW(t) per reactor x three reactors per site = 3000 MW(t) per site
- GT-MHR – 600 MW(t) per reactor x four reactors per site = 2400 MW(t) per site
- PBMR – 400 MW(t) per reactor x eight reactors per site = 3200 MW(t) per site.

As shown above, single-unit ABWR and ESBWR plants exceed the 3800-MW(t) condition in 10 CFR 51.52(a)(1). In addition, the twin-reactor ACR-700 and AP1000 site exceed the core thermal power condition.

G.1.2.2 Reactor Fuel Form

All of the advanced LWRs (i.e., the ABWR, ESBWR, surrogate AP1000, IRIS, and ACR-700) use sintered UO_2 fuel pellets encapsulated in zircaloy rods. The average enrichment for the ACR-700 fuel is about 2 percent, which is well within the 10 CFR 51.52(a)(2) condition. The average enrichments for the other advanced LWR fuels exceed the 4 percent uranium-235 by weight condition in 10 CFR 51.52(a)(2) (INEEL 2003).

The gas-cooled reactors (i.e., the GT-MHR and PBMR) have substantially different fuel forms than those described in 10 CFR 51.52(a)(2). The fuel forms for these reactors are coated uranium oxycarbide fuel kernels (GT-MHR) or coated uranium dioxide fuel kernels (PBMR). The fuel kernels are coated with layers of pyrolitic carbon and silicone carbide. Thus, these fuel forms are not the same as those specified in 10 CFR 51.52(a)(2). Furthermore, the equilibrium enrichments for these fuels are 12.9 percent (PBMR) and 19.8 percent (GT-MHR).

G.1.2.3 Shipping Mode

Trucks are used to ship unirradiated fuel to the various sites for all the reactor types (INEEL 2003).

G.1.2.4 WASH-1238 and Table S–4 of 10 CFR 51.52(c)

The condition specified in Table S–4 that applies to shipment of unirradiated fuel limits the number of shipments of fuel and waste to and from a commercial nuclear power plant to less than one per day. Table G-1 summarizes the number of truck shipments of unirradiated fuel required for each reactor type. The numbers of shipments are normalized to the net electrical generation output for the reference reactor in WASH-1238 (AEC 1972) or 880 MW(e) (1100-MW(e)) plant operating at 80-percent annual capacity factor.

As shown in Table G-1, the ACR-700, PBMR, and GT-MHR advanced reactor types exceed the number of truck shipments estimated for the reference LWR in WASH-1238 (AEC 1972). The largest number of shipments, in excess of 700 shipments over 40 years, is for the GT-MHR. However, the combined number of unirradiated fuel, spent fuel, and radioactive waste

Table G-1. Numbers of Truck Shipments of Unirradiated Fuel for Each Advanced Reactor Type

Reactor Type	Number of Shipments per Unit			Unit Electric Generation, MW(e)[c]	Capacity Factor[c]	Normalized, Shipments per 1100 MW(e)[d]
	Initial Core[a]	Annual Reload	Total[b]			
Reference LWR (WASH-1238)	18	6	252	1100	0.8	252
ABWR/ESBWR[e]	30	6.1	267	1500[f]	0.95	165
Surrogate AP1000	14	3.8	161	1150[f]	0.95	130
ACR-700[e]	30	15.4	628	1462[g]	0.9	420
IRIS	34	4.3	201	1005[h]	0.96	184
GT-MHR	51	20	831	1140[I]	0.88	729
PBMR	44	20	824	1320[j]	0.95	579

(a) Shipments of the initial core have been rounded up to the next highest whole number.

(b) Total shipments of unirradiated fuel over a 40-year plant lifetime (i.e., initial core load plus 39 years of average annual reload quantities).

(c) Unit capacities and capacity factors were taken from INEEL (2003).

(d) Normalized to net electric output for WASH-1238 reference LWR (i.e., 1100 MW(e) reactor at 80 percent or net electrical output of 880 MW(e)).

(e) Ranges of capacities are given in INEEL (2003) for these reactor unirradiated fuel shipments. The unirradiated fuel shipment data for these reactors were derived using the upper limits of the ranges.

(f) The ABWR/ESBWR unit includes one reactor at 1500 MW(e), and the surrogate AP1000 unit includes one reactor at 1150 MW(e).

(g) The ACR-700 unit includes two reactors at 731 MW(e) per reactor.

(h) The IRIS unit includes three reactors at 335 MW(e) per reactor.

(I) The GT-MHR unit includes four reactors at 285 MW(e) per reactor.

(j) The PBMR unit includes eight reactors at 165 MW(e) per reactor.

Note: The reference LWR shipment values have all been normalized to 880 MW(e) net electrical generation.

shipments per day equate to far less than one truck shipment per day for all reactor types. Consequently, the numbers of shipments for all the advanced reactor types are within the conditions specified in Table S–4 of 10 CFR 51.52. Table S–4 includes a condition that the truck shipments not exceed 33,100 kg (73,000 lb) as governed by Federal or State gross vehicle weight restrictions. All of the advanced reactors were indicated in INEEL (2003) to be capable of meeting this restriction for unirradiated fuel shipments.

Finally, Table S–4 includes conditions related to radiological doses to transport workers and members of the public along transport routes. These doses are a function of the radiation dose rate emitted from the unirradiated fuel shipments, the number of exposed individuals and their locations relative to the shipment, the time in transit (including travel time and stop time), and the number of shipments to which the individuals are exposed. The radiological dose impacts of the transportation of unirradiated fuel were calculated using the RADTRAN 5 computer code (Neuhauser et al. 2003). The RADTRAN 5 calculations were performed to develop estimates of the worker and public doses associated with annual unirradiated fuel shipments to the ESP sites.

One of the key assumptions in WASH-1238 (AEC 1972) for the reference light water reactor (LWR) unirradiated fuel shipments is that the radiation dose rate at 1 m (3 ft) from the transport vehicle is about 0.001 mSv/hr (0.1 mrem/hr). This assumption was also used in the analysis of advanced reactor unirradiated fuel shipments. This assumption is reasonable for all the advanced reactor fuel types because the fuel materials will be low-dose-rate uranium radionuclides and will be packaged similarly (i.e., inside a metal container that provides little radiation shielding). The numbers of shipments per year were obtained by dividing the normalized shipments in Table G-1 by 40 years of operation. Other key input parameters used in the radiation dose analysis for unirradiated fuel are shown in Table G-2.

The RADTRAN 5 results for this "generic" unirradiated fuel shipment are as follows:

- Worker dose: 1.71×10^{-5} person-Sv/shipment (1.71×10^{-3} person-rem/shipment)

- General public dose (onlookers/persons at stops and sharing the highway): 6.65×10^{-5} person-Sv/shipment (6.65×10^{-3} person-rem/shipment)

- General public dose (along route - persons living near a highway): 1.61×10^{-6} person-Sv/shipment (1.61×10^{-4} person-rem/shipment).

Table G-2. RADTRAN 5 Input Parameters for Unirradiated Fuel Shipments

Parameter	RADTRAN 5 Input Value	Source
Shipping distance, km	3200	AEC (1972)[a]
Travel fraction – rural	0.90	NRC (1977a)
Travel fraction – suburban	0.05	
Travel fraction – urban	0.05	
Population density – rural, persons/km^2	10	DOE (2002a)
Population density – suburban, persons/km^2	349	
Population density – urban, persons/km^2	2260	
Vehicle speed – rural, km/hr	88.49	Based on average speed in rural areas given in
Vehicle speed – suburban, km/hr	88.49	DOE (2002a)
Vehicle speed – urban, km/hr	88.49	
Traffic count – rural, vehicles/hr	530	DOE (2002a)
Traffic count – suburban, vehicles/hr	760	
Traffic count – urban, vehicles/hr	2400	
Dose rate at 1 m from vehicle, mSv/hr	0.001	AEC (1972)
Packaging length, m	7.3	Approximate length of two LWR fuel element packages placed on end
Number of truck crew	2	AEC (1972), NRC (1977a), and DOE (2002a)
Stop time, hr/trip	4.5	Based on 0.0014-hour stop time per km (Hostick et al. 1992)
Population density at stops, persons/km^2	64,300	Based on 20 people in annular ring extending from 1 to 10 m (3.3 to 33 ft) from the vehicle.

(a) AEC (1972) provides a range of shipping distances between 40 km (25 mi) and 4800 km (3000 mi) for unirradiated fuel shipments. A 3200-km (2000-mi) "average" shipping distance was assumed here.

These values were combined with the average annual shipments of unirradiated fuel for each advanced reactor type (see Table G-1) normalized to the WASH-1238 (AEC 1972) reference LWR electric output (880 MW(e)) to calculate annual doses to the public and workers. The results are compared to Table S–4 conditions and are shown in Table G-3. As demonstrated, the calculated radiation doses for shipping unirradiated fuel to advanced reactor sites are within the conditions shown in Table S–4.

Although radiation may cause cancers at high doses and high dose rates, currently there are no data that unequivocally establish the occurrence of cancer following exposures to low doses below about 100 mSv (10,000 mrem) and at low dose rates. However, radiation protection experts conservatively assume that any amount of radiation exposure may pose some risk of causing cancer or a severe hereditary effect and that the risk is higher for higher radiation exposures. Therefore, a linear, no-threshold dose response model is used to describe the relationship between radiation dose and detriments such as cancer induction. A recent report by the National Research Council (2006), the BEIR VII report, supports the linear, no-threshold dose response theory. Simply stated, any increase in dose, no matter how small, results in an

Table G-3. Radiological Impacts of Transporting Unirradiated Fuel to ESP Sites

Plant Type	Normalized Average Annual Shipments	Cumulative Annual Dose, person-Sv/yr[a] per 1100 MW(e)		
		Workers	Public – Onlookers	Public – Along Route
Reference LWR (WASH-1238 (AEC 1972))	6.3	1.1×10^{-4}	4.2×10^{-4}	1.0×10^{-5}
ABWR/ESBWR	4.1	7.1×10^{-5}	2.7×10^{-4}	6.6×10^{-6}
Surrogate AP1000	3.3	5.6×10^{-5}	2.2×10^{-4}	5.2×10^{-6}
ACR-700	10.5	1.8×10^{-4}	7.0×10^{-4}	1.7×10^{-5}
IRIS	4.6	7.9×10^{-5}	3.1×10^{-4}	7.4×10^{-6}
GT-MHR	18.2	3.1×10^{-4}	1.2×10^{-3}	2.9×10^{-5}
PBMR	14.5	2.5×10^{-4}	9.6×10^{-4}	2.3×10^{-5}
10 CFR 51.52, Table S-4 Condition	<1 per day	4×10^{-2}	3.0×10^{-2}	3.0×10^{-2}

(a) Person-Sv = person-sievert; multiply person-Sv/yr times 100 to obtain dose in person-rem/yr.

incremental increase in health risk. This theory is accepted by the NRC as a conservative model for estimating health risks from radiation exposure, recognizing that the model probably overestimates those risks.

Based on this model, the staff estimates the risk to the public from radiation exposure using the nominal probability coefficient for total detriment (730 fatal cancers, nonfatal cancers, and severe hereditary effects per 10,000 person-Sv (1,000,000 person-rem)) from International Commission on Radiological Protection (ICRP) Publication 60 (ICRP 1991). All the public doses presented in Table G-3 are less than or equal to 0.0012 person-Sv/yr (0.12 person-rem/yr); therefore, the total detriment estimates associated with these doses would all be less than 1×10^{-4} fatal cancers, nonfatal cancers, and severe hereditary effects per year. These risks are very small compared to the fatal cancers, nonfatal cancers, and severe hereditary effects that would be expected to occur annually to the same population from exposure to natural sources of radiation, based on the same risk model.

G.1.3 Transportation Accidents

Accidents involving unirradiated fuel shipments are also addressed in Table S–4. Accident risks are the product of accident frequency times consequence. Accident frequencies are likely to be lower than those used in the analysis in WASH-1238 (AEC 1972) because traffic accident, injury, and fatality rates have fallen over the past 30 years. Consequences of accidents that are severe enough to result in a release of unirradiated fuel particles are not significantly different for advanced LWRs because the fuel form, cladding, and packaging are similar to those analyzed in WASH-1238. Consequently, the impacts of accidents during transport of unirradiated fuel to advanced LWR sites would be smaller than the WASH-1238 results that formed the basis for Table S–4.

With respect to the advanced gas-cooled reactors, accident rates (accidents per unit distance) and associated accident frequencies (accidents per year) would follow the same trends as for LWRs (i.e., overall reduction relative to the accident rates used in WASH-1238). The consequences of accidents involving gas-cooled reactor unirradiated fuel, however, are more uncertain. A literature search was conducted to identify publicly available documents that describe the effects of accidents (i.e., exposure of unirradiated gas-cooled reactor fuel to structural and thermal transients). No definitive references were found. Consequently, it was assumed that the gas-cooled reactor unirradiated fuel shipments would have the same abilities as LWR unirradiated fuel to maintain functional integrity following a traffic accident. This assumption is judged to be conservative because gas-cooled reactor fuel operates at significantly higher temperatures and thus maintains integrity under more severe thermal conditions than LWR fuel. Detailed information about the behavior of the gas-cooled reactor fuel under impact conditions was not available. However, packaging systems for unirradiated gas-cooled reactor fuel will be required to meet the same performance requirements as unirradiated LWR fuel packages including fissile material controls to prevent criticality under normal and accident conditions. Consequently, packaging systems for unirradiated gas-cooled reactor fuels are expected to provide protection equivalent to those designed for unirradiated LWR fuels. In addition, the fuel forms for the gas-cooled reactors are similar to those for LWRs (i.e., uranium oxide for the PBMR and uranium oxycarbide for the GT-MHR versus uranium oxide for LWRs); thus, the inherent failure resistance provided by unirradiated gas-cooled reactor fuels is expected to be similar to that provided by LWR fuels. Based on the assumption that unirradiated gas-cooled and LWR fuels and associated packaging systems provide similar resistance to various environmental conditions, the staff concluded that the impacts of accidents involving unirradiated gas-cooled reactor fuel are not expected to be significantly different than those for unirradiated LWR fuel.

G.2 Spent Fuel Shipping

This section discusses the impact of transporting irradiated or spent advanced reactor fuel from ESP sites to a potential high-level waste repository at Yucca Mountain, Nevada. The section is divided into two parts. The first part considers incident-free transportation, and the second part considers transportation accidents.

The analysis is based on shipment of spent fuel by legal-weight trucks in casks with characteristics similar to casks currently available (i.e., massive, heavily shielded, cylindrical metal pressure vessels). Each shipment is assumed to consist of a single shipping cask loaded onto a modified trailer. These assumptions are consistent with assumptions made in the evaluation of the environmental impacts of transportation of spent fuel presented in Addendum I to NUREG-1437 (NRC 1999). As discussed in Addendum I, these assumptions are conservative because the alternative assumptions involve rail transportation or heavy-haul trucks, which would reduce the number of spent-fuel shipments.

Environmental impacts of the transportation of spent fuel were calculated using the RADTRAN 5 computer code (Neuhauser et al. 2003). Routing and population data for input to RADTRAN 5 for shipment by truck were obtained from the TRAGIS routing code (Johnson and Michelhaugh 2000). The population data in the TRAGIS code is based on the 2000 U.S. Census.

G.2.1 Incident-Free Transportation of Spent Fuel

"Incident-free" transportation refers to transportation activities in which the shipments of radioactive material reach their destination without releasing any radioactive cargo to the environment. The vast majority of radioactive shipments are expected to reach their destination without experiencing an accident or incident or releasing any cargo. The "incident-free" impacts from these normal, routine shipments arise from the low levels of radiation that penetrate the heavily shielded spent fuel shipping cask. Although Federal regulations in 10 CFR Part 71 and 49 CFR Part 173 impose constraints on radioactive material shipments, some radiation penetrates the shipping container and exposes nearby persons to low levels of radiation.

Incident-free, legal-weight truck transportation of spent fuel has been evaluated by considering shipments from 11 representative reactor sites to the proposed high-level waste repository at Yucca Mountain, Nevada, (referred to here as the proposed Yucca Mountain Repository) for disposal. This assumption is conservative because it tends to maximize the shipping distance from the East Coast and Midwest, where most of the reactors are assumed to be located. Therefore, shipment to one or more other potential sites, such as a monitored retrievable storage facility, would reduce the impacts.

Environmental impacts from these shipments will occur to persons residing along the transportation corridors between the potential advanced reactor sites and the proposed repository; to persons in vehicles passing the spent-fuel shipment; to persons at vehicle stops for refueling, rest, and vehicle inspections; and to transportation crew members. The impacts to these exposed population groups were quantified using the RADTRAN 5 computer code (Neuhauser et al. 2003).

This analysis assumes that all spent nuclear fuel will be transported to the proposed Yucca Mountain Repository because Congress has directed (Nuclear Waste Policy Act of 1982, as amended) the U.S. Department of Energy to study only Yucca Mountain for the proposed repository.

The characteristics of specific shipping routes (e.g., population densities and shipping distances) influence the normal radiological exposures. To address the differences that arise from the specific reactor site from which the spent fuel shipment originates, each advanced reactor design was assumed to be located at all of the primary and alternative ESP sites. These sites are:

- Primary Sites
    - North Anna Power Station, Virginia
    - Clinton Nuclear Power Station, Illinois
    - Grand Gulf Nuclear Power Station, Mississippi

- Alternative Sites[(a)]
    - Savannah River Site (SRS), South Carolina
    - Portsmouth Gaseous Diffusion Plant (PGDP), Ohio
    - FitzPatrick Nuclear Power Plant, New York
    - Pilgrim Nuclear Power Station, Massachusetts
    - Zion Nuclear Power Station, Illinois
    - Quad Cities Nuclear Power Station, Iowa
    - Braidwood Nuclear Power Station, Illinois
    - Surry Power Station, Virginia

Input to RADTRAN 5 includes the total shipping distance between the origin and destination sites and the population distributions along the routes. This information was obtained by running the TRAGIS computer code (Johnson and Michelhaugh 2000) for the origin-destination combinations of interest for legal-weight trucks. The resulting route characteristics information is shown in Table G-4. Note that for truck shipments, all the spent fuel is assumed to be shipped to the proposed Yucca Mountain Repository over designated highway route controlled quantity (HRCQ) routes. The routes used here are the same as those used in the Yucca Mountain Environmental Impact Statement (DOE 2002b).

Shipping casks have not been designed for advanced reactor spent fuel. Although some of the advanced reactor fuel designs are similar to current LWR fuel, no attempt has been made to optimize the cargo capacities of shipping casks for advanced LWR fuels. For the non-LWR fuel types (i.e., the GT-MHR and PBMR), there is little information on even a conceptual basis that would provide a defensible technical basis for shipping-cask capacities. The shipping-cask

---

(a) Impacts were not calculated for the River Bend site because the analysis is bounded by the impacts calculated for Grand Gulf. Impacts were not calculated for the Dresden and LaSalle sites because they are bounded by the Braidwood analysis.

Table G-4. Transportation Route Information for Shipments from ESP Sites to the Proposed High-Level Waste Repository at Yucca Mountain

ESP Site	One-Way Shipping Distance, km				Population Density, persons/km²			Stop Time per Trip, hr
	Total	Rural	Suburban	Urban	Rural	Suburban	Urban	
Primary Site								
North Anna	4409.5	3498	812.4	99.1	11.3	319	2310.6	5
Clinton	3076.3	2626.3	398.3	51.7	9.4	306.1	2372.2	3.5
Grand Gulf[a]	3718.3	3030.4	581.3	106.6	9.2	339.4	2429.4	4
Alternative Site								
Savannah River Site	4263	3260	881	122	11	331.5	2311.2	5
Portsmouth Gaseous Diffusion Plant	3902.2	3166.9	647.2	88.1	10.7	316.4	2339.7	4.5
FitzPatrick	4212.2	3228.6	875.4	108.2	11.4	312.4	2348.7	5
Pilgrim	4682.3	3469.3	1091.7	121.3	11.8	312.3	2377.2	5.5
Zion	3138.9	2629.6	441.3	68	9.5	323.8	2360.3	3.5
Quad Cities	2853.1	2451	352.6	49.5	9.1	310.2	2391.3	3
Braidwood[b]	3034.5	2604.4	378.7	51.4	9.4	308.9	2377.2	3.5
Surry	4555.4	3590.7	863.9	100.8	11.4	317.6	2301.6	5

(a) The River Bend alternative site can be assumed to be bounded by the Grand Gulf values because of the proximity of the sites.

(b) Dresden and LaSalle can be assumed to be bounded by the Braidwood values because of the proximity of the sites.

capacity data in the *Early Site Permit Environmental Report Sections and Supporting Documentation* (INEEL 2003) is summarized as follows:

- ABWR – The ABWR fuel is not significantly different from existing LWR fuel designs; thus, the number of ABWR assemblies that can be transported in a legal-weight truck shipment (i.e., 23 MT [25-ton] shipping cask) is not expected to be different from current cargo capacities.

- ESBWR – The ESBWR fuel is similar to the ABWR fuel.

- Surrogate AP1000 – The surrogate AP1000 fuel assemblies are similar to current-generation PWR fuel. No information was provided in INEEL (2003) on shipping cask capacities for surrogate AP1000 spent nuclear fuel.

- ACR-700 – The ACR-700 fuel is somewhat different from the current and advanced LWR fuel designs. INEEL (2003) estimated that an ACR-700 rail cask would hold about 10 MTU of spent fuel. This value is nearly identical to the cargo capacities of current rail cask designs; thus, it was assumed that the truck cask capacity for ACR-700 and current-generation LWRs would also be about the same (i.e., 1.8 MTU/shipment).

- IRIS – The IRIS fuel is similar to current-generation PWR fuel. No information was provided in INEEL (2003) on shipping-cask capacities for IRIS spent nuclear fuel.

- GT-MHR – The GT-MHR fuel is a spherical coated-particle fuel with a uranium oxycarbide fuel kernel loaded into graphite fuel assemblies. This fuel concept is significantly different from current and advanced LWR fuels (sintered UO_2 pellets loaded into zircaloy tubes). According to INEEL (2003), six spent fuel assemblies containing 0.023 MTU of spent fuel is assumed to be transported in a legal weight truck cask.

- PBMR – The PBMR fuel is also a spherical coated-particle fuel with uranium oxide fuel kernels. INEEL (2003) estimated that 0.495 MTU of spent PBMR fuel can be transported in a single legal-weight truck shipment.

These shipping cask capacities are approximations based on current shipping cask designs. Actual shipping cask capacities in the future may be significantly different. Applicants must account for changes in shipping cask capacities in applications at the construction permit or combined operating license stage.

Incident-free radiation doses are a function of many variables. The most important of these variables are presented in Table G-5. Most of these variables, which are extracted from the literature, are considered to be "standard" values used in many RADTRAN 5 applications, including environmental impact statements and regulatory analyses.

For purposes of this Appendix G analysis, the transportation crew for spent fuel shipments delivered by truck is assumed to consist of two drivers. Escorts were considered, but they were not included because their distance from the shipping cask would reduce the dose rates to levels well below the dose rates experienced by the drivers. Stop times were assumed to accrue at the rate of 30 minutes per 4-hour driving time. TRAGIS outputs were used to determine the number of stops for each origin-destination.

Doses to the public at truck stops have been significant contributors to the doses calculated in previous RADTRAN 5 analyses. For this Appendix G analysis, stop doses are the sum of the doses to individuals located in two annular rings centered at the stopped vehicle, as illustrated in Figure G-1. The inner ring represents persons who may be at the truck stop at the same

Table G-5. RADTRAN 5 Incident-Free Exposure Parameters

Parameter	RADTRAN 5 Input Value	Source
Vehicle speed – rural, km/hr	88.49	Based on average speed in rural areas given in DOE (2002a). Because most travel is on interstate highways, the same vehicle speed is assumed in rural, suburban, and urban areas. No speed reductions were assumed for travel at rush hour.
Vehicle speed – suburban, km/hr	88.49	
Vehicle speed – urban, km/hr	88.49	
Traffic count – rural, vehicles/hr	530	DOE (2002a)
Traffic count – suburban, vehicles/hr	760	
Traffic count – urban, vehicles/hr	2400	
Dose rate at 1 m from vehicle, mSv/hr	0.14	Approximate dose rate at 1 m (3 ft) that is equivalent to maximum dose rate allowed by the U.S. Department of Transportation and NRC regulations (i.e., 0.1 mSv/hr at 2 m (~7 ft) from the side of a transport vehicle) (DOE 2002b)
Packaging dimensions, m	Length – 5.2 Diameter – 1.0	DOE (2002b)
Number of truck crew	2	(AEC 1972; NRC 1977a; DOE 2002a)
Stop time, hr/trip	Route-specific	See Table H-6.
Population density at stops, persons/km^2	30,000	Sprung et al. (2000)
Min/max radii of annular area around vehicle at stops, m	1 to 10	Sprung et al. (2000)
Shielding factor applied to annular area surrounding vehicle at stops	1 (no shielding)	Sprung et al. (2000)
Population density surrounding truck stops, persons/km^2	340	Sprung et al. (2000)
Min/max radius of annular area surrounding truck stop, m	10 to 800	Sprung et al. (2000)
Shielding factor applied to annular area surrounding truck stop	0.2	Sprung et al. (2000)

time as a spent fuel shipment and extends 1 to 10 m from the edge of the vehicle. The outer ring represents persons who reside near a truck stop and extends from 10 to 800 m from the vehicle. This scheme is the same as that used in Sprung et al. (2000).

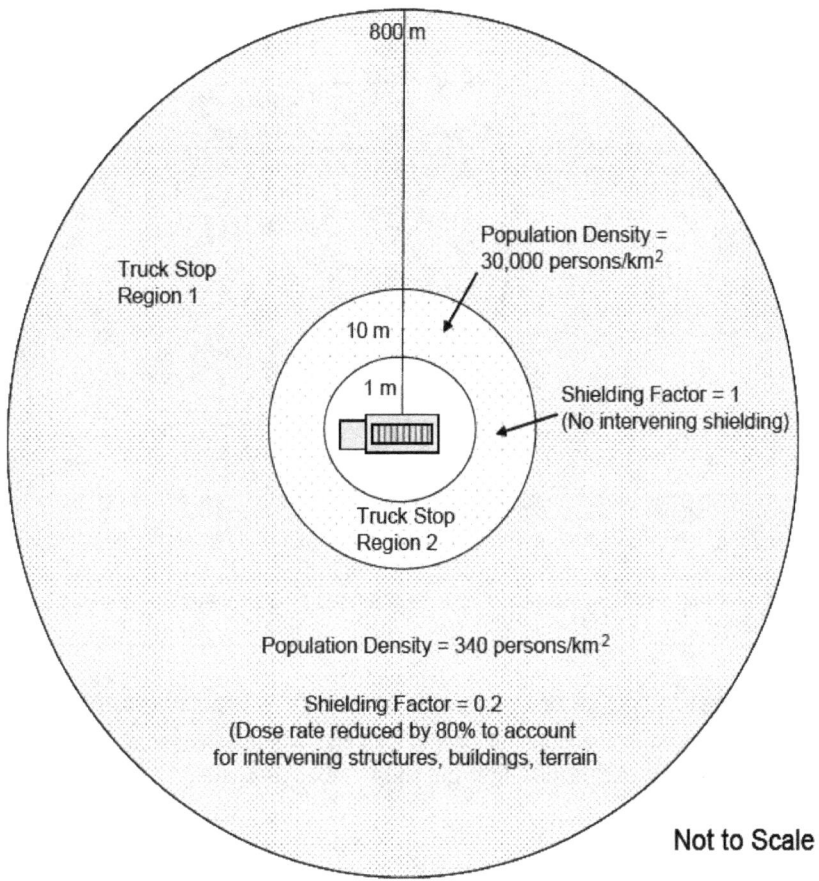

Figure G-1. Illustration of Truck Stop Model (Sprung et al. 2000)

Population densities and shielding factors were also taken from Sprung et al. (2000) and were based on the observations of Griego et al. (1996).

The results of these routine (incident-free) exposure calculations are shown in Table G-6 for spent fuel shipments from all 11 primary and alternative sites to the proposed Yucca Mountain Repository. Population dose estimates are given for workers (i.e., truck crew members), onlookers (doses to persons at truck stops and persons and on highways exposed to the spent fuel shipments), and along the route (persons living near the highway).

Table G-6. Routine (Incident-Free) Radiation Doses to Transport Workers and the Public from Shipping Spent Fuel from Potential ESP Sites to the Proposed High-Level Waste Repository at Yucca Mountain

Reactor Site	Population Dose, person-Sv/shipment[a]		
	Crew	Onlookers	Along Route
Braidwood[b]	7.1×10^{-4}	2.4×10^{-3}	4.4×10^{-5}
Clinton	7.2×10^{-4}	2.5×10^{-3}	4.5×10^{-5}
FitzPatrick	9.8×10^{-4}	3.5×10^{-3}	9.5×10^{-5}
Grand Gulf[c]	8.7×10^{-4}	2.8×10^{-3}	7.0×10^{-5}
North Anna	1.0×10^{-3}	3.5×10^{-3}	9.2×10^{-5}
Pilgrim	1.1×10^{-3}	3.9×10^{-3}	1.2×10^{-4}
Portsmouth	9.1×10^{-4}	3.2×10^{-3}	7.3×10^{-5}
Quad Cities	6.7×10^{-4}	2.1×10^{-3}	4.1×10^{-5}
Savannah River	9.9×10^{-4}	3.5×10^{-3}	1.0×10^{-4}
Surry	1.1×10^{-3}	3.5×10^{-3}	9.7×10^{-5}
Zion	7.3×10^{-4}	2.5×10^{-3}	5.2×10^{-5}

(a) Multiply person-Sv/shipment by 100 to obtain doses in person-rem/shipment.

(b) Dresden and LaSalle can be assumed to be bounded by the Braidwood values because of the proximity of the sites.

(c) The River Bend alternative site can be assumed to be bounded by the Grand Gulf values because of the proximity of the sites.

This discussion addresses whether or not the environmental effects of incident-free advanced reactor spent fuel shipments are within the guidelines established in Table S–4. The bounding cumulative doses to the exposed population given in Table S–4 are:

- Transport workers   0.04 person-Sv (4 person-rem) per reference reactor year.

- General public (onlookers and along route)   0.03 person-Sv (3 person-rem) per reference reactor year.

Calculation of the cumulative doses entailed converting the per-shipment risks given in Table G-6 to estimates of environmental effects per reference reactor year of operation. The per-shipment results, which are independent of reactor type (i.e., the doses are dependent on the assumed external radiation dose rate emitted from the cask, which is fixed at the regulatory maximum limit for all of the advanced reactor types), are given in terms of the population dose per shipment of spent fuel. To develop estimates of the annual environmental impacts, the following assumptions were made:

- The basis for the annual number of shipments of spent fuel from the reference LWR in WASH-1238 (AEC 1972) will be used. In WASH-1238, it was assumed that 60 shipments per year would be made, each shipment carrying 0.5 MTU of spent fuel.

This equates to shipping 30 MTU of spent fuel per year. This is equivalent to the annual refueling requirements for the reference LWR. It was assumed that the other reactor types would also ship spent fuel at a rate equal to their annual refueling requirements.

- Shipping cask capacities that were used to calculate annual spent fuel shipments for the advanced LWRs were assumed to be the same as for the reference LWR (i.e., approximately 0.5 MTU per truck shipment).

- The annual numbers of spent fuel shipments from the advanced gas-cooled reactors were taken directly from INEEL (2003). These estimates were 34 shipments per year from a GT-MHR site and 12 shipments per year from the PBMR site.

Table G-7 provides the estimated annual population doses from routine (incident-free) transportation of spent fuel from ESP sites to the proposed Yucca Mountain Repository. The results in Table G-9 have been normalized to the WASH-1238 (AEC 1972) net electrical generation (i.e., 880 MW(e)). Although radiation may cause cancers at high doses and high dose rates, currently there are no data that unequivocally establish the occurrence of cancer following exposure to low doses below about 100 mSv (10,000 mrem) and at low dose rates. However, radiation protection experts conservatively assume that any amount of radiation exposure may pose some risk of causing cancer or a severe hereditary effect and that the risk is higher for higher radiation exposures. Therefore, a linear, no-threshold dose response model is used to describe the relationship between radiation dose and detriments such as cancer induction. A recent report by the National Research Council (2006), the BEIR VII report, supports the linear, no-threshold dose response theory. Simply stated, any increase in dose, no matter how small, results in an incremental increase in health risk. NRC accepts this theory as a conservative model for estimating health risks from radiation exposure, recognizing that the model probably over-estimates those risks.

Based on this model, the staff estimates the risk to the public from radiation exposure using the nominal probability coefficient for total detriment (730 fatal cancers, nonfatal cancers, and severe hereditary effects per 10,000 person-Sv [1,000,000 person-rem]) from ICRP Publication 60 (ICRP 1991). All the population doses presented in Table G-7 are less than one person-Sv/yr (100 person-rem/yr); therefore, the total detriment estimates associated with these population doses would all be less than 0.1 fatal cancers, nonfatal cancers, and severe hereditary effects per year. These risks are very small compared to the fatal cancers, nonfatal cancers, and severe hereditary effects that would occur annually in the same population from exposure to natural sources of radiation.

As shown in Table G-7, some of the estimated population doses are higher than the Table S–4 conditions. Two key reasons for the higher population doses relative to Table S–4 are the higher number of spent fuel shipments estimated for some of the reactor technologies and the

Table G-7. Routine (Incident-Free) Population Doses from Spent Fuel Transportation, Normalized to Reference LWR Net Electrical Generation

Reactor Type	Reference LWR (WASH-1238)			ABWR/ESBWR			Surrogate AP1000			ACR-700		
No. Shipments per year	60			41			40			90		
	Environmental Effects, person-Sv per reference reactor year[a]											
Reactor Site	Crew	Onlookers	Along Route	Crew	Onlookers	Along Route	Crew	Onlookers	Along Route	Crew	Onlookers	Along Route
Braidwood[b]	4.2×10^{-2}	1.5×10^{-1}	2.6×10^{-3}	2.9×10^{-2}	1.0×10^{-1}	1.8×10^{-3}	2.8×10^{-2}	9.7×10^{-2}	1.7×10^{-3}	6.3×10^{-2}	2.2×10^{-1}	3.9×10^{-3}
Clinton	4.3×10^{-2}	1.5×10^{-1}	2.7×10^{-3}	2.9×10^{-2}	1.0×10^{-1}	1.8×10^{-3}	2.8×10^{-2}	9.7×10^{-2}	1.8×10^{-3}	6.4×10^{-2}	2.2×10^{-1}	4.1×10^{-3}
FitzPatrick	5.9×10^{-2}	2.1×10^{-1}	5.7×10^{-3}	4.0×10^{-2}	1.4×10^{-1}	3.9×10^{-3}	3.9×10^{-2}	1.4×10^{-1}	3.8×10^{-3}	8.8×10^{-2}	3.1×10^{-1}	8.5×10^{-3}
Grand Gulf[c]	5.2×10^{-2}	1.7×10^{-1}	4.2×10^{-3}	3.5×10^{-2}	1.2×10^{-1}	2.8×10^{-3}	3.4×10^{-2}	1.1×10^{-1}	2.7×10^{-3}	7.8×10^{-2}	2.5×10^{-1}	6.2×10^{-3}
North Anna	6.2×10^{-2}	2.1×10^{-1}	5.5×10^{-3}	4.2×10^{-2}	1.4×10^{-1}	3.7×10^{-3}	4.1×10^{-2}	1.4×10^{-1}	3.6×10^{-3}	9.2×10^{-2}	3.2×10^{-1}	8.2×10^{-3}
Pilgrim	6.5×10^{-2}	2.3×10^{-1}	7.0×10^{-3}	4.4×10^{-2}	1.6×10^{-1}	4.8×10^{-3}	4.3×10^{-2}	1.5×10^{-1}	4.6×10^{-3}	9.8×10^{-2}	3.5×10^{-1}	1.0×10^{-2}
Portsmouth	5.5×10^{-2}	1.9×10^{-1}	4.4×10^{-3}	3.7×10^{-2}	1.3×10^{-1}	3.0×10^{-3}	3.6×10^{-2}	1.2×10^{-1}	2.9×10^{-3}	8.1×10^{-2}	2.8×10^{-1}	6.6×10^{-3}
Quad Cities	4.0×10^{-2}	1.3×10^{-1}	2.4×10^{-3}	2.7×10^{-2}	8.6×10^{-2}	1.7×10^{-3}	2.6×10^{-2}	8.4×10^{-2}	1.6×10^{-3}	6.0×10^{-2}	1.9×10^{-1}	3.6×10^{-3}
Savannah River	6.0×10^{-2}	2.1×10^{-1}	6.0×10^{-3}	4.0×10^{-2}	1.4×10^{-1}	4.1×10^{-3}	3.9×10^{-2}	1.4×10^{-1}	4.0×10^{-3}	8.9×10^{-2}	3.2×10^{-1}	9.0×10^{-3}
Surry	6.4×10^{-2}	2.1×10^{-1}	5.8×10^{-3}	4.3×10^{-2}	1.4×10^{-1}	3.9×10^{-3}	4.2×10^{-2}	1.4×10^{-1}	3.8×10^{-3}	9.5×10^{-2}	3.2×10^{-1}	8.7×10^{-3}
Zion	4.4×10^{-2}	1.5×10^{-1}	3.1×10^{-3}	3.0×10^{-2}	1.0×10^{-1}	2.1×10^{-3}	2.9×10^{-2}	9.7×10^{-2}	2.0×10^{-3}	6.5×10^{-2}	2.2×10^{-1}	4.6×10^{-3}

Table G-7. (contd)

Reactor Type	IRIS			GT-MHR			PBMR		
No. Shipments per year	35			34			12		
	Environmental Effects, person-rem per reference reactor year[a]								
Reactor Site	Crew	Onlookers	Along Route	Crew	Onlookers	Along Route	Crew	Onlookers	Along Route
Braidwood	2.5×10^{-2}	8.5×10^{-2}	1.5×10^{-3}	2.4×10^{-2}	8.2×10^{-2}	1.5×10^{-3}	7.9×10^{-3}	2.7×10^{-2}	4.9×10^{-4}
Clinton	2.5×10^{-2}	8.5×10^{-2}	1.6×10^{-3}	2.4×10^{-2}	8.2×10^{-2}	1.5×10^{-3}	8.0×10^{-3}	2.8×10^{-2}	5.1×10^{-4}
FitzPatrick	3.4×10^{-2}	1.2×10^{-1}	3.3×10^{-3}	3.3×10^{-2}	1.2×10^{-1}	3.2×10^{-3}	1.1×10^{-2}	3.9×10^{-2}	1.1×10^{-3}
Grand Gulf	3.0×10^{-2}	9.8×10^{-2}	2.4×10^{-3}	2.9×10^{-2}	9.4×10^{-2}	2.3×10^{-3}	9.7×10^{-3}	3.2×10^{-2}	7.8×10^{-4}
North Anna	3.6×10^{-2}	1.2×10^{-1}	3.2×10^{-3}	3.4×10^{-2}	1.2×10^{-1}	3.1×10^{-3}	1.2×10^{-2}	4.0×10^{-2}	1.0×10^{-3}
Pilgrim	3.8×10^{-2}	1.3×10^{-1}	4.0×10^{-3}	3.6×10^{-2}	1.3×10^{-1}	3.9×10^{-3}	1.2×10^{-2}	4.3×10^{-2}	1.3×10^{-3}
Portsmouth	3.1×10^{-2}	1.1×10^{-1}	2.5×10^{-3}	3.0×10^{-2}	1.1×10^{-1}	2.4×10^{-3}	1.0×10^{-2}	3.6×10^{-2}	8.2×10^{-4}
Quad Cities	2.3×10^{-2}	7.4×10^{-2}	1.4×10^{-3}	2.2×10^{-2}	7.1×10^{-2}	1.4×10^{-3}	7.5×10^{-3}	2.4×10^{-2}	4.6×10^{-4}
Savannah River	3.4×10^{-2}	1.2×10^{-1}	3.5×10^{-3}	3.3×10^{-2}	1.2×10^{-1}	3.3×10^{-3}	1.1×10^{-2}	3.9×10^{-2}	1.1×10^{-3}
Surry	3.7×10^{-2}	1.2×10^{-1}	3.3×10^{-3}	3.5×10^{-2}	1.2×10^{-1}	3.2×10^{-3}	1.2×10^{-2}	4.0×10^{-2}	1.1×10^{-3}
Zion	2.5×10^{-2}	8.5×10^{-2}	1.8×10^{-3}	2.4×10^{-2}	8.2×10^{-2}	1.7×10^{-3}	8.2×10^{-3}	2.8×10^{-2}	5.8×10^{-4}

(a) Multiply person-Sv/yr by 100 to obtain doses in person-rem/yr.
(b) The River Bend alternative site can be assumed to be bounded by the Grand Gulf values because of the proximity of the sites.
(c) Dresden and LaSalle can be assumed to be bounded by the Braidwood values because of the proximity of the sites.

longer shipping distances used in this assessment than were used in WASH-1238 (AEC 1972).
WASH-1238 used a "typical" distance for a spent fuel shipment of 1600 km (1000 mi), whereas
the shipping distances used in this assessment ranged from about 2900 km (1800 mi) to
4700 km (2900 mi). The higher numbers of shipments are based on spent fuel shipping-casks
designed to transport short-cooled fuel (150 days out of the reactor). It was assumed in this
analysis that the shipping-cask capacities are 0.5 MTU/shipment, roughly equivalent to one
PWR or two BWR spent fuel assemblies per shipment. Newer designs are based on
longer-cooled spent fuel (5 years out of reactor) and have larger capacities than those used in
this assessment. DOE (2002b) spent fuel shipping-cask capacities were approximately
1.8 MTU/shipment, or up to four PWR or nine BWR fuel assemblies per shipment. Use of the
newer shipping-cask designs will reduce the number of spent fuel shipments and the associated
environmental impacts. If the population doses are adjusted for the shipping distance (a factor
of 2 to 3) and shipping cask capacity (a factor of 4), the routine population doses from spent fuel
shipments from all reactor types and all sites fall within the Table S–4 conditions.

Most of the stops made for actual spent fuel shipments are short duration stops
(i.e., 10 minutes) for brief visual inspections of the cargo (e.g., checking the cask tie-downs).
These stops typically occur in areas devoid of people, such as overpasses or freeway ramps in
unpopulated areas. Therefore, doses to residents surrounding these types of stops are
negligible. In DOE (2002b), close-proximity exposures (i.e., from 1 to 15.8 m from the cask)
were not assumed to occur at the short-duration inspection stops. In this analysis, for the
purpose of developing bounding estimates of environmental effects, close-proximity (1 to 10 m
from cask) exposures at all truck stops were included in the RADTRAN 5 calculations. Because
the numbers of stops in this analysis are effectively doubled relative to DOE (2002b), truck stop
doses are also doubled. The doses to residents would also be lower; however, because doses
to residents are two to three orders of magnitude (i.e., a factor of 100 to 1000) less than the
calculated close-proximity doses, this reduction does not affect the total stop dose.

The number of exposed persons at stops is higher in this analysis by about a factor of 1.5
relative to DOE (2002b) assumptions (6.9 persons in DOE 2002b versus 10 persons assumed
in this analysis). Thus, the bounding doses calculated in this analysis are also a factor of 1.5
(10 divided by 6.9) greater than those given in DOE (2002b). Furthermore, empirical data
provided in Griego et al. (1996) indicate that a 30-minute stop is toward the high end of the stop
time distribution. Average stop times for food and refueling observed by Griego et al. (1996) are
on the order of 18 minutes. This amounts to another factor of 1.5 increase in stop doses
calculated here relative to DOE (2002b).

Based on these observations, the staff concluded that the stop model used in this study overestimates public doses at stops by approximately a factor of four (factor of two for close-proximity exposure time at stops, a factor of 1.5 for average stop time at food and refueling stops, and a factor of 1.5 for the number of people in proximity to the shipping cask). Coupled with the factor of two reduction in shipping cask dose rates that result from fuel aging, the doses to onlookers at stops could be reduced to about one-eighth of the doses shown in Table G-7 [1/(2 x 1.5 x 1.5 x 2) ≈ 0.12] to reflect more realistic truck shipping conditions. Based on the previous discussion, use of more realistic dose rates, shipping cask capacities, and truck stop model assumptions in the RADTRAN 5 calculations could substantially reduce the environmental effects presented in Table G-7.

Table G-8 provides a comparison between the radiological incident-free doses calculated in NUREG-0170 (NRC 1977a) and those calculated here. The table also summarizes the key incident-free input parameters used in NUREG-0170 and in this study. Comparisons are also made between the doses for spent fuel shipments in NUREG-0170 and doses calculated for a shipment from the Quad Cities, Iowa, to the proposed Yucca Mountain Repository because the shipping distances are comparable (2530 km in NUREG-0170 versus 2853 km for Quad Cities to Yucca Mountain). As shown in the table, many parameters have changed over the years and the technical bases for them have improved. For example, the work of Griego et al. (1996) has improved the basis for assumptions about stop times and persons exposed at truck stops, and the TRAGIS computer code has improved the basis for shipping distances and population distributions along highway routes.

The incident-free impacts at truck stops shown in the table have been adjusted, as discussed above, to reflect more realistic conditions than assumed in the bounding analysis. Adjustments were not made to the onlookers, along route, and crew doses shown in Table G-7. As shown, the adjusted doses in Table G-10 for spent fuel shipments from the Quad Cities to the proposed Yucca Mountain Repository are about a factor of two lower than the per-shipment doses from NUREG-0170 when the doses to and doses associated with in-transit storage from NUREG-0170 are excluded. Storage doses were excluded from this analysis because spent fuel shipments proceed directly from the reactor site to Yucca Mountain with no intermediate storage involved. Handler doses were excluded from this Appendix G analysis because doses to workers who load the spent fuel cask at reactors and unload them at the proposed repository are treated as facility doses, not transportation doses.

Table G-8. Comparison of Incident-Free Doses from NUREG-0170 (NRC 1977a) Spent Fuel Shipments and Spent Fuel Shipment from Quad-Cities to the Proposed High-Level Waste Repository at Yucca Mountain

Incident-Free Exposure Parameter	NUREG-0170 (NRC 1977a)	This Study (Quad Cities to Yucca Mountain)[a]
One-way shipping distance, km	2530	2853
Travel fraction		
Urban	0.05	0.02
Suburban	0.05	0.12
Rural	0.9	0.86
Population density along highway, persons per km^2		
Urban	3861	2391.3
Suburban	719	310.2
Rural	6	9.1
Speed, km/hr		
Urban	24	88
Suburban	40	88
Rural	88	88
Traffic count, vehicles/hr		
Urban	2800	2400
Suburban	780	760
Rural	470	530
Shipment dose Rate, mSv/hr at 2m	0.1	0.1
Crew dose rate, mSv/hr	0.02	Calculated (7.4 m from package)
Stop time, hr per trip		
Urban	2	3 hours per trip (30 minutes per
Suburban	5	4 hours driving time)
Rural	1	
Population density at stops (per km^2)		
Urban	3861	Distribution: 1 to 10 m - 30,000;
Suburban	719	10 to 800 m - 340 (see Figure G-1)
Rural	6	
Person-Sv/shipment		
Crew	1.2×10^{-3}	4.8×10^{-4}
Off-link	1.5×10^{-4}	3.1×10^{-4}
On-link	7.4×10^{-5}	1.7×10^{-4}
Stops	1.9×10^{-4}	$1.7 \times 10^{-4(b)}$
Total	1.6×10^{-3}	8.5×10^{-4}
Handlers + Storage	2.1×10^{-3}	Not calculated
Grand Total	**3.7×10^{-3}**	**8.5×10^{-4}**

(a) Tables G-7 and G-9 provide the basis for these input parameters.
(b) Stop doses have been adjusted as described in the text to reflect more realistic assumptions than were used in the bounding analysis (Table G-9).

G.2.2 Transportation Accident Impacts

RADTRAN 5 assesses accident risk by calculating a risk value, which is the product of probabilities and the consequences of accidents. RADTRAN 5 considers a spectrum of potential transportation accidents, ranging from those with high frequencies and low consequences (e.g., "fender-benders") to those with low frequencies and high consequences (e.g., accidents in which the shipping container is exposed to severe mechanical and thermal conditions).

Radionuclide inventories are important parameters in the calculation of accident risks. The radionuclide inventories used in this analysis were taken directly from the *Early Site Permit Environmental Report Sections and Supporting Documentation* (INEEL 2003). The report included hundreds of radionuclides for each advanced reactor type. A screening analysis was conducted to select the dominant contributors to accident risks to simplify the RADTRAN 5 calculations. The screening identifies the radionuclides that will contribute more than 99.999 percent of the dose from inhalation.

A sum-of-fractions approach was used for this screening. First, the inventory of each radionuclide was multiplied by its respective inhalation dose conversion factor, taken from Federal Guidance Report 13 (EPA 2002). These values were then summed. Then, each inventory-conversion factor product was divided by the sum of the products to obtain the fraction of the total inhalation dose for each radionuclide. The resulting fractions were then sorted from largest to smallest, their cumulative contributions were calculated, and those that contributed to 99.999 percent of the inhalation-dose potential were selected. Two gases, krypton-85 and iodine-129, were added to the list because they are more easily released than the solid and semi-volatile species contained in the fuel.

The inventories of radionuclides used in this study are shown in Table G-9. Note that the list of radionuclides provided in the table includes all of the radionuclides that were included in the analysis conducted by Sprung et al. (2000), which validates the screening process used in this EIS. Also note that INEEL (2003) did not provide radionuclide source terms for radioactive material deposited on the external surfaces of LWR spent fuel rods, which is commonly referred to as "crud." In addition, data on activation products was provided for only the ABWR. The ABWR spent fuel transportation risks were calculated assuming the entire Co-60 inventory is in the form of crud. This is very conservative as the source term used here is about two orders of magnitude greater than that given in Sprung et al. (2000). Because crud is deposited from corrosion products generated elsewhere in the reactor cooling system and the complete reactor design and operating parameters are uncertain, the quantities and characteristics of crud deposited on advanced reactor spent fuel are unknown at this time. Consequently, the impacts of crud and activation products on spent fuel transportation accident risks will need to be examined at the construction permit or combined operating license stage.

Table G-9 shows that the dominant radionuclides are approximately the same regardless of fuel type. The table does not show radionuclide inventory data for the ACR-700 and IRIS advanced reactors, as those were not given in INEEL (2003). Nor were they provided in WASH-1238 (AEC 1972) for the reference LWR. Consequently, accident risks were not quantified for these reactor types.

Table G-9. Radionuclide Inventories Used in the Transportation Accident Risk Calculations for Each Advanced Reactor Type

Radionuclide	ABWR and ESBWR Inventory, Bq/MTU[a]	Surrogate AP1000 Inventory, Bq/MTU	GT-MHR Inventory, Bq/MTU	PBMR Inventory, Bq/MTU
Am-241	4.96×10^{13}	2.69×10^{13}	8.18×10^{13}	7.55×10^{13}
Am-242m	1.24×10^{12}	4.85×10^{11}	5.03×10^{11}	8.51×10^{11}
Am-243	1.20×10^{12}	1.24×10^{12}	5.14×10^{11}	4.77×10^{12}
Ce-144	4.22×10^{14}	3.28×10^{14}	2.15×10^{15}	1.19×10^{15}
Cm-242	2.04×10^{12}	1.05×10^{12}	1.51×10^{12}	2.78×10^{12}
Cm-243	1.37×10^{12}	1.14×10^{12}	2.02×10^{11}	1.96×10^{12}
Cm-244	1.80×10^{14}	2.87×10^{14}	2.83×10^{13}	5.48×10^{14}
Cm-245	2.43×10^{10}	4.48×10^{10}	1.65×10^{8}	5.29×10^{10}
Co-60	1.01×10^{14}	--[b]	--[b]	--[b]
Cs-134	1.78×10^{15}	1.78×10^{15}	2.21×10^{15}	4.03×10^{15}
Cs-137	4.59×10^{15}	3.44×10^{15}	1.08×10^{16}	1.41×10^{16}
Eu-154	3.81×10^{14}	3.38×10^{14}	3.23×10^{14}	3.74×10^{14}
Eu-155	1.93×10^{14}	1.71×10^{14}	8.77×10^{13}	1.08×10^{14}
I-129	1.55×10^{9}	1.55×10^{9}	1.55×10^{9}	1.55×10^{9}
Kr-85	3.29×10^{14}	3.29×10^{14}	3.29×10^{14}	3.29×10^{14}
Pm-147	1.25×10^{15}	6.51×10^{14}	6.92×10^{15}	5.07×10^{15}
Pu-238	2.27×10^{14}	2.25×10^{14}	1.17×10^{14}	4.55×10^{14}
Pu-239	1.43×10^{13}	9.44×10^{12}	2.25×10^{13}	1.11×10^{13}
Pu-240	2.28×10^{13}	2.01×10^{13}	3.96×10^{13}	3.32×10^{13}
Pu-241	4.51×10^{15}	2.58×10^{15}	8.33×10^{15}	7.18×10^{15}
Pu-242	8.29×10^{10}	6.73×10^{10}	1.56×10^{11}	4.51×10^{11}
Ru-106	6.07×10^{14}	5.74×10^{14}	1.48×10^{15}	1.68×10^{15}
Sb-125	1.99×10^{14}	1.42×10^{14}	2.21×10^{14}	2.51×10^{14}
Sr-90	3.27×10^{15}	2.29×10^{15}	8.95×10^{15}	1.08×10^{16}
Y-90	3.27×10^{15}	2.29×10^{15}	8.95×10^{15}	1.08×10^{16}

(a) To convert Bq/MTU to Ci/MTU, divide the value by 3.7×10^{10}.

(b) Co-60 is an activation product. Only the ABWR/ESBWR submittal in INEEL (2003) provided inventory data for activation products.

Robust shipping casks are used to transport spent fuel because of the heavy radiation shielding and accident resistance required by 10 CFR Part 71. Spent fuel shipping casks must be certified Type B packaging systems, which means they must withstand a series of severe hypothetical accident conditions with essentially no loss of containment or shielding capability. These casks are also designed with fissile material controls to ensure that the spent fuel remains subcritical under both normal and accident conditions. The tests include a 9-m (30-ft) free drop onto an unyielding surface, a drop onto a puncture probe, an exposure to an engulfing 800°C fire for 30 minutes, and an underwater immersion. According to Sprung et al. (2000), the probability of encountering accident conditions more severe than these tests that could lead to shipping cask failure are less than 0.01 percent of all accidents (i.e., more than 99.99 percent of all accidents would not result in a release of radioactive material from the shipping cask). It was assumed that shipping casks for advanced reactor spent fuels will provide equivalent mechanical and thermal protection of the spent fuel cargo.

The RADTRAN 5 accident risk calculations were performed using unit radionuclide inventories (Bq/MTU) for the spent fuel shipments from the various reactor types. The resulting risk estimates were then multiplied by assumed annual spent fuel shipments (MTU/yr) to derive estimates of the annual accident risks associated with spent fuel shipments from each potential ESP site. As was done for routine exposures, it was assumed that the numbers of shipments of spent fuel per year are equivalent to the annual discharge quantities: 32.76 MTU/yr for the ABWR and ESBWR; 24.4 MTU/yr for a single-reactor surrogate AP1000 site; 6.8 MTU/yr for the four-reactor GT-MHR site; and 8.3 MTU/yr for the eight-reactor PBMR site. These data were taken from INEEL (2003) and have not been normalized to the reference LWR net electrical generation.

Route-specific accident rates (accidents per km) were derived for the RADTRAN 5 accident risk analysis. The approach used to develop accident rates for spent fuel shipments is as follows. The TRAGIS data provide estimates of the distance traveled in each state along a route and the type of highway (interstate, state highway, or other). Saricks and Tompkins (1999) provide accident rates for each state that are a function of highway type. The approach taken to estimate route-specific accident rates was to multiply the state-level accident or fatality rates by the distances traveled in each state on the corresponding highway type and then sum over all the states on each route. For example, for interstate highways, the interstate distances and interstate accident rates were used. For non-interstate highway travel, either the "Primary" or "Other" accident rates given by Saricks and Tompkins (1999) were used. This approach allowed computation of route-specific accident rates.

Transportation accident risk analysis in RADTRAN 5 is performed using an accident severity and package release model. The user can define up to 30 severity categories, with each category increasing in magnitude. Severity categories are related to fire, puncture, crush, and immersion environments created in vehicular accidents. For this analysis, the 19 severity categories defined by Sprung et al. (2000) were adopted.

Each severity category has an assigned conditional probability (or the probability, given an accident occurs, that it will be of the specified severity). The accident scenarios are further defined by allowing the user to input release fractions and aerosol and respirable fractions for each severity category. These fractions are a function of the physical-chemical properties of the materials being transported as well as the mechanical and thermal accident conditions that define the severity categories. The severity and release fractions used here are presented in Table G-10.

The severity categories and release fractions published by Sprung et al. (2000) were designed specifically to address accidents involving current generation LWR fuel and the current generation of spent fuel shipping casks. While some of the advanced reactor fuel designs are similar to current-generation reactor fuel designs (e.g., the ABWR, ESBWR, Surrogate

Table G-10. Severity and Release Fractions Used to Model Spent Fuel Transportation Accidents (Sprung et al. 2000)

Severity Category	Severity Fraction[b]	Release Fractions[a]				
		Gas	Cesium	Ruthenium	Particulates	Corrosion Products
1	1.53×10^{-8}	0.8	2.4×10^{-8}	6.0×10^{-7}	6.0×10^{-7}	2.0×10^{-3}
2	5.88×10^{-5}	0.14	4.1×10^{-9}	1.0×10^{-7}	1.0×10^{-7}	1.4×10^{-3}
3	1.81×10^{-6}	0.18	5.4×10^{-9}	1.3×10^{-7}	1.3×10^{-7}	1.8×10^{-3}
4	7.49×10^{-8}	0.84	3.6×10^{-5}	3.8×10^{-6}	3.8×10^{-6}	3.2×10^{-3}
5	4.65×10^{-7}	0.43	1.3×10^{-8}	3.2×10^{-7}	3.2×10^{-7}	1.8×10^{-3}
6	3.31×10^{-9}	0.49	1.5×10^{-8}	3.7×10^{-7}	3.7×10^{-7}	2.1×10^{-3}
7	0	0.85	2.7×10^{-5}	2.1×10^{-6}	2.1×10^{-6}	3.1×10^{-3}
8	1.13×10^{-8}	0.82	2.4×10^{-8}	6.1×10^{-7}	6.1×10^{-7}	2.0×10^{-2}
9	8.03×10^{-11}	0.89	2.7×10^{-8}	6.7×10^{-7}	6.7×10^{-7}	2.2×10^{-3}
10	0	0.91	5.9×10^{-6}	6.8×10^{-7}	6.8×10^{-7}	2.5×10^{-3}
11	1.44×10^{-10}	0.82	2.4×10^{-8}	6.1×10^{-7}	6.1×10^{-7}	2.0×10^{-3}
12	1.02×10^{-12}	0.89	2.7×10^{-8}	6.7×10^{-7}	6.7×10^{-7}	2.2×10^{-3}
13	0	0.91	5.9×10^{-6}	6.8×10^{-7}	6.8×10^{-7}	2.5×10^{-3}
14	7.49×10^{-11}	0.84	9.6×10^{-5}	8.4×10^{-5}	1.8×10^{-5}	6.4×10^{-3}
15	0	0.85	5.5×10^{-5}	5.0×10^{-5}	9.0×10^{-6}	5.9×10^{-3}
16	0	0.91	5.9×10^{-6}	6.4×10^{-6}	6.8×10^{-7}	3.3×10^{-3}
17	0	0.91	5.9×10^{-6}	6.4×10^{-6}	6.8×10^{-7}	3.3×10^{-3}
18	5.86×10^{-6}	0.84	1.7×10^{-5}	6.7×10^{-8}	6.7×10^{-8}	2.5×10^{-3}
19	0.99993	0	0	0	0	0

(a) RADTRAN 5 also models the fraction of the released particulate material that is small enough to be dispersible in prevailing wind conditions and the fraction that is respirable. For this analysis, these parameters were set to 1.0 (i.e., 100 percent dispersible and 100 percent respirable).

(b) Severity fractions are the conditional probabilities, given the occurrence of an accident, that the mechanical and thermal conditions experienced by a spent fuel shipping cask are within the conditions defined by the Severity Category. See Sprung et al. (2000) for detailed information about the derivation of these data. Generic steel-depleted uranium-steel cask designs were assumed for the severity fractions.

AP1000, ACR-700, and IRIS), others are significantly different, including the GT-MHR and PBMR. Extrapolating the current generation of LWR fuel and shipping casks to advanced LWR fuels and shipping casks is expected to be relatively straightforward because the fuel form, cladding, and physical and mechanical properties are similar. Furthermore, substantial experimental data exist to develop technically defensible release fractions for various radionuclide groups (e.g., gases, semi-volatiles such as cesium and ruthenium, and particulates). However, because detailed experimental studies of releases from GT-MHR and PBMR fuels have not been this approach is bounding. However, gas-cooled reactors operate at much higher temperatures than LWRs; thus, high-temperature conditions anticipated in transportation accident fires are expected to have less effect on radionuclide releases than they would for LWR fuels. Consequently, smaller release fractions are anticipated for advanced gas-cooled reactor fuels than for LWR fuels subjected to thermal transients.

For accidents that result in a release of radioactive material, RADTRAN 5 assumes the material is dispersed into the environment according to standard Gaussian diffusion models. The code allows the user to choose two different methods for modeling the atmospheric transport of radionuclides after a potential accident. The user can input either Pasquill atmospheric-stability category data or averaged time-integrated concentrations. In this analysis, the default standard cloud option (using time-integrated concentrations) was used.

RADTRAN 5 was used to calculate the population dose from the released radioactive material for four of five[a] possible exposure pathways:

- External dose from exposure to the passing cloud of radioactive material (cloudshine).

- External dose from radionuclides deposited on the ground by the passing plume (groundshine). This analysis included the radiation exposures from this pathway even though the area surrounding a potential accidental release would be evacuated and decontaminated, thus preventing long-term exposures from this pathway.

- Internal dose from inhalation of airborne radioactive contaminants (inhalation).

- Internal dose from radioactive materials that were deposited on the ground and then resuspended (resuspension). This analysis included the radiation exposures from this pathway even though evacuation and decontamination of the area surrounding a potential accidental release would prevent long-term exposures.

(a) The internal dose from ingestion of contaminated food was not considered, as the staff assumed evacuation and subsequent interdiction of foodstuffs following a potential transportation accident.

A sixth pathway, external doses arising from increased radiation fields surrounding a shipping cask with damaged shielding, was considered but not included in the analysis. It is possible that shielding materials incorporated into the cask structures could become damaged as a result of an accident. For example, casks with lead shielding could undergo a slumping phenomenon in which impact or fire causes gaps to form in the lead. Radiation would penetrate through the gaps in the shielding at higher intensities, leading to higher radiation dose rates. These events, which are commonly referred to as "loss of shielding events," were not included in this assessment because their contribution to spent fuel transportation risks is much smaller than the dispersal accident risks.

Standard radionuclide uptake and dosimetry models are incorporated into RADTRAN 5. The computer code combines the accident consequences and frequencies of each severity category, sums up the severity categories, and then integrates across all the shipments. Accident-risk impacts are provided in the form of a collective population dose (person-rem over the entire shipping campaign).

The shipping distances and population distribution information for the routes used for the evaluation of the impacts of incident-free transportation (see Table G-4) were also used to calculate transportation accident impacts. Representative shipping casks described above were assumed.

Table G-11 presents unit (per MTU) accident risks associated with transportation of spent fuel from each potential ESP site to the proposed Yucca Mountain Repository.

Projected annual accident risks, normalized to the WASH-1238 (AEC 1972) reference LWR net electrical generation (i.e., 880 MW(e)) are presented in Table G-11. As expected, accident risks are highest for the longest shipments. Also, consistent with past spent fuel transportation risk assessments, the routine impacts are several orders of magnitude greater than accident impacts.

Considering the small magnitude of the risks presented in Table G-10 and the conservative computational methods and data used to address uncertainties, the overall transportation accident risks associated with ABWR, ESBWR, Surrogate AP1000, GT-MHR, and PBMR spent fuel shipments are judged to be small. Although likely to also be small, accident risks associated with IRIS and ACR-700 spent fuel shipments could not be analyzed because of the lack of radionuclide source-term data. Additional analyses are necessary to quantify the impacts of IRIS and ACR-700 spent fuel shipments.

Table G-11. Unit Spent Fuel Transportation Accident Risks for Advanced Reactors

Site	Advanced Reactor Type			
	ABWR/ ESBWR	Surrogate AP1000	GT-MHR	PBMR
Population Dose, person-Sv/MTU[a]				
Braidwood[b]	1.0×10^{-7}	1.0×10^{-8}	1.5×10^{-8}	2.5×10^{-8}
Clinton	1.1×10^{-7}	1.0×10^{-8}	1.5×10^{-8}	2.6×10^{-8}
FitzPatrick	1.9×10^{-7}	1.7×10^{-8}	2.5×10^{-8}	4.3×10^{-8}
Grand Gulf[c]	2.0×10^{-7}	1.9×10^{-8}	2.8×10^{-8}	4.7×10^{-8}
North Anna	2.3×10^{-7}	2.1×10^{-8}	3.2×10^{-8}	5.4×10^{-8}
Pilgrim	4.0×10^{-7}	3.7×10^{-8}	5.5×10^{-8}	9.3×10^{-8}
Portsmouth	2.3×10^{-7}	2.1×10^{-8}	3.1×10^{-8}	5.2×10^{-8}
Quad Cities	1.0×10^{-7}	9.4×10^{-9}	1.4×10^{-8}	1.4×10^{-8}
Savannah River	2.6×10^{-7}	2.4×10^{-8}	3.6×10^{-8}	6.1×10^{-8}
Surry	2.4×10^{-7}	2.2×10^{-8}	3.3×10^{-8}	5.6×10^{-8}
Zion	1.5×10^{-7}	1.4×10^{-8}	2.1×10^{-8}	3.5×10^{-8}

(a) To convert to person-rem, multiply person-Sv by 100.
(b) Dresden and LaSalle can be assumed to be bounded by the Braidwood values because of the proximity of the sites.
(c) The River Bend alternative site can be assumed to bounded by the Grand Gulf values because of the proximity of the sites.

Table G-12 presents the environmental consequences of transportation accidents when shipping spent fuel from the proposed ESP sites and alternative sites to the proposed Yucca Mountain Repository. The shipping distances and population distribution information for the routes were the same as those used for the normal "incident-free" conditions. The table presents estimates of population dose (person-Sv/reference reactor year) for several of the advanced reactor designs. These values are normalized to the WASH-1238 reference reactor (880-MW(e)) net electrical generation, 1100-MW(e) reactor operating at 80 percent capacity).

Although radiation may cause cancers at high doses and high dose rates, currently there are no data that unequivocally establish the occurrence of cancer following exposure to low doses below about 100 mSv (10,000 mrem) and low dose rates. However, radiation protection experts conservatively assume that any amount of radiation exposure may pose some risk of causing cancer or a severe hereditary effect and that the risk is higher for higher radiation exposures. Therefore, a linear, no-threshold dose response model is used to describe the relationship between radiation dose and detriments such as cancer induction. A recent report, the BEIR VII report (National Research Council 2006), supports the linear, no-threshold dose

Table G-12. Annual Spent Fuel Transportation Accident Impacts for Advanced Reactors, Normalized to Reference LWR Net Electrical Generation

	Advanced Reactor Type			
	ABWR/ESBWR	Surrogate AP1000	GT-MHR	PBMR
MTU/reference reactor year	20.3	19.7	6.0	5.8
Population Dose, person-Sv per reference reactor year[a]				
Braidwood[b]	2.1×10^{-6}	2.0×10^{-7}	8.9×10^{-8}	1.5×10^{-7}
Clinton	2.3×10^{-6}	2.0×10^{-7}	9.1×10^{-8}	1.5×10^{-7}
FitzPatrick	3.8×10^{-6}	3.3×10^{-7}	1.5×10^{-7}	2.5×10^{-7}
Grand Gulf[c]	4.1×10^{-6}	3.7×10^{-7}	1.7×10^{-7}	2.7×10^{-7}
North Anna	4.7×10^{-6}	4.2×10^{-7}	1.9×10^{-7}	3.1×10^{-7}
Pilgrim	8.1×10^{-6}	7.2×10^{-7}	3.3×10^{-7}	5.4×10^{-7}
Portsmouth	4.6×10^{-6}	4.0×10^{-7}	1.8×10^{-7}	3.0×10^{-7}
Quad Cities	2.1×10^{-6}	1.8×10^{-7}	8.4×10^{-8}	8.1×10^{-8}
Savannah River	5.3×10^{-6}	4.7×10^{-7}	2.2×10^{-7}	3.5×10^{-7}
Surry	4.9×10^{-6}	4.3×10^{-7}	2.0×10^{-7}	3.2×10^{-7}
Zion	3.0×10^{-6}	2.7×10^{-7}	1.2×10^{-7}	2.0×10^{-7}

(a) Multiply person-Sv/reference reactor year by 100 to obtain doses in person-rem/reference reactor year.
(b) Dresden and LaSalle can be assumed to be bounded by the Braidwood values because of the proximity of the sites.
(c) The River Bend alternative site can be assumed to be bounded by the Grand Gulf values because of the proximity of the sites.

response theory. The theory states that any increase in dose, no matter how small, results in an incremental increase in health risk. This theory is accepted by the NRC as a conservative model for estimating health risks from radiation exposure, recognizing that the model probably over-estimates those risks.

Based on this model, the staff estimates the risk to the public from radiation exposure using the nominal probability coefficient for total detriment (730 fatal cancers, nonfatal cancers, and severe hereditary effects per 10,000 person-Sv [1,000,000 person-rem]) from ICRP Publication 60 (ICRP 1991). All the population doses presented in Table G-12 are less than 1.0×10^{-5} person-Sv (1.0×10^{-3} person-rem) per reference reactor year; therefore, the total detriment estimates associated with these population doses would all be less than 1.0×10^{-6} fatal cancers, nonfatal cancers, and severe hereditary effects per reference reactor year. These risks are quite small compared to the fatal cancers, nonfatal cancers, and severe hereditary effects that would be expected to occur annually in the same population from exposure to natural sources of radiation using the same risk model.

G.3 Shipment of Radioactive Waste

This section discusses the environmental effects of transporting radioactive waste from advanced reactor sites. The environmental conditions listed in 10 CFR 51.52 that apply to shipments of radioactive waste are as follows:

- Radioactive waste (except spent fuel) is packaged and in a solid form [51.52(a)(4)]

- Radioactive waste (except spent fuel) is shipped from the reactor by truck or rail [51.52(a)(5)].

INEEL (2003) indicates that all of the advanced reactors will transport their radioactive waste by truck. Furthermore, INEEL (2003) indicates that all of the advanced reactors plan to solidify and package their radioactive waste. In addition, all of the advanced reactors will be subject to NRC (10 CFR Part 71) and U.S. Department of Transportation regulations for the shipment of radioactive material (49 CFR Parts 171, 172, 173, 178).

Table S–4 also specifies the following conditions that apply to shipments of radioactive waste:

- Weight – less than 33,100 kg (73,000 lb) per truck or 100 tons per cask per rail car
- Traffic density – less than one truck shipment per day or three rail cars per month.

The advanced reactors are assumed to be capable of shipping their radioactive wastes in compliance with Federal or State weight restrictions. With respect to the traffic density, all of the advanced reactor vendors provided radioactive waste generation estimates. Table G-13 provides these estimates, in addition to the radioactive waste generation estimates for the reference LWR in WASH-1238 (AEC 1972).

As shown in the table, only the PBMR generates a larger volume of radioactive waste than the reference LWR in WASH-1238. However, the GT-MHR and PBMR information in INEEL (2003) assumed these advanced reactors would ship wastes using two different packaging systems: one that hauls 28 m^3/shipment (1000 ft^3 per shipment) and one that hauls 5.7 m^3/shipment (200 ft^3/per shipment). Under those conditions, the number of shipments of radioactive waste per year, normalized to 1100 MW(e) electric generation capacity, would be about six shipments/year per 1100 MW(e) (880 net MW(e)) for the GT-MHR and seven shipments/year per 1100 MW(e) for the PBMR. These estimates are well below the reference LWR (42 shipments per 1100 MW(e)). In any event, all the estimates are well below the one truck shipment per day condition given in 10 CFR 51.52, Table S–4. Doubling the shipment estimates to account for empty return shipments is still well below the one shipment per day condition.

Table G-13. Summary of Radioactive Waste Shipments for Advanced Reactors

Reactor Type	INEEL (2003) Waste Generation Information	Annual Waste Volume, m³/yr per Unit	Electrical Output, MW(e) per Unit	Normalized Rate, m³/1100 MW(e) Reactor (880 MW(e) Net)[a]	Shipments/ 1100 MW(e) (880 MW(e) Net) Electrical Output[b]
Reference LWR (WASH-1238)	100 m³/yr per unit	108	1100	108	46
ABWR	100 m³/yr per unit	100	1500[c]	62	27
ESBWR	100 m³/yr per unit	100	1500[c]	62	27
Surrogate AP1000	55 m³/yr per unit	110 (2 units)	2300[c]	45	20
ACR-700	47.5 m³/yr per unit	95 (2 units)	1462[d]	64	28
IRIS	25 m³/yr per unit	75 (3 units)	1005[e]	67	29
GT-MHR	98 m³/yr (4-unit plant)	98 (4 units)	1140[f]	86	37[h]
PBMR	100 drums/yr per unit	168 (8 units)	1320[g]	118	51[h]

(a) Capacity factors used to normalize the waste generation rates to an equivalent electrical generation output are given in Table 6-3 for each reactor type. All are normalized to 880 MW(e) net electrical output (1100-MW(e) plant with an 80 percent capacity factor).

(b) The number of shipments per 1100 MW(e) was calculated assuming the WASH-1238 average waste shipment capacity of 2.34 m³ per shipment (108 m³/yr divided by 46 shipments).

(c) The ABWR and ESBWR units include one reactor at 1500 MW(e) and the surrogate AP1000 unit includes two reactors at 1150 MW(e).

(d) The ACR-700 unit includes two reactors at 731 MW(e) per reactor.

(e) The IRIS unit includes three reactors at 335 MW(e) per reactor.

(f) The GT-MHR unit includes four reactors at 285 MW(e) per reactor.

(g) The PBMR unit includes eight reactors at 165 MW(e) per reactor.

(h) INEEL (2003) states that 90 percent of the waste could be shipped on trucks carrying 28 m³ (1000 ft³) of waste and the remaining 10 percent in shipments carrying 5.7 m³ (200 ft³) of radioactive waste. This would result in five to six shipments per year after normalization to the reference LWR electrical output.

Conversions: 1 m³ = 35.31 ft³. Drum volume = 210 liters (0.21 m³).

G.4 References

10 CFR Part 51. Code of Federal Regulations. Title 10, *Energy,* Part 51, "Environmental Protection Regulations for Domestic Licensing and Related Regulatory Functions."

10 CFR Part 52. Code of Federal Regulations. Title 10, *Energy,* Part 52, "Early Site Permits, Standard Design Certifications, and Combined Licenses for Nuclear Power Plants."

10 CFR Part 71. Code of Federal Regulations Title 10, *Energy*, "Packaging and Transportation of Radioactive Material."

49 CFR Part 171. Code of Federal Regulations. Title 49, *Transportation,* Part 171, "General Information, Regulations, and Definitions."

49 CFR Part 172. Code of Federal Regulations. Title 49, *Transportation,* Part 172, "Hazardous Materials Table, Special Provisions, Hazardous Materials Communications, Emergency Response Information, and Training Requirements."

49 CFR Part 173. Code of Federal Regulations. Title 49, *Transportation,* Part 173, "Shippers - General Requirements for Shipments and Packagings."

49 CFR Part 178. Code of Federal Regulations. Title 49, *Transportation,* Part 178, "Specifications for Packagings."

Griego, N.R., J.D. Smith, and K.S. Neuhauser. 1996. "Investigation of RADTRAN Stop Model Input Parameters for Truck Stops" in *Conference Proceedings B Waste Management 96*, CONF-960212-44. Tucson, Arizona.

Hostick, C.J., J.C. Lavender, and B.H. Wakeman. 1992. *Time/Motion Observations and Dose Analysis of Reactor Loading, Transportation, and Dry Unloading of an Overweight Truck Spent Fuel Shipment.* PNL-7206, Pacific Northwest Laboratory, Richland, Washington.

Idaho National Engineering and Environmental Laboratory (INEEL). 2003. *Early Site Permit Environmental Report Sections and Supporting Documentation.* Office of Nuclear Energy, Science, and Technology, U.S. Department of Energy, Washington, D.C.

International Commission on Radiological Protection (ICRP). 1991. *1990 Recommendations of the International Commission on Radiological Protection.* ICRP Publication 60, November 1990, Pergamon Press, New York.

Johnson, P.E., and R.D. Michelhaugh. 2000. *Transportation Routing Analysis Geographic Information System (WebTRAGIS) User's Manual.* ORNL/TM-2000/86, Oak Ridge National Laboratory, Oak Ridge, Tennessee. Accessed on the Internet on February 7, 2005 at http://www.ornl.gov/~webworks/cpr/v823/rpt/106749.pdf.

National Research Council. 2006. Health Risks for Exposure to Low Levels of Ionizing Radiation: BEIR VII - Phase 2. Committee to Assess Health Risks from Exposure to Low Levels of Ionizing Radiation, National Research Council, National Academies Press, Washington, D.C.

Neuhauser, K.S., F.L. Kanipe, and R.F. Weiner. 2003. *RADTRAN 5 User Guide.* SAND2003-2354, Sandia National Laboratories, Albuquerque, New Mexico. Accessed on the Internet on February 7, 2005 at http://infoserve.sandia.gov/sand_doc/2003/032354.pdf.

Nuclear Waste Policy Act of 1982, as amended. 42 USC 10101, et seq.

Saricks, C.L., and M.M. Tompkins. 1999. *State-Level Accident Rates of Surface Freight Transportation: A Reexamination.* ANL/ESD/TM-150, Argonne National Laboratory, Argonne, Illinois.

Sprung, J.L., D.J. Ammerman, N.L. Breivik, R.J. Dukart, F.L. Kanipe, J.A. Koski, G.S. Mills, K.S. Neuhauser, H.D. Radloff, R.F. Weiner, and H.R. Yoshimura. 2000. *Reexamination of Spent Fuel Shipment Risk Estimates.* NUREG/CR-6672, U.S. Nuclear Regulatory Commission, Washington, D.C.

U.S. Atomic Energy Commission (AEC). 1972. *Environmental Survey of Transportation of Radioactive Materials To and From Nuclear Power Plants.* WASH-1238, AEC, Washington, D.C.

U.S. Department of Energy (DOE). 2002a. *A Resource Handbook on DOE Transportation Risk Assessment.* DOE/EM/NTP/HB-01, DOE, Washington, D.C.

U.S. Department of Energy (DOE). 2002b. *Final Environmental Impact Statement for a Geologic Repository for the Disposal of Spent Nuclear Fuel and High-Level Radioactive Waste at Yucca Mountain, Nye County, Nevada.* DOE/EIS-0250, Office of Civilian Radioactive Waste Management, DOE, Washington, D.C.

U.S. Environmental Protection Agency (EPA). 2002. *Cancer Risk Coefficients for Environmental Exposure to Radionuclides* Federal Guidance Report No. 13, EPA, Washington, D.C.

U.S. Nuclear Regulatory Commission (NRC). 1977. *Final Environmental Statement on Transportation of Radioactive Material by Air and Other Modes.* NUREG-0170, Vol.1, NRC, Washington, D.C.

U.S. Nuclear Regulatory Commission (NRC). 1999. *Generic Environmental Impact Statement for License Renewal of Nuclear Plants.* NUREG-1437, NRC, Washington, D.C.

Appendix H

Supporting Documentation on Radiological Dose Assessment

Appendix H

Supporting Documentation on Radiological Dose Assessment

The staff performed an independent radiological dose assessment on the radiological impacts of normal operation for a new nuclear unit at the Exelon Generation Company, LLC (Exelon) early site permit (ESP) site. Results of this assessment are presented in this appendix and are compared to Exelon's results found in Section 5.9 ("Radiological Impacts of Normal Operation.") The appendix is divided into three sections: (1) dose estimates to the public from liquid effluents, (2) dose estimates to the public from gaseous effluents, and (3) dose estimates to the biota from both the liquid and gaseous effluents.

For comparative purposes with Exelon's estimates, all doses and radioactivity levels are reported in millirem (mrem) and curies (Ci), respectively.

H.1 Dose Estimates to the Public from Liquid Effluents

The staff used the LADTAP II code (Strenge et al. 1986) and input parameters supplied by Exelon in its Environmental Report (ER) (Exelon 2006) to estimate doses to the maximally exposed individual from the liquid effluent pathway. Population doses were not calculated for radioactive liquid effluents.

H.1.1 Scope

Doses to the maximally exposed individual were calculated for the following:

- *Total Body* – Dose was the total for all pathways (i.e., fish consumption, shoreline usage, swimming exposure, boating) with the highest value for either the adult, teen, child, or infant compared to the 0.03 mSv/yr (3 mrem/yr) per reactor design objective in Title 10 of the Code of Federal Regulations (CFR), Part 50, Appendix I.

- *Organ* – Dose was the total for each organ for all pathways (i.e., fish consumption, shoreline usage, swimming exposure, boating) with the highest value for either the adult, teen, child, or infant compared to the 0.1 mSv/yr (10 mrem/yr) per reactor design objective in 10 CFR Part 50, Appendix I.

The staff reviewed the input parameters used by Exelon for appropriateness. Default values from Regulatory Guide 1.109 (NRC 1977) were used when input parameters were not available. The staff concluded that all the input parameters used by Exelon were appropriate.

Population doses were not calculated because (1) there are no municipal or industrial water intakes within 80 km (50 mi) downstream of the ESP site, and (2) no commercial fishing is allowed on Salt Creek. The only possible aquatic pathway would be sport fishing on Clinton Lake; however, detailed dilution and statistics on the number of fish caught by sport fishermen were not available (Exelon 2006).

H.1.2 Resources Used

The staff used a personal computer version of the LADTAP II code entitled NRCDOSE, Version 2.3.5 (Bland 2000) obtained through the Oak Ridge Radiation Safety Information Computational Center (RSICC) to calculate doses to the public from liquid effluents.

H.1.3 Input Parameters

Table H-1 provides a listing of the major parameters used in calculating dose to the public from liquid effluent releases during normal operation. The values used by the applicant and the staff for each parameter are listed along with the appropriateness of the value.

Table H-1. Parameters Used in Calculating Dose to the Public from Liquid Effluent Releases

Parameter	Exelon Value		Staff Value		Comments (Appropriateness of Value)
Source term (Ci/yr)[a]	Table 3.5-1 of Exelon (2006) (modified as discussed in "Comments" column)		Table 3.5-1 of Exelon (2006)		The source term used in the Exelon calculation differs from Table 3.5-1 for the radionuclides where ACR-700 releases were
	H-3	3.1×10^3	H-3	3.1×10^3	bounding to include C-14,
	Na-24	3.26×10^{-3}	Na-24	3.26×10^{-3}	Cr-51, Fe-59, Co-60,
	Cr-51	9.73×10^{-3}	Cr-51	7.7×10^{-3}	Zr-95, Nb-95, and Sb-124.
	Mn-54	2.6×10^{-3}	Mn-54	2.6×10^{-3}	These releases were
	Mn-56	3.81×10^{-3}	Mn-56	3.81×10^{-3}	higher in the Exelon
	Fe-55	5.81×10^{-3}	Fe-55	5.81×10^{-3}	source term. After
	Fe-59	5.08×10^{-4}	Fe-59	4.0×10^{-4}	calculations were initially
	Co-57	7.19×10^{-5}	Co-57	7.19×10^{-5}	performed by Exelon, the
	Co-58	6.72×10^{-3}	Co-58	6.72×10^{-3}	ACR-700 liquid effluent

Table H-1. (contd)

Parameter	Exelon Value		Staff Value		Comments (appropriateness of value)
Source term (Ci/yr)[a] (contd)	Co-60	1.35×10^{-2}	Co-60	9.11×10^{-3}	releases were revised; however, because initial calculations were bounding, Exelon chose not to recalculate doses. Therefore, Exelon doses may be slightly higher than the staff's.
	Zn-65	8.2×10^{-4}	Zn-65	8.2×10^{-4}	
	Rb-88	5.4×10^{-4}	Rb-88	5.4×10^{-4}	
	Sr-89	2.0×10^{-4}	Sr-89	2.0×10^{-4}	
	Sr-90	3.51×10^{-5}	Sr-90	3.51×10^{-5}	
	Nb-95	1.95×10^{-2}	Nb-95	1.91×10^{-3}	
	Zr-95	9.19×10^{-3}	Zr-95	1.04×10^{-3}	
	Mo-99	1.14×10^{-3}	Mo-99	1.14×10^{-3}	
	Tc-99m	1.1×10^{-3}	Tc-99m	1.1×10^{-3}	
	Ru-103	9.86×10^{-3}	Ru-103	9.86×10^{-3}	Both the Exelon and NRC staff used the LADTAP II code to calculate doses. The code accepts only 35 radionuclides; therefore, the 35 radionuclides listed here represent those that contribute significantly to dose.
	Ru-106	1.47×10^{-1}	Ru-106	1.47×10^{-1}	
	Ag-110m	2.1×10^{-3}	Ag-110m	2.1×10^{-3}	
	Sb-124	1.78×10^{-3}	Sb-124	6.79×10^{-4}	
	I-131	2.83×10^{-2}	I-131	2.83×10^{-2}	
	I-132	3.28×10^{-3}	I-132	3.28×10^{-3}	
	I-133	1.34×10^{-2}	I-133	1.34×10^{-2}	
	I-134	1.7×10^{-3}	I-134	1.7×10^{-3}	
	I-135	9.94×10^{-3}	I-135	9.94×10^{-3}	
	Cs-134	1.99×10^{-2}	Cs-134	1.99×10^{-2}	
	Cs-136	1.26×10^{-3}	Cs-136	1.26×10^{-3}	
	Cs-137	2.66×10^{-2}	Cs-137	2.66×10^{-2}	
	Cs-138	1.9×10^{-4}	Cs-138	1.9×10^{-4}	
	Ba-140	1.1×10^{-2}	Ba-140	1.1×10^{-2}	
	La-140	1.49×10^{-2}	La-140	1.49×10^{-2}	
	Ce-141	1.8×10^{-4}	Ce-141	1.8×10^{-4}	
	Ce-144	6.32×10^{-4}	Ce-144	6.32×10^{-4}	
Discharge flow rate m³/s (ft³/s)	0.152 (5.35)		0.152 (5.35)		Site-specific value from Exelon (2006) - Table 5.4-1
Source term multiplier	1		1		Site-specific value from Exelon (2006)
Site type	Fresh water		Fresh water		Site-specific value from Exelon (2006)
Reconcentration model	Partially mixed		Partially mixed		Site-specific value from Exelon (2006)
Effluent discharge rate from impoundment system to receiving water body m³/s (ft³/s)	5.61 (198)		5.61 (198)		Site-specific value from Exelon (2006)

Table H-1. (contd)

Parameter	Exelon Value	Staff Value	Comments (appropriateness of value)
Impoundment total volume m³ (ft³)	9.1×10^7 (3.2×10^9)	9.1×10^7 (3.2×10^9)	Site-specific value from Exelon (2006) - Table 5.4-1
Shore width factor	0.3	0.3	Site-specific value from Exelon (2006) - appropriate per guidance in NRC (1977)
Dilution factors for aquatic food and boating, shoreline and swimming, and drinking water	1	1	Site-specific value from Exelon (2006) - conservative value
Transit time (hr)	0	0	Site-specific value from Exelon (2006) - conservative value
Consumption and usage factors for adults, teens, children, and infants	Values from Table 5.4-2 of Exelon (2006)	Values from Table 5.4-2 of Exelon (2006)	Default values from Regulatory Guide 1.109 (NRC 1977)

(a) To convert Ci/yr to Bq/yr, multiply the value by 3.7×10^{10}.

H.1.4 Comparison of Results

Table H-2 compares the applicant's results with those performed by the staff. Doses calculated were similar.

The LADTAPI code will accept only 35 radionuclides. The staff used the 35 primary radionuclides listed in Table H-1 in their calculations. Another computer run was made with the remaining radionuclides listed in Table 3.5-1 of Exelon (2006). The results are shown in Table H-3 and confirmed that the remaining radionuclides contribute insignificantly (less than 1 percent) to the dose.

Table H-2. Comparison of Doses to the Public from Liquid Effluent Releases

Type of Dose	Exelon's ER (Exelon 2006)[a]	Staff's Calculation[a]	Percent Difference
Total Body (mrem/yr)	0.95 (adult)	0.95 (adult)	0
Organ Dose (mrem/yr)	1.33 (teen liver)	1.32 (teen liver)	-0.8

(a) To convert mrem/yr to mSv/yr divide by 100.

Table H-3. Impact on Dose from Remaining Radionuclides in Liquid Effluent Source Term

Type of Dose	Dose from Remaining Radionuclides[a]	Dose from Primary Radionuclides[a]	Percent Contribution to Dose from Remaining Radionuclides
Total Body (mrem/yr)	1.46×10^{-4} (adult)	0.95 (adult)	0.015
Organ Dose (mrem/yr)	1.69×10^{-4} (teen liver)	1.32 (teen liver)	0.013

(a) To convert mrem/yr to mSv/yr, divide by 100.

H.2 Dose Estimates to the Public from Gaseous Effluents

The staff used the GASPAR II code (Strenge et al. 1987) and input parameters supplied by Exelon in its ER (Exelon 2006) to estimate doses to the maximally exposed individual and to the population within an 80-km (50-mile) radius of the ESP site from the gaseous effluent pathway.

H.2.1 Scope

The staff and Exelon calculated gamma air dose, beta air dose, total body dose, and skin dose from noble gases at the exclusion area boundary located 1.0 km (0.64 mi) north-northeast of the ESP site. Dose to the maximally exposed individual was also calculated for the following locations:

- Nearest residence (plume and inhalation)
- Nearest garden (vegetable)
- Nearest meat cow (meat)
- Nearest milk cow
- Nearest milk goat.

The input parameters used by the applicant were found in Exelon (2006) or the applicant's supporting calculation sheets. These parameters were reviewed for appropriateness. Default values from Regulatory Guide 1.109 (NRC 1977) were used when site-specific input parameters were not available. The staff concluded that all the input parameters used by Exelon were appropriate. These parameters were used by the staff in its independent calculations using GASPAR.

Population doses were calculated for the following pathways (plume, ground, inhalation, vegetable ingestion, cow milk ingestion, and meat ingestion) using the GASPAR II code.

H.2.2 Resources Used

The staff used a personal computer version of GASPAR II code entitled NRCDOSE Version 2.3.5 (Bland 2000), obtained through the Oak Ridge RSICC to calculate doses to the public from gaseous effluents.

H.2.3 Input Parameters

Table H-4 provides a list of the major parameters used in calculating dose to the public from gaseous effluent releases during normal operation. The values used by the applicant and the staff for each parameter are listed along with comments regarding the appropriateness of the value.

Table H-4. Parameters Used in Calculating Dose to Public from Gaseous Effluent Releases

Parameter	Exelon Value		Staff Value		Comments (Appropriateness of Value)
Source term for calculating noble gas dose at exclusion area boundary (Ci/yr)[a]	Table 3.5-3 of Exelon (2006)		Table 3.5-3 of Exelon (2006)		These are bounding plant parameter envelope (PPE) values and are appropriate.
	Ar-41	4.0×10^2	Ar-41	4.0×10^2	
	Kr-85	8.2×10^3	Kr-85	8.2×10^3	
	Kr-85m	7.2×10^1	Kr-85m	7.2×10^1	
	Kr-87	3.0×10^1	Kr-87	3.0×10^1	
	Kr-88	9.2×10^1	Kr-88	9.2×10^1	
	Kr-89	2.41×10^2	Kr-89	2.41×10^2	
	Xe-131m	3.6×10^3	Xe-131m	3.6×10^3	
	Xe-133	9.2×10^3	Xe-133	9.2×10^3	
	Xe-133m	1.74×10^2	Xe-133m	1.74×10^2	
	Xe-135	6.6×10^2	Xe-135	6.6×10^2	
	Xe-135m	4.05×10^2	Xe-135m	4.05×10^2	
	Xe-137	5.14×10^2	Xe-137	5.14×10^2	
	Xe-138	4.32×10^2	Xe-138	4.32×10^2	
Source term for calculating dose to the maximally exposed individual (Ci/yr)[a]	Ar-41	4.0×10^2	Ar-41	4.0×10^2	These are bounding PPE values and are appropriate.
	Kr-85	8.2×10^3	Kr-85	8.2×10^3	
	Kr-85m	7.2×10^1	Kr-85m	7.2×10^1	
	Kr-87	3.0×10^1	Kr-87	3.0×10^1	
	Kr-88	9.2×10^1	Kr-88	9.2×10^1	The GASPAR II code
	Kr-89	2.41×10^2	Kr-89	2.41×10^2	accepts only 33

Table H-4. (contd)

Parameter	Exelon Value		Staff Value		Comments (Appropriateness of Value)
	Xe-131m	3.6×10^3	Xe-131m	3.6×10^3	radio nuclides; therefore,
	Xe-133	9.2×10^3	Xe-133	9.2×10^3	the radionuclides listed
	Xe-135	6.6×10^2	Xe-135	6.6×10^2	here represent those that
	Xe-135m	4.05×10^2	Xe-135m	4.05×10^2	significantly contribute to
	Xe-137	5.14×10^2	Xe-137	5.14×10^2	the dose from gaseous
	Xe-138	4.32×10^2	Xe-138	4.32×10^2	effluents.
	I-131	2.59×10^{-1}	I-131	2.59×10^{-1}	
	I-132	2.19×10^0	I-132	2.19×10^0	
	I-133	1.7×10^0	I-133	1.7×10^0	
	I-134	3.78×10^0	I-134	3.78×10^0	
	I-135	2.41×10^0	I-135	2.41×10^0	
	H-3	3.53×10^3	H-3	3.53×10^3	
	C-14	1.46×10^1	C-14	1.46×10^1	
	Mn-54	5.41×10^{-3}	Mn-54	5.41×10^{-3}	
	Fe-55	6.49×10^{-3}	Fe-55	6.49×10^{-3}	
	Co-58	4.6×10^{-2}	Co-58	4.6×10^{-2}	
	Co-60	1.74×10^{-2}	Co-60	1.74×10^{-2}	
	Fe-59	8.11×10^{-4}	Fe-59	8.11×10^{-4}	
	Zn-65	1.1×10^{-2}	Zn-65	1.1×10^{-2}	
	Sr-89	6.0×10^{-3}	Sr-89	6.0×10^{-3}	
	Sr-90	2.4×10^{-3}	Sr-90	2.4×10^{-3}	
	Zr-95	2.0×10^{-3}	Zr-95	2.0×10^{-3}	
	Nb-95	8.38×10^{-3}	Nb-95	8.38×10^{-3}	
	Ru-103	3.51×10^{-3}	Ru-103	3.51×10^{-3}	
	Sb-124	1.81×10^{-4}	Sb-124	1.81×10^{-4}	
	Cs-134	6.22×10^{-3}	Cs-134	6.22×10^{-3}	
	Cs-137	9.46×10^{-3}	Cs-137	9.46×10^{-3}	
Population distribution	Used data from Exelon's supporting documentation (equivalent to data found in Tables 2.5-1 and 2.5-3 of Exelon [2006])		Used data from Exelon's supporting documentation (equivalent to data found in Tables 2.5-1 and 2.5-3 of Exelon [2006])		Site-specific data - appropriate for use
Atmospheric dispersion factors (sec/m³)	Used data from Exelon's supporting documentation (equivalent to Tables 2.7-53, 2.7-55, and 2.7-56 of Exelon [2006])		Used data from Exelon's supporting documentation (equivalent to Tables 2.7-53, 2.7-55, and 2.7-56 of Exelon [2006])		Site-specific data - appropriate for use

Appendix H

Table H-4. (contd)

Parameter	Exelon Value	Staff Value	Comments (Appropriateness of Value)
Ground deposition factors (m^{-2})	Used data from Exelon's supporting documentation (equivalent to Table 2.7-54 of Exelon [2006])	Used data from Exelon's supporting documentation (equivalent to Table 2.7-54 of Exelon [2006])	Site-specific data - appropriate for use
Mi k production rate within 80 km (50 mi) (L/yr)	Used data from Exelon's supporting documentation	Used data from Exelon's supporting documentation	Site-specific data - appropriate for use
Meat production rate within 80 km (50 mi) (kg/yr)	Used data from Exelon's supporting documentation	Used data from Exelon's supporting documentation	Site-specific data - appropriate for use
Vegetable/fruit production rate within 80 km (50 mi) (kg/yr)	Used data from Exelon's supporting documentation	Used data from Exelon's supporting documentation	Site-specific data - appropriate for use
Pathway receptor locations (direction, distance, and atmospheric dispersion factors)- nearest site boundary, vegetable garden, residence, meat animal	Used data from Exelon's supporting documentation (equivalent to Table 2.7-53 of Exelon [2006])	Used data from Exelon's supporting documentation (equivalent to Table 2.7-53 of Exelon [2006])	Site-specific data - appropriate for use
Consumption factors for leafy vegetables, meat, mi k, and vegetable/fruit	Table 5.4-4 of Exelon (2006)	Table 5.4-4 of Exelon (2006)	Site-specific data - appropriate for use
Fraction of year leafy vegetables are grown	0.33	0.33	Site-specific data - appropriate for use
Fraction of year that mi k cows are on pasture	0.58	0.58	Site-specific data - appropriate for use
Fraction of milk-cow intake that is from pasture while on pasture	1.0	1.0	Default value of GASPAR II code

Table H-4. (contd)

Parameter	Exelon Value	Staff Value	Comments (Appropriateness of Value)
Average absolute humidity over the growing season	8.0 g/m^3	8.0 g/m^3	Default value of GASPAR II code
Average temperature over the growing season(°F)	0	0	Default value of GASPAR II code
Fraction of year goats are on pasture	0.67	0.67	Site-specific data - appropriate for use
Fraction of year beef-cattle are on pasture	0.58	0.58	Site-specific data - appropriate for use
Fraction of beef-cattle intake that is from pasture while on pasture	1.0	1.0	Default value of GASPAR II code

(a) To convert Ci/yr to Bq/yr, multiply the value by 3.7 x 10^{10}.

H.2.4 Comparison of Doses to the Public from Gaseous Effluent Releases

Table H-5 compares Exelon's results for doses from noble gases at the exclusion area boundary with the results calculated by the staff. Doses calculated were similar.

Table H-5. Comparison of Doses to the Public from Noble Gas Releases

Type of Dose	Exelon's ER (Exelon 2006)	Staff's Calculation	Percent Difference
Gamma air dose at exclusion area boundary – noble gases only (mrad/yr)[a]	1.35	1.35	0
Beta air dose at exclusion area boundary – noble gases only (mrad/yr)[a]	2.89	2.91	0.7
Total body dose at exclusion area boundary – noble gases only (mrem/yr)[a]	0.875	0.877	0.2
Skin dose at exclusion area boundary – noble gases only (mrem/yr)[a]	2.94	2.95	0.3

(a) To convert from mrad/yr or mrem/yr to mGy/yr or mSv/yr, divide by 100.

Table H-6 compares doses to the maximally exposed individual calculated by Exelon and the staff. Doses to the maximally exposed individual were calculated at the nearest residence, nearest garden, nearest meat cow, and nearest milk cow. Doses calculated were similar.

Table H-6. Comparison of Doses to the Maximally Exposed Individual from Gaseous Effluent Releases

Location	Pathway	Total Body Dose (mrem/yr)[a, b]	Skin Dose (mrem/yr)[a, b]	Thyroid Dose (mrem/yr)[a, b]
Nearest residence, 1.2 km (0.73 mi) SW	Plume	0.39 (0.39)	1.4 (1.4)	--
Nearest residence, 1.2 km (0.73 mi) SW	Inhalation			
	Adult	0.12 (0.12)	--	0.48 (0.48)
	Teen	0.12 (0.12)	--	0.60 (0.60)
	Child	0.11 (0.11)	--	0.70 (0.70)
	Infant	0.063 (0.063)	--	0.60 (0.60)
Nearest garden, 1.5 km (0.93 mi) N	Vegetable			
	Adult	0.27 (0.27)	--	2.6 (2.6)
	Teen	0.36 (0.36)	--	3.6 (3.6)
	Child	0.68 (0.68)	--	7.0 (7.0)
Nearest meat animal, 1.5 km (0.93 mi) N	Meat			
	Adult	0.061 (0.061)	--	--
	Teen	0.045 (0.045)	--	--
	Child	0.073 (0.073)	--	--
Nearest milk cow,[c] 8.1 km (5.0 mi) N	Cow Milk			
	Adult	0.0097 (0.0097)	--	0.15 (0.15)
	Teen	0.014 (0.0.14)	--	0.24 (0.24)
	Child	0.027 (0.027)	--	0.47 (0.47)
	Infant	0.050 (0.050)	--	1.1 (1.1)
Nearest milk goat,[d] 7.1 km (4.4 mi) SE	Goat Milk			
	Adult	0.015 (0.015)	--	0.17 (0.17)
	Teen	0.02 (0.02)	--	0.27 (0.27)
	Child	0.034 (0.034)	--	0.54 (0.54)
	Infant	0.059 (0.059)	--	1.3 (1.3)

(a) Values in parentheses represent the values that the staff calculated. The Exelon values (those not in parentheses were taken from Table 5.4-6 of Exelon (2006).

(b) To convert from mrem/yr to mSv/yr, divide by 100.

(c) This distance and direction from the ESP site represent the location of the nearest cow producing milk for human consumption.

(d) This distance and direction from the ESP site represent the location of the nearest milk goat. In Table 2.7-54 of the ER (Exelon 2006), the largest relative deposition factor for the nearest milk goat is listed at a distance of 8 km (5 mi) north-northeast of the ESP site. This relative deposition factor is approximately 20 percent greater than the relative deposition factor used in Exelon's calculation; however, it would not result in a significant increase in the dose to the maximally exposed individual.

H.2.5 Comparison of Results - Population Doses

Table H-7 compares the Exelon's population dose estimates taken from Table 5.4-11 of
Exelon (2006) with the staff's estimate. Doses calculated were similar.

Table H-7. Comparison of Population Doses from Gaseous Effluent Releases

Pathway	Applicant's Estimate	Staff's Estimate	Percent Difference
TOTAL BODY (person-rem/yr)[a]			
Plume	0.403	0.403	0%
Ground	0.145	0.145	0%
Inhalation	0.480	0.48	0%
Vegetable ingestion	0.108	0.108	0%
Cow-milk ingestion	0.392	0.391	0.3%
Meat ingestion	0.298	0.298	0%
Total	1.830	1.82	-0.6%
THYROID (WORST CASE ORGAN) (person-rem/yr)[a]			
Plume	0.403	0.403	0%
Ground	0.145	0.145	0%
Inhalation	1.530	1.52	-0.7%
Vegetable ingestion	0.109	0.109	0%
Cow-milk ingestion	3.350	3.16	-5.7%
Meat ingestion	0.42	0.415	-1.2%
Total	5.95	5.75	-3.4%

(a) To convert from person-rem/yr to person-Sv/yr, divide by 100.

H.3 Dose Estimates to the Biota from Liquid and Gaseous Effluents

To estimate doses to the biota from the liquid and gaseous effluent pathways, the staff used the
LADTAP II code (Strenge et al. 1986) and the GASPAR II code (Strenge et al. 1987) and input
parameters supplied by Exelon as part of its ER (Exelon 2006).

H.3.1 Scope

Doses to both terrestrial and aquatic biota were calculated using the LADTAP II code. Aquatic
biota include fish, invertebrates, and algae. Terrestrial biota include muskrat, raccoon, heron,
and duck. The LAPTAP II code calculates an internal dose component and an external dose
component and sums them for a total body dose. The staff reviewed the input parameters used
by Exelon for appropriateness. Default values from Regulatory Guide 1.109 (NRC 1977) were

used when input parameters were not available. The staff concluded that all of the input parameters used by Exelon were appropriate. These parameters were used by the staff in its independent calculations using LADTAP.

The LADTAP II code calculates only biota dose from the liquid effluent pathway. Terrestrial biota could also be exposed via the gaseous effluent pathway. These values would be the same as those for the maximally exposed individual calculated using the GASPAR II code. Exelon (2006) used the maximally exposed individual doses at the exclusion area boundary (1 km [0.64 mi] from the plant) to estimate these doses. The maximally exposed individual calculation for the biota assumed a ground deposition factor twice that used in the maximally exposed individual calculation for a member of the public. Gaseous doses are not significant compared to the liquid pathway.

H.3.2 Resources Used

To calculate doses to the biota, the staff used a personal computer version of the LADTAP II and GASPAR II computer codes entitled NRCDOSE Version 2.3.5 (Bland 2000) obtained through the Oak Ridge RSICC.

H.3.3 Input Parameters

Most of the LADTAP II input parameters are specified in Section H.1.3 to include the source term, discharge flow rate, reconcentration model, effluent discharge rate from the impoundment system to the receiving water body, impoundment total volume, and shore width factor. Parameters unique to the biota dose calculation were taken from Table 5.4-15 (terrestrial biota parameters), Table 5.4-16 (shoreline and swimming exposures), and Table 5.4-17 of the ER (Exelon 2006). These parameters were default values used in the LADTAP II code (Strenge et al. 1986) and are appropriate values to use in calculating biota dose.

H.3.4 Comparison of Results

Table H-8 compares Exelon's biota dose estimates from liquid effluents taken from Table 5.4-18 of Exelon (2006) with the staff's estimate. Dose estimates were similar.

Table H-9 compares Exelon's biota dose estimates for gaseous effluents taken from Table 5.4-18 of Exelon (2006) with the staff's estimate. Dose estimates were similar except for the staff's dose estimate to the heron, which were approximately twice that of the applicant. The difference is likely due to the applicant's considering the heron to be present at the impact site only 50 percent of the time, which is a reasonable assumption.

Table H-8. Comparison of Dose Estimates to Biota from Liquid Effluents

Biota	Type of Dose	Exelon's ER (mrad/yr)[a]	Staff's Calculation (mrad/yr)[a]	Percent Difference
Fish	Internal	2.43	2.42	-0.4
	External	3.82	3.81	-0.3
Invertebrates	Internal	6.11	6.75	10.5
	External	7.63	7.61	-0.3
Algae	Internal	27.8	30.9	11
	External	0.00718	0.00701	-2.4
Muskrat	Internal	13.4	15.1	12.7
	External	2.55	2.54	-0.4
Raccoon	Internal	4.57	5.16	12.9
	External	1.91	1.9	-0.5
Heron	Internal	66.3	75.1	13.3
	External	2.55	2.54	-0.4
Duck	Internal	12.0	13.5	12.5
	External	3.82	3.81	-0.3

(a) To convert from mrad/yr to mGy/yr, divide by 100.

Table H-9. Comparison of Dose Estimates to Biota from Gaseous Effluents

Biota	Type of Dose	Exelon's ER (mrad/yr)[a]	Staff's Calculation (mrad/yr)[a]	Percent Difference
Fish	Internal	--	--	--
	External	--	--	--
Invertebrates	Internal	--	--	--
	External	--	--	--
Algae	Internal	--	--	--
	External	--	--	--
Muskrat	Internal	0.166	0.166	0
	External	1.06	1.44[c]	36
Raccoon	Internal	0.166	0.166	0
	External	1.44	1.44[c]	0
Heron	Internal	0.083	0.166	100[b]
	External	0.627	1.44[c]	130[b]
Duck	Internal	0.166	0.166	0
	External	1.16	1.44[c]	36

(a) To convert from mrad/yr to mGy/yr divide by 100.
(b) Difference is likely due to the applicant considering the heron to be present at the impact site only 50 percent of the time. This is a reasonable assumption.
(c) This dose is equal to the sum of the total body dose from the plume and twice the ground deposition dose at the exclusion area boundary (1 km [0.64 mi] from the plant): 0.875 mrad+ 2 (0.284 mrad) = 1.44 mrad.

H.4 References

10 CFR Part 50. Code of Federal Regulations, Title 10, *Energy*, Part 50, "Domestic Licensing of Production and Utilization Facilities."

Bland, J.S. 2000. NRCDOSE for Windows. Radiation Safety Information Computational Center, Oak Ridge, Tennessee.

Exelon Generation Company, LLC (Exelon). 2006. *Exelon Generation Company, LLC, Early Site Permit: Environmental Report, Rev. 4*. Exelon Nuclear, Kennett Square, Pennsylvania.

Strenge, D.L., R.A. Peloquin, and G. Whelan. 1986. *LADTAP II – Technical Reference and User Guide*. NUREG/CR-4013, Pacific Northwest Laboratory, Richland, Washington.

Strenge, D.L., T.J. Bander, and J.K. Soldat. 1987. *GASPAR II – Technical Reference and User Guide*. NUREG/CR-4653, Pacific Northwest Laboratory, Richland, Washington.

U.S. Nuclear Regulatory Commission (NRC). 1977. *Calculation of Annual Doses to Man from Routine Releases of Reactor Effluents for the Purpose of Evaluating Compliance with 10 CFR Part 50, Appendix I*. Regulatory Guide 1.109, NRC, Washington, D.C.

Appendix I

Authorizations and Consultations

Appendix I

Authorizations and Consultations

This appendix contains a list of the environmental-related authorizations, permits, and certifications potentially required by Federal, Sate, regional, local, and affected Native American tribal agencies related to the construction and operation of the potential new nuclear unit at the Exelon ESP site, reproduced from Table 1.2-1 of the Environmental Report.

Table I-1. Federal, State, and Local Authorizations

Agency	Authority	Requirement	License/ Permit No.	Expiration Date	Authorization Granted
U.S. Nuclear Regulatory Commission (USNRC)	10 CFR 40	Source Material License	—[a]	—[a]	Possession of source material
USNRC	Atomic Energy Act of 1954 (AEA), 10 CFR 51	ER	—[a]	—[a]	Site approval for a nuclear power station separate from an application for a standard design certification or combined operating license (COL)
USNRC	10 CFR 52	COL	—[a]	—[a]	Construction and Operation Safety Review for a nuclear power station
USNRC	10 CFR 70	Special Nuclear Materials License	—[a]	—[a]	Possession of fuel
USNRC	10 CFR 30	By-product License	—[a]	—[a]	Possession of special nuclear materials
U.S. Fish and Wildlife Services (USFWS)	Threatened and Endangered Species Act	Letter of Compliance	—[a]	—[a]	Compliance with Threatened and Endangered Species Act
Federal Aviation Administration (FAA)	49 USC 1501	Construction Notice	—[a]	—[a]	Construction of structures affecting air navigation

Table I-1. (contd)

Agency	Authority	Requirement	License/ Permit No.	Expiration Date	Authorization Granted
U.S. Environmental Protection Agency (USEPA)	Clean Water Act (CWA)	Storm Water Pollution Prevention Plan (SWP3)	—[a]	—[a]	Discharge of storm water associated with construction activities
U.S. Army Corps of Engineers (USACOE)	CWA	Section 404 Permit	—[a]	—[a]	Disturbance of the crossing of a navigable stream
USACOE	Section 404 Conditional Permit	Walleye Spawning Areas Permit	—[a]	—[a]	Disturbances of walleye spawning areas
USACOE	33 CFR 209	Dredge and Fill Discharge Permit	—[a]	—[a]	Construction/ modification of the discharge to Salt Creek
State Historic Preservation Office (SHPO)	36 CFR 800	Cultural Resources Review	—[a]	—[a]	Confirmation that site and transmission line right-of-way are not considered historic preservation areas
Illinois Commerce Commission	Illinois Public Utilities Act	Certification of Public Convenience and Necessity	—[a]	—[a]	Construction and operation of plant
Illinois Department of Transportation (IDOT)	Illinois Rev. Stat. 1971	Construction Permit	—[a]	—[a]	Construct lift crane
(IDOT)	Illinois Rev. Stat. 1971	Construction Permit	—[a]	—[a]	Construct dome lighting mast
IDOT	Illinois Commerce Act 1911	Construction Permit	—[a]	—[a]	Construction/ modification of discharge structures on Salt Creek

Table I-1. (contd)

Agency	Authority	Requirement	License/ Permit No.	Expiration Date	Authorization Granted
IDOT	Illinois Commerce Act 1911	Construction Permit[b]	_[a]	_[a]	Construction of transmission lines crossing waterways
IDOT	Illinois Commerce Act 1911	Construction Permit[b]	_[a]	_[a]	Construction of transmission lines crossing state highways
Illinois Environmental Protection Agency (IEPA)	Resource Conservation and Recovery Act (RCRA)	Development (DE), Operating (OP), and Supplemental Permits	_[a]	_[a]	Storage and transportation of hazardous materials
IEPA	17 IL Adm. Code Part 120	Surface Water Withdrawal Permit	_[a]	_[a]	Withdrawal of water from a public surface water source
IEPA	CWA	IEPA Section 401 Water Quality Certification	_[a]	_[a]	Certification that activities will comply with water quality standards of the State
IEPA	General permit for discharges associated with construction activities	Notice of Intent (NOI) for Construction	_[a]	_[a]	Discharge of storm water from site during construction
IEPA	General permit for discharges associated with construction activities	Notice of Termination (NOT) for Construction	_[a]	_[a]	Termination of coverage under the general permit for storm water discharge associated with construction site activities
IEPA	CWA	NPDES Permit	_[a]	_[a]	Discharges to surface water

Table I-1. (contd)

Agency	Authority	Requirement	License/ Permit No.	Expiration Date	Authorization Granted
IEPA	CAA	Minor Source Construction Permit	—[a]	—[a]	Construction and operation of facilities generating air emissions
IEPA	Title V	Title V Operating Permit	—[a]	—[a]	Operation of facility generating air emissions
IEPA	General Storm Water Permit	Notice of Termination (NOT) for Industrial Activities	—[a]	—[a]	Termination of coverage under the general permit for storm water discharge associated with operations activities
IEPA	Environmental Protection Act (415 ILCS 5)	Sanitary Waste Water Hauling Permit	—[a]	—[a]	Transportation of sanitary waste water
IEPA	Environmental Protection Act (415 ILCS 5)	Sludge Disposal Operating Permit	—[a]	—[a]	Disposal of sludge
IEPA	Environmental Protection Act (415 ILCS 5)	Non-Hazardous Domestic Waste Water or Sludge Transporting Permit	—[a]	—[a]	Transportation of non-hazardous waste water or sludge
IEPA	IL Adm. Code, Part 170	Emergency Petroleum Storage Tank Permit	—[a]	—[a]	Implementation of storage tanks containing petroleum products
IEPA	Environmental Protection Act (415 ILCS 5)	Open Burning Permit	—[a]	—[a]	Open burning of petroleum products for backup generators

Table I-1. (contd)

Agency	Authority	Requirement	License/ Permit No.	Expiration Date	Authorization Granted
IEPA	Environmental Protection Act (415 ILCS 5)	Supplemental Waste Stream Permit	—[a]	—[a]	Disposal of waste from additional waste streams
IEPA	N/A	Refrigerant Recovery/ Recycling Equipment Certifications	—[a]	—[a]	Recovery and recycling of refrigerants
IEPA	Environmental Protection Act (415 ILCS 5)	Construction Permit	—[a]	—[a]	Construction of waste treatment facilities
IEPA	Environmental Protection Act (415 ILCS 5)	Construction Permit	—[a]	—[a]	Construction of temporary sewage treatment unit for construction phase only
IEPA	Environmental Protection Act (415 ILCS 5)	Operating Permit	—[a]	—[a]	Operation of temporary sewage treatment unit for construction phase only
IEPA	Environmental Protection Act (415 ILCS 5)	Operating Permit	—[a]	—[a]	Treatment of waste water discharge
DeWitt County Zoning Board of Appeals	Illinois Zoning Act	Approvals	—[a]	—[a]	Construction of the plant

Table I-1. (contd)

Agency	Authority	Requirement	License/ Permit No.	Expiration Date	Authorization Granted
Circuit Court of DeWitt County	Eminent Domain Act	Petition for Condemnation	—[a]	—[a]	Exercise right of eminent domain

(a)   Data not available. Applicable permits may not be applied for until the COL phase. Applications for permits will be made before the beginning of construction, as required. Some permits may be combined with existing CPS permits.

(b)   To be obtained by the Regional Transmission Operator.

Notes: All permits will be applied for before the beginning of construction. Some permits may not be obtained since the area may be combined with some existing CPS permits.

Appendix J

Plant Parameter Envelope Values

Appendix J

Plant Parameter Envelope Values

This appendix contains the Exelon Plant Parameter Envelope reproduced from Section 1.4, "Plant Parameter Envelope" of the *Site Safety Analysis Report*, Rev. 3.

Appendix J

Table J-1. Plant Parameter Envelope (PPE) Values

PPE	Section	PPE Value	Site Characteristic Value	Usage
1.	Structure			
1.1	Building Characteristics			
1.1.1	Height	234 ft above grade	Not Applicable	ER
1.1.2	Foundation Embedment	140 ft below grade	Not Applicable	ER
1.2	Precipitation (for Roof Design)			
1.2.1	Maximum Rainfall Rate	(a)	18.15 in./hr (6.08 in./5 min)	SSAR
1.2.2	Snow Load	(a)	40 b/ft^2	SSAR
1.3	Safe Shutdown Earthquake (SSE)			
1.3.1	Design Response Spectra	(a)	Site Specific Determination: Figure 2.5-12	SSAR
1.3.2	Peak Ground Acceleration	(a)	0.35 g	SSAR
1.3.3	Time History	(a)	NUREG/CR-6728	SSAR
1.3.4	Capable Tectonic Structures or Sources	(a)	No active faults: <25 mi Possible faults: >25 mi <200 mi	SSAR
1.4	Site Water Level (Allowable)			
1.4.1	Maximum Flood (or Tsunami)	(a)	26.1 ft below grade	SSAR
1.4.2	Maximum Ground Water	(a)	1.5 ft below grade	SSAR
1.5	Soil Properties Design Bases			
1.5.1	Liquefaction	(a)	None at site below 60 ft below ground surface (bgs) Soils above 60 ft bgs to be replaced or improved	SSAR
1.5.2	Minimum Bearing Capacity (Static)	(a)	50,000 bs/ft^2	SSAR
1.5.3	Minimum Shear Wave Velocity	(a)	0-51 ft = 820 fps 50-285 ft = 1090 fps 285-310 ft = 2580 fps	SSAR
1.6	Tornado (Design Bases)			
1.6.1	Maximum Pressure Drop	(a)	2.0 psi	SSAR
1.6.2	Maximum Rotational Speed	(a)	240 mph	SSAR
1.6.3	Maximum Translational Speed	(a)	60 mph	SSAR
1.6.4	Maximum Wind Speed	(a)	300 mph	SSAR
1.6.6	Radius of Maximum Rotational Speed	(a)	150 ft	SSAR
1.6.7	Rate of Pressure Drop	(a)	1.2 psi/sec	SSAR

Table J-1. (contd)

PPE	Section	PPE Value	Site Characteristic Value	Usage
1.7	Wind			
1.7.1	Basic Wind Speed or	(a)	75 mph	SSAR
	3-second gust	(a)	96 mph	SSAR
1.7.2	Importance Factors	1.11 (Safety Related)	Not Applicable	SSAR
2.	Normal Plant Heat Sink			
2.1	Ambient Air Temperatures			
2.1.1	Normal Shutdown Ambient Temperature (1% exceedance)	(a)	91°F	SSAR
2.1.2	Normal Shutdown Max Wet Bulb Temperature (1% exceedance)	(a)	78°F	SSAR
2.1.3	Normal Shutdown Min Ambient Temperature (1% exceedance)	(a)	0°F	SSAR
2.1.4	Rx Thermal Power Max Ambient Temperature (0% exceedance)	(a)	117°F	SSAR
2.1.5	Rx Thermal Power Max Wet Bulb Temperature (0% exceedance)	(a)	86°F	SSAR
2.1.6	Rx Thermal Power Min Ambient Temperature (0% exceedance)	(a)	-36°F	SSAR
2.3	Condenser			
2.3.2	Condenser/Heat Exchanger Duty	15.08 E+09 Btu/hr	Not Applicable	SSAR ER
2.4	Mechanical Draft Cooling Towers			
2.4.1	Acreage	50 ac	Not Applicable	ER
2.4.3	Blowdown Constituents and Concentrations	See Table 1.4-2	Not Applicable	SSAR ER
2.4.4	Blowdown Flow Rate	12,000 gpm (49,000 gpm max.)	Not Applicable	SSAR ER
2.4.5	Blowdown Temperature	100°F	Not Applicable	SSAR ER
2.4.7	Evaporation Rate	31,500 gpm (b)	Not Applicable	SSAR
2.4.8	Height	60 ft	Not Applicable	ER
2.4.9	Makeup Flow Rate	42,000 gpm	Not Applicable	ER
2.4.10	Noise	55 dBa @ 1000 ft	Not Applicable	ER
2.4.12	Cooling Water Flow Rate	1,200,000 gpm	Not Applicable	SSAR
2.4.13	Heat Rejection Rate (Blowdown)	12,000 gpm (49,000 gpm max.) @ 100°F	Not Applicable	ER
2.4.14	Maximum Consumption of Raw Water	60,000 gpm	Not Applicable	ER
2.5	Natural Draft Cooling Towers			

Appendix J

Table J-1. (contd)

PPE	Section	PPE Value	Site Characteristic Value	Usage
2.5.1	Acreage	34.5 ac total (with 3 x 2.75 ac per reactor basin, 8.25 ac total for basins)	Not Applicable	ER
2.5.3	Blowdown Constituents and Concentrations	See Table 1.4-2	Not Applicable	ER
2.5.4	Blowdown Flow Rate	12,000 gpm (49,000 gpm max.)	Not Applicable	SSAR ER
2.5.5	Blowdown Temperature	100°F	Not Applicable	SSAR ER
2.5.7	Evaporation Rate	31,500 gpm (b)	Not Applicable	SSAR
2.5.8	Height	550 ft	Not Applicable	ER
2.5.9	Makeup Flow Rate	42,000 gpm	Not Applicable	ER
2.5.10	Noise	55 dBa @ 1000 ft	Not Applicable	ER
2.5.12	Cooling Water Flow Rate	1200,000 gpm	Not Applicable	SSAR ER
2.5.13	Heat Rejection Rate (Blowdown)	12,000 gpm normal (49,000 gpm max.) @ 100°F	Not Applicable	ER
2.5.14	Maximum Consumption of Raw Water	60,000 gpm	Not Applicable	ER
3.	Ultimate Heat Sink			
3.1	Ambient Air Requirements			
3.1.1	Maximum Ambient Temperature (0% exceedance)	(a)	117°F	SSAR
3.1.2	Maximum Web Bulb Temperature (0% exceedance)	(a)	86°F	SSAR
3.1.3	Minimum Ambient Temperature (0% exceedence)	(a)	-36°F	SSAR
3.1.4	Maximum 30-day Average Web Bulb Temperature	(a)	74.7°F	SSAR
3.1.5	Coincident 30-day Average Dry Bulb Temperature	(a)	82°F	SSAR
3.1.6	Maximum 1-day Average Web Bulb Temperature	(a)	81°F	SSAR
3.1.7	Coincident 1-day Average Dry Bu b Temperature	(a)	87.6°F	SSAR
3.1.8	Maximum 5-day Average Wet Bulb Temperature	(a)	79.7°F	SSAR
3.1.9	Coincident 5-day Average Dry Bu b Temperature	(a)	86.2°F	SSAR

Table J-1. (contd)

PPE	Section	PPE Value	Site Characteristic Value	Usage
3.1.10	Maximum Cumulative Degree-Days Below Freezing	(a)	1141.5 degree-days	SSAR
3.1.11	Maximum Ambient Temperature (1% exceedance)	(a)	91°F	SSAR
3.1.12	Maximum Wet Bulb Temperature (1% exceedance)	(a)	78°F	SSAR
3.1.13	Minimum Ambient Temperature (1% exceedance)	(a)	0°F	SSAR
3.2	CCW Heat Exchanger			
3.2.1	Maximum Inlet Temp. to CCW Heat Exchanger	95°F	Not Applicable	SSAR
3.2.2	CCW Heat Exchanger Duty	225 E+06 Btu/hr 411.4E+06 Btu/hr (Shutdown)	Not Applicable	ER
3.3	Mechanical Draft Cooling Towers			
3.3.1	Acreage	0.5 ac	Not Applicable	ER
3.3.3	Blowdown Constituents and Concentrations	See Table 1.4-2	Not Applicable	ER
3.3.4	Blowdown Flow Rate	144 gpm expected (700 gpm max.)	Not Applicable	SSAR ER
3.3.5	Blowdown Temperature	95°F	Not Applicable	SSAR ER
3.3.7	Evaporation Rate	411 gpm (700 gpm max.)	Not Applicable	SSAR ER
3.3.8	Height	60 ft	Not Applicable	ER
3.3.9	Makeup Flow Rate	555 gpm (1400 gpm max)	Not Applicable	ER ER
3.3.10	Noise	55 dBa @ 1000 ft	Not Applicable	ER
3.3.12	Cooling Water Flow Rate	26,125 gpm normal (52,250 gpm shutdown)	Not Applicable	SSAR ER
3.3.13	Heat Rejection Rate (blowdown)	144 gpm expected (700 max. gpm) @ 95°F	Not Applicable	ER
4.	Containment Heat Removal System (Post-Accident)			
4.1	Ambient Air Requirements			
4.1.1	Maximum Ambient Air Temperature (0% exceedance)	(a)	117°F	SSAR
4.1.2	Minimum Ambient Air Temperature (0% exceedance)	(a)	-36°F	SSAR

Table J-1. (contd)

PPE	Section	PPE Value	Site Characteristic Value	Usage
5.	Potable Water/Sanitary Waste System			
5.1	Discharge to Site Water Bodies			
5.1.1	Flow Rate	60 gpm expected (198 max gpm)	Not Applicable	SSAR ER
5.2	Raw Water Requirements			
5.2.1	Maximum Use	198 gpm	Not Applicable	ER
5.2.2	Monthly Average Use	90 gpm	Not Applicable	SSAR ER
6.	Demineralized Water System			
6.1	Discharge to Site Water Bodies			
6.1.1	Flow Rate	110 gpm expected	Not Applicable	ER
6.2	Raw Water Requirements			
6.2.1	Maximum Use	720 gpm	Not Applicable	ER
6.2.2	Monthly Average Use	550 gpm	Not Applicable	SSAR ER
7.	Fire Protection System			
7.1	Raw Water Requirements			
7.1.1	Maximum Use	2500 gpm	Not Applicable	ER
7.1.2	Monthly Average Use	10 gpm	Not Applicable	SSAR ER
8.	Miscellaneous Drain			
8.1	Discharge to Site Water Bodies			
8.1.1	Flow Rate	75 gpm total (150 gpm max)	Not Applicable	ER
9.	Unit Vent/Airborne Effluent Release Point			
9.1	Atmospheric Dispersion (χ/Q) (Accident)			
9.1.1	0-2 hr @ EAB (sec/m^3)	(a)(c)	2.52E-04 (5%) 3.56E-05 (50%)	SSAR ER
9.1.2	0-8 hr @ LPZ (sec/m^3)	(a)(c)	3.00E-05 (5%) 3.40E-06 (50%)	SSAR ER
9.1.3	8-24 hr @ LPZ (sec/m^3)	(a)(c)	2.02E-05 (5%) 2.85E-06 (50%)	SSAR ER

Table J-1. (contd)

PPE	Section	PPE Value	Site Characteristic Value	Usage
9.1.4	1-4 day @ LPZ (sec/m^3)	(a)(c)	8.53E-06 (5%)	SSAR
			1.85E-06 (50%)	ER
9.1.5	4-30 day @ LPZ (sec/m^3)	(a)(c)	2.48E-06 (5%)	SSAR
			1.00E-06 (50%)	ER
9.2	Atmospheric Dispersion (χ/Q)(Annual Average)	(a)	2.04E-06 sec/m^3 @ EAB[d]	SSAR
				ER
9.3	Dose Consequences	(a)		
9.3.1	Normal	(a)	10 CFR 20, 10 CFR 50 Appendix I, and 40 CFR 190 dose limits. Refer to SSAR 3.1.1 and 3.1.1.2 and ER 5.4	SSAR
				ER
9.3.2	Post-Accident	(a)	10 CFR 50.34(a)(1) and 10 CFR 100 dose limits. Refer to SSAR 3.3 and ER 7.1	SSAR
				ER
9.4	Release Point	(a)		
9.4.2	Elevation (Normal)	(a)	Ground Level	SSAR
9.4.3	Elevation (Post-Accident)	(a)	Ground Level	SSAR
9.4.4	Minimum Distance to Site Boundary	(a)(c)	1025 m (3362 ft)	SSAR
9.4.7	Minimum Distance to LPZ	(a)	4018 m (2.5 mi)	SSAR
9.5	Source Term			
9.5.1	Gaseous (Normal)	See Table 1.4-3 for isotopic breakdown.	Not Applicable	SSAR
				ER
9.5.2	Gaseous (Post-Accident)	Based on limiting DBAs[f]. (Refer to SSAR 3.3)	Not Applicable	SSAR
9.5.3	Tritium (Normal)	See Table 1.4-3	Not Applicable	SSAR
				ER
10.	Liquid Radwaste System			
10.1	Dose Consequences			
10.1.1	Normal	(a)	10 CFR 20, 10 CFR 50 Appendix I, 40 CFR 190 dose limits. Refer to SSAR 3.1.2 and 3.1.2.2 and ER 5.4	SSAR
10.2	Release Point			

Table J-1. (contd)

PPE	Section	PPE Value	Site Characteristic Value	Usage
10.2.1	Flow Rate	Average daily discharge for 292 days per year with dilution flow of 2400 gpm	Not Applicable	SSAR
				ER
10.3	Source Term			
10.3.1	Liquid	See Table 1.4-4 for isotopic listing.	Not Applicable	SSAR
				ER
10.3.2	Tritium	See Table 1.4-4	Not Applicable	SSAR
				ER
11.	Solid Radwaste System			
11.2	Solid Radwaste			
11.2.1	Activity	See Table 1.4-5	Not Applicable	SSAR
				ER
11.2.2	Principal Radionuclides	See Table 1.4-5	Not Applicable	SSAR
				ER
11.2.3	Volume	15,087 ft^3/yr avg.	Not Applicable	SSAR
				ER
13.	Auxiliary Boiler System			
13.1	Exhaust Elevation	110 ft above grade	Not Applicable	ER
13.2	Flue Gas Effluents	See Table 1.4-6	Not Applicable	ER
14.	Heating, Ventilating, and Air Conditioning System			
14.1	Ambient Air Requirements			
14.1.1	Non-safety HVAC Max Ambient Temperature (1% exceedance)	(a)	91°F	SSAR
14.1.2	Non-safety HVAC Min Ambient Temperature (1% exceedance)	(a)	0°F	SSAR
14.1.3	Safety HVAC Max Ambient Temperature (0% exceedance)	(a)	117°F	SSAR
14.1.4	Safety HVAC Min Ambient (0% exceedance)	(a)	-36°F	SSAR
15.	Onsite/Offsite Electrical Power System			
15.1	Acreage			
15.1.1	Switchyard	15 ac	Not Applicable	ER

Table J-1. (contd)

PPE	Section	PPE Value	Site Characteristic Value	Usage
16.	Standby Power System			
16.1	Diesel			
16.1.2	Diesel Exhaust Elevation	30 ft above grade	Not Applicable	ER
16.1.3	Diesel Flue Gas Effluents	See Table 1.4-7	Not Applicable	ER
16.2	Gas-Turbine			
16.2.2	Gas-Turbine Exhaust Elevation	60 ft	Not Applicable	ER
16.2.3	Gas-Turbine Flue Gas Effluents	See Table 1.4-8	Not Applicable	ER
16.2.5	Gas-Turbine Fuel Type	Distillate	Not Applicable	ER
17.	Plant Characteristics			
17.3	Megawatts Thermal	6800 MW(t)	Not Applicable	SSAR & ER
17.4	Plant Design Life	60 years	Not Applicable	ER
17.5	Plant Population			
17.5.1	Operation	580 people	Not Applicable	ER & EP
17.5.2	Refueling/Major Maintenance	1000 people	Not Applicable	EP ER
18.	Construction			
18.2	Acreage			
18.2.1	Laydown Area	29 ac	Not Applicable	ER
18.2.2	Temporary Construction Facilities	52 ac	Not Applicable	ER
18.3	Construction			
18.3.1	Noise	76-101 dBa at 50 ft	Not Applicable	ER
18.4	Plant Population			
18.4.1	Construction	3150 people (max.)	Not Applicable	ER
18.5	Site Preparation Duration	18 months	Not Applicable	ER

(a) Surrogate PPE value not used since actual site characteristic value is available.
(b) 5 percent margin added to vendor supplied PPE quantity to establish value.
(c) Re-evaluated site accident 5% χ/Qs using 36 months of data for the period 1-1-2000 to 12-31-2002 and a minimum distance of 805 m. Also shown are the 50% Chi/Qs used in the ER accident assessments.
(d) LPZ = low population zone
(e) EAB = exclusion area boundary
(f) DBA = design basis accident

Appendix K

**Key Statements Made in the Environmental Report
Considered in the NRC Staff's Environmental Review**

Appendix K

Key Statements Made in the Environmental Report
Considered in the NRC Staff's Environmental Review

Throughout the Environmental Report (ER) supporting the Exelon ESP application, Exelon provides

(1) commitments to address certain issues in the design, construction, and operation of the facility

(2) statements of planned compliance with current laws, regulations, and requirements

(3) commitments to future activities and actions that it will take should it decide to apply for a construction permit (CP) or combined operating license (COL)

(4) descriptions of Exelon's estimate of the environmental impacts resulting from the construction and operation of a new nuclear unit on the ESP site

(5) descriptions of Exelon's estimates of future activities and actions of others and the likely environmental impacts of those activities and actions that would be expected should an applicant holding an Exelon ESP decide to apply for a CP or COL.

Those statements are discussed throughout this environmental impact statement (EIS) and are listed in this Appendix.[a] Some of those statements considered by the staff in determining the level of impacts to a resource are related to matters that are within Exelon's control. Table K-1 lists those matters that were considered in the staff's evaluation of the environmental impacts related to the construction and operation of a new nuclear unit at the Exelon ESP site. The table shows the section and page number where the matter is addressed in the ER, Exelon's statement that addresses the matter, and the location in the EIS where the item was considered in the staff's evaluation. Table K-2 lists those matters that are identified in the ER, but were not directly considered by the staff in its evaluation. Table K-3 lists statements related to likely activities and actions of others that were considered by the staff.

In some cases the same statement or similar statements are made in more than one place in the ER. Where statements contain essentially the same information, the location of the more comprehensive statements are listed first in the table, and the text provided is the text from that location. Locations of similar statements and information are listed, but the text is not included.

(a) The listings are not intended to be a complete list of the commitments described in the ER.

Table K-1. Key Statements Made in the Environmental Report Related to Future Actions and Activities by Exelon and the Impacts of Those Activities Considered in the NRC Staff's Environmental Analysis

Environmental Report			Environmental Impact Statement Sections
Section	**Page**	**Environmental Report Statement**	**Sections**
1.1.4 3.4.2.3	1.1-2 3.4-3	The approach velocity to the intake will be limited to a maximum velocity of 0.50 feet per second (fps) at the normal lake elevation of 690 ft above mean sea level (msl). The intake water for the facility will pass through bar racks or similar devices in order to remove large debris. In addition, it will also pass through traveling screens in order to remove smaller debris before entering the pump suction chamber.	3.2.2.2, 5.4.2.1, 7.5
2.1	2.1-1	The EGC ESP Facility will be colocated on the site of the existing facility and adjacent to the CPS 4,895-ac man-made cooling reservoir; Clinton Lake (IDNR, 2002). The EGC ESP Facility will be located just south of the CPS Facility.	2.1, 4.1.1
2.2.1 2.5.2.2	2.2-2 2.5-5	The EGC ESP Site will not conflict with the proposed zoning for the site, since the facility will be constructed within the CPS Site, which is already designated for transportation and utilities.	2.2, 4.1.1, 5.1, 7.1
2.4.1.3.1 4.3.2.4.1.1 5.3.3.3.1.1 5.6.2.1.1 5.10.3.12.2.1 6.5.2.1.1.1 6.5.2.2.1.1	2.4-4 4.3-5 5.3-8 5.6-4 5.10-18 6.5-2 6.5-5	Federal wildlife agencies will be formally contacted at a date closer to the facility construction to confirm the absence of federal listed threatened and endangered species, since confirmation letters are valid for only one year after issuance.	2.7, 4.4, 5.4, 7.4, 8.5, 8.6
2.4.1.3.2 4.3.1.4.1.2 4.3.2.4.1.2 5.3.3.3.1.2 5.6.2.1.2 5.10.3.12.2.1	2.4-5 4.3-2 4.3-5 5.3-8 5.6-4 5.10-18	State wildlife agencies will be formally contacted at a date closer to the facility construction to confirm the absence of state-listed threatened and endangered species, since conformation letters are valid for only two years after issuance.	2.7, 4.4, 5.4, 7.4, 8.5, 8.6
2.4.2.3.1 5.3.3.3.1.1 5.6.2.1.1	2.4-9 5.3-8 5.6-4	Applicable federal agencies, including the National Marine Fisheries Service and the USFWS will be formally contacted in order to confirm the presence or absence of any federally-listed (or proposed for listing) threatened or endangered fish or other aquatic species.	2.7, 4.4, 5.4, 7.4, 8.5, 8.6
2.5.3	2.5-12	If additional area within the EGC ESP Site will be required, further evaluation will be performed to determine if additional archaeological review is required.	2.9, 4.6, 5.6, 7.6, 8.5, 8.6
2.6	2.6-2	Excavated material will be disposed either on site or off site. Normal methods will be used to mitigate the potential for erosion of material at the disposal site, such as reseeding and drainage control. Excavated slopes or soil surfaces exposed during construction will be protected from erosion.	4.8.1

Table K-1. (contd)

Environmental Report			Environmental Impact Statement Sections
Section	Page	Environmental Report Statement	
2.7.5.1	2.7-17	[A] new [meteorologica] system is being designed to be fully compliant with Regulatory Guide 1.23.	2.3.3
3.1.4	3.1-3	Any visual impacts from the visible plumes from the EGC ESP Facility will be similar to those associated with the CPS. There is the potential that an additional visible plume will result from the heat dissipation system. The viewshed of the EGC ESP Facility is limited to a few residences and recreational users in the vicinity. Based on the fact that the EGC ESP Site will have similar visual impacts as the CPS (with the exception of the new plume from the heat dissipation system assumed for the EGC ESP Facility), the EGC ESP Site will have a minor impact on aesthetic quality for nearby residences and recreational users of Clinton Lake. Therefore, no mitigation will be provided.	4.5.3.4, 5.5.1, 7.6
3.4	3.4-1	Details regarding the design of intake and discharge structures and cooling system comparison tables for the proposed reactor cooling systems will be presented at the COL phase.	4.4, 5.4, 7.4
3.5	3.5-1	Detailed information regarding the description of the liquid and gaseous radioactive waste management and effluent control systems; process/instrumentation diagrams; system process flow diagrams of the liquid and gaseous radioactive waste management and effluent control systems; identification of principal release points; identification of sources of radioactive liquid and gaseous waste materials to the environment; and identification of direct radiation sources stored on site as solid waste will be provided at the COL phase.	3.2.3, 5.9, 6.1, 7.8, 8.11.8
3.5.1	3.5-1	The process systems will be designed to minimize the releases to, and impact on, the aquatic environment. Discharges will be via the existing discharge plume of the CPS.	5.9
3.6	3.6-1	Detailed information regarding the description of the nonradioactive waste management and effluent control systems, process/ instrumentation diagrams, and system process flow diagrams will be provided at the COL phase.	3.2.4, 5.8, 7.7, 8.11.7
3.6.2	3.6-2	Sanitary systems installed for preconstruction and construction activities will include the use of portable toilets, which are supplied and serviced by an off-site vendor. Sanitary system wastes that are anticipated to be discharged to Clinton Lake during actual station operations include discharges from the potable and sanitary water treatment system.... As with the CPS, these discharges will be controlled in compliance with an approved NPDES permit for the EGC ESP Facility, to be issued by IEPA.	4.8, 5.8
3.7.1.1	3.7-1	EGC plans to develop a merchant generator facility at the site; the proposed site will be set aside for a unit that generates power for sale on the open wholesale market. The facility owner will not be responsible for building transmission lines. Rather, it will interconnect with the transmission system owner.	4.4.1, 5.4.1, Chapters 2, 3, 8

Table K-1. (contd)

Environmental Report			Environmental Impact Statement Sections
Section	Page	Environmental Report Statement	
3.7.1.2	3.7-2	To the extent that new transmission lines are needed, they would be interconnected to the Brokaw, Oreana, or Latham substations.	3.3, 4.1.1, 5.1.1
3.7.2	3.7-2	The existing transmission system was sized for a larger capacity than currently used and would be able to carry some new generation. However, in order to accommodate the bounding case of an output of 2,180 MWe, new lines will be required, as there is insufficient capacity on the existing system to carry the load, and the existing structures were not designed for additional circuits. Parallel lines are required in each direction because a single line can not carry the full output of both the EGC ESP Facility and CPS. Four new transmission lines will be required to connect the EGC ESP Facility to the existing transmission grid in southern Illinois. Two parallel, double circuit transmission lines will depart the station north to an interconnect point at the Brokaw substation near Bloomington, Illinois, approximately 15 mi from the site (see Figure 2.2-4). A second pair of parallel double circuit lines will depart the station south to an interconnect point on Illinois Power Company's Latham-Rising 345-kV line (Number 4571) approximately 9 mi from the site (see Figure 2.2-4). As discussed above, it is assumed that any new transmission lines related to this project would be 345 kV.	3.3, 4.1.1, 5.1.1
3.8.1	3.8-3	The LWR technologies being considered will use either Zircaloy or ZIRLO rods and therefore meet this subsequent evaluation condition.	Chapter 6, Appendix G
3.8.1	3.8-3	The LWR technologies being considered will have average burnup of less than or equal to 62,000 MWd/MTU for the peak rod and therefore meet this subsequent evaluation condition.	Chapter 6, Appendix G
3.8.1	3.8-3	The LWR technologies being considered will solidify and package their radioactive waste.	Chapter 6, Appendix G
3.8.1	3.8-4	10 CFR 51.52(a)(5) allows for truck, rail, or barge transport of irradiated fuel. The LWR technologies being considered will comply with this transport mode requirement.	Chapter 6.0, Appendix G
3.8.2.1	3.8-8	The gas-cooled technologies being considered will solidify and package their radioactive waste.	Chapter 6.0, Appendix G
4.0 4.1.1.1	4-1 4-1	It is estimated that site preparation activities (preconstruction) will take up to eighteen months to complete. Based on estimates provided by the reactor vendors, assuming that appropriate licenses are obtained, actual construction is expected to take from three to five years. The construction laydown area will be approximately 29 ac with an additional 52 ac needed for temporary construction facilities, and another 15 ac for a substation (see SSAR Table 1.4-1). To the extent possible, the CPS roads will be used for construction traffic. The site has at least one access road that can be used to transport heavy construction equipment. Construction of the EGC ESP Facility will occur at a location approximately 700 ft to the south of the CPS.	2.2, 3.2.1, 4.1, 4.4.1, 5.1, 7.1, 8.5

Table K-1. (contd)

Environmental Report			Environmental Impact Statement Sections
Section	Page	Environmental Report Statement	
4.1.1.1	4.1-1	No construction activities within the site will take place within a floodplain (IDNR 1986), coastal zone (USGS 1990), or wild and scenic river (USFWS 2002). There are four minor areas (less than 1 ac) within the site boundary that have been identified as wetland areas. They are all palustrine unconsolidated bottom (IDNR 1987). None are within the power block footprint, cooling tower footprint, or intake areas of the EGC ESP Facility, and therefore will not be impacted by construction. Additionally, care will be taken so that these areas are not impacted by other construction activities such as construction laydown, and disposal of fill material. As defined by ESRP Section 4.1.1, since the expected disturbance of construction is less than 1236 ac and does not have any special resources that will be affected, "it may be concluded that the expected impacts of construction on land use are not a major significance and there are no land use changes that will influence the decision on a construction permit" (USNRC 1999).	Chapters 2, 3, 4, 7
4.1.1.2	4.1-2	Normal recreational practices near the site will not be altered during construction. Access to the lake and camp areas will still be afforded to the recreational public.	4.1, 4.5, 8.5
4.1.1.2	4.1-2	In Section 2.2.1, Figure 2.2-3 shows the highways, RR, and utilities that cross the site and the vicinity. None of these facilities will be physically impacted by construction. Approximately 3,200 additional worktrips and 100 truck deliveries during peak hours will occur on the roads and highways during construction, but the roads and highways will not be unduly congested, except for brief periods (10 to 15 minutes) during the beginning and end of shifts.	4.1, 4.5, 7.6
4.1.1.4	4.1-3	Mitigation measures, designed to lessen the impact of construction activities, will be specific to erosion control, controlled access roads for personnel and vehicle traffic, and restricted construction zones. The site preparation work will be completed in two stages. The first stage will consist of stripping, excavating, and backfilling the areas occupied by the structure and roadways. The second stage will consist of developing the site with the necessary facilities to support construction, such as construction offices, warehouses, trackwork, large unloading facilities, water wells, construction power, construction drainage, etc. In addition, structures will be razed and holes will be filled. Grading and drainage will be designed to avoid erosion during the construction period. Action will be taken to restore areas consistent with existing and natural vegetation. A total of approximately 96 ac will be required for construction facilities including permanent facility structures and laydown. To the extent possible, CPS roads will be used for construction traffic. If necessary, temporary stone roads will be installed along with site grading and drainage facilities. This will permit an all weather use of the site for travel and storage of materials and equipment during construction.	2.2, 3.1, 4.1, 4.4.1, 7.1, 8.5, 8.6

Table K-1. (contd)

Environmental Report			Environmental Impact Statement Sections
Section	Page	Environmental Report Statement	
4.1.2.1	4.1-4	In both normal and special condition construction, the methods used will be selected to minimize the impact on the local environment.	4.1, 4.4.1, 7.1
4.1.3	4.1-9	If additional areas within the EGC ESP Site will be required for development, further evaluation will be performed to determine if additional archaeological review is necessary.	2.9, 4.6, 5.6, 7.6, 8.5, 8.6
4.2	4.2-1	The construction will be confined to the station site and the existing transmission corridor. Proper mitigation and management methods implemented during construction will limit the potential water quantity and quality impacts to the surface water (e.g., Clinton Lake, stream crossings, and intermittent drainage ways) and adjacent groundwater.	4.4.2, 4.3
4.2.1.2.1	4.2-4	The construction area will be temporarily isolated from the lake by cofferdams, or similar structures, and dewatered. The water will be pumped to a sedimentation basin if necessary and allowed to drain back into the lake at a location away from the CPS intake structure. Construction of the intake structure will be designed to control shoreline and bank erosion and minimize impacts on Clinton Lake, the UHS, and the CPS intake structure. Special erosion and siltation control measures will be incorporated with lakeshore construction to minimize these impacts. Any sediment deposition in the vicinity of the intake structure will be removed following construction. This work will be bounded by the requirements of the stormwater pollution prevention plans (SWPPP). Appropriate USACOE Section 404, IEPA 401 Water Quality Certification, and NPDES permits will be obtained for these activities.	4.4.2, 4.3
4.2.1.2.2	4.2-4	Comprehensive construction erosion control measures will be employed to minimize the effects of the runoff and minimize siltation in the adjacent drainage ways and Clinton Lake. Runoff from construction areas will be diverted to the south or to the discharge side of the Clinton Lake cooling system in order to avoid impacts to the CPS intake and cooling system. A limited amount of silt deposition in the drainage ways and Clinton Lake will be unavoidable; however, erosion will be monitored and control measures implemented to minimize the potential for additional sediment deposition during the construction period. Proper safeguards (such as sediment basins, silt fencing, and revegetation of disturbed areas) will be used to minimize sediment and nutrient transport to Clinton Lake in order to prevent long-term effects on downstream habitats. Surface disturbance due to construction of overhead transmission lines is expected to be limited to temporary disturbance from removal of trees and shrubs, movement of construction equipment, and excavation for the foundation of the transmission line towers. This disturbance is expected to be minimal, as the disturbances will be short-term or isolated at individual tower pads. The appropriate erosion control measures will be incorporated	4.4.2, 4.3, 4.4, 7.3, 7.4, 7.5

Table K-1. (contd)

Environmental Report			Environmental Impact Statement Sections
Section	Page	Environmental Report Statement	
		into the design contract documents to minimize the impacts of disturbances that occur near the lake or other surface waters. Ground disturbance will be minimized and native ground vegetation will be reestablished following construction in order to minimize erosion.	
4.3.1.2	4.3-1	As previously discussed, transmission system improvements will be required to support the EGC ESP Facility. These modifications will be located within or immediately adjacent to the existing substation at the CPS and along the existing transmission corridor. The proposed transmission line improvements will be sited within the existing utility rights-of-way to the greatest extent poss ble. Construction of the proposed transmission line improvements will temporarily impact habitats within the existing rights-of-way; however, the agricultural and open field areas will be allowed to revegetate to preconstruction conditions. There will be no significant loss of agricultural or open field habitats resulting from construction of the transmission systems. Where right-of-way expansion is required in forested lands, clearing will be required. Forested habitats do not make up a significant amount of the proposed utility corridor; therefore, significant impacts to forested lands are not anticipated.	4.4.1, 5.4.1
4.3.1.4.2.4 Table 10.1-1	4.3-4 10 T-1	The wetlands and floodplains will be restored and there will be no net loss of wetland resources. It is assumed that any pole placement will occur outside of the designated wetland areas. Therefore, the project is not anticipated to adversely affect any wetlands or floodplains within the site or vicinity.	4.4.1, 5.4.1, 7.4, 7.5, 8.5, 8.6
4.4.1.1 4.6.3.2 Table 10.1-1	4.4-1 4.6-2 10 T-1	Noise levels will be controlled by using the following criteria: • The Occupational Safety and Health Administration (OSHA) noise exposure limit to workers and workers' annoyance that are determined through consideration of acceptable noise levels for offices, control rooms, etc. (29 CFR 1910); • Federal (40 CFR 204) noise pollution control regulations; and • State regulation or local (35 Illinois Administrative Code [IAC] Subtitle H, 1987) noise pollution control rules. ... activities with significant noise impacts, such as blasting, will be limited to normal weekday business hours.	4.8, 7.7
4.4.1.3 4.6.1.3	4.4-2 4.6-2	Some recreational users of Clinton Lake will be able to view the construction areas. However, the construction area will not visually impact most recreational users and areas of the Clinton Lake. Therefore, overall aesthetic impacts during construction are minimal. Mitigation measures designed to lessen the minor visual impact of construction activities include restricting construction laydown to as small of an area as possible, and removing construction debris from the site in a timely and suitable manner.	4.5.3.4, 7.6, 8.5, 8.6

Table K-1. (contd)

Environmental Report			Environmental Impact Statement Sections
Section	Page	Environmental Report Statement	
4.4.2.7	4.4-4	Also, since private security guards will be used at the site, dependence on local police forces will not be required.	4.5
4.5.2.	4.5-1	During the construction of the EGC ESP Facility, the construction workers will be exposed to direct radiation and to the radioactive effluents emanating from the routine operation of the CPS. The direct radiation exposure has two principal sources: (1) the cycled condensate storage tank located on the northern boundary of the protected area adjacent to the existing switchyard; and (2) the skyshine from the N-16 activity present in the reactor steam in the high pressure and low pressure turbines, the intercept valves, and the associated piping located on the main floor of the turbine building. The design basis radiation source term for the cycled condensate storage tank is listed in the CPS USAR Table 12.2-8 (CPS 2002). The N-16 activity that is present in the reactor steam in the primary steam lines, turbines, and moisture separators provides an air-scattered radiation dose contribution to locations outside the CPS plant structure. The design basis radiation source inventory in these pieces of equipment is listed in the CPS USAR Table 12.2-7 (CPS 2002). To reduce the turbine skyshine doses, radiation shielding has been provided. The CPS Facility releases airborne effluents via two gaseous effluent release points to the environment. These are the common station heating, ventilating, and air conditioning stack and the standby gas treatment system vent. The expected radiation sources in the gaseous effluents are listed in the CPS USAR Table 11.3-8 (CPS 2002). The CPS Facility has achieved zero liquid radioactivity release from the plant in the past nine years. Therefore, the radiation sources expected to be present in liquid effluents in the future are considered negligible.	4.9
4.6.3.2	4.6-3	Procedures and a hearing conservation program will be developed at the construction site for any employees exposed to excessive noise, which is defined as an 8-hr exposure of 85 dB or more.	4.8, 7.7

Table K-1. (contd)

Environmental Report			Environmental Impact Statement Sections
Section	Page	Environmental Report Statement	
4.6.3.3 10.3.1 Table 10.1-1	4.6-3 10.3-1 10 T-1	During construction, a number of controls will be imposed to mitigate air emissions from construction sources including good drainage and dry weather wetting. In addition, the most traveled construction roads will be paved in order to reduce dust generated by vehicular traffic. Bare areas will be seeded to provide ground cover, where necessary. Applicable air pollution control regulations will be adhered to as they relate to open burning or the operation of fuel burning equipment. Permits and operating certificates will be secured where required. Fuel burning equipment will be maintained in good mechanical order to reduce excessive emissions. Reasonable precautions will be taken to prevent accidental brush or forest fires. The concrete facility will be equipped with dust control systems to avoid excessive releases of cement dust. ... Nevertheless, dust emissions will be mitigated to the extent practical and will be in compliance with local, state, and federal air emissions standards.	4.2, 4.8, 7.2, 7.7
4.6.3.4	4.6-3	If construction activities are not properly controlled and monitored, erosion from improperly graded or excavated areas will lead to the runoff of large amounts of sediments to nearby areas or surface waters. Therefore, the construction activities at the EGC ESP Site will conform to the following goals and criteria, as applicable. • Erosion and sedimentation controls will comply with the requirements specified in this section and, if appropriate, with a stormwater pollution prevention plan. • Implement erosion and sediment controls during construction in order to retain sediment on site to the greatest extent practicable. • Select, install, and maintain control measures in accordance with the manufacturer's specifications and good engineering practices. If periodic inspections or other information indicate that a particular erosion control measure is ineffective, the control measure will be modified or replaced as necessary. • If practical and if required, remove off-site accumulations of sediment in order to minimize the off-site impacts in the event that sediment escapes the construction site. • Routinely remove sediment from sediment traps or sedimentation routinely. • Implement construction practices that prevent litter, construction debris, and construction chemicals exposed to stormwater from becoming pollutant sources for stormwater discharges. • Control erosion and sediment runoff through the use of structural and/or stabilization practices. Structural control practices may include the use of straw bales, silt fences, earth dikes, drainage swales, sediment traps, and sediment basins. Sediment traps an basins will be designed to accommodate the large potential load from the deep	4.3.3, 4.4.1.1, 4.4.1.2, 4.4.2

Table K-1. (contd)

Environmental Report			Environmental Impact Statement Sections
Section	Page	Environmental Report Statement	
		excavation dewatering operations. Stabilization practices may include temporary seeding, permanent seeding, mulching, geotextiles, sod stabilization, vegetative buffer strips, protection of trees, and preservation of mature vegetation. Several different structural controls may be used to regulate the quality of the stormwater running off the construction site. Table 4.6-1 lists the controls that may be instituted during construction activities. Based on site conditions, the final location of these controls will be determined just prior to the commencement of construction. Stabilization practices that may be implemented are listed in Table 4.6-2. Final stabilization will consist of grading and revegetation areas in which potential pollutant sources are used.	
4.6.3.4	4.6-4	In addition, the following general erosion control requirements will be implemented during construction activities, as appropriate: • Where practical, disturbed soil areas will be reseeded with maintenance seed (if activities are temporary) or permanent seed mix (for permanent or final cover) as soon as possible after redress activities are either temporarily or permanently stopped. • Where practical, excelsior blankets will be mulched or installed and slopes greater than 3:1 will be reseeded, depending on the length, exposure, and texture of the soils on the slope. Mulch may be natural and consist of slash, brush, manure, and vegetation previously chipped and stockpiled; clean straw, free from noxious weed seed, mold, and other harmful elements; or wood cellulose fiber. Mulch will be applied as soon as possible after seeding to reduce runoff and promote vegetation. • Sidehill slopes will be furrow-contoured as practical. Otherwise the final grading will be performed in a manner that will result in tracks and depressions contoured across the slope instead of down the "fall-line." This will not only minimize wind erosion, but will also "roughen" the earth to provide a microclimate of wind protection for new plants, and will help conserve precipitation for use in growth of new seed. This results in a reduction of sediment erosion. • The time that bare soil is exposed before stabilized will be minimized. • The disturbance to existing vegetation will be minimized. • Where slope cuts have developed from erosion (particularly along the faces of flood detention structures), loose material will be removed, and the area will be filled with suitable soils to the original profile of the bank or slightly above the original profile. If the cut is not completely filled, the steeper area at the brow of the cut will encourage erosion and may cause redevelopment of the cut. The area upstream from the cut will be carefully inspected to determine if there is an irregularity in the ground profile that will cause stormwater to concentrate and erode	4.2.1, 4.3.3, 4.4.1.1, 4.4.1.2, 4.4.2

Table K-1. (contd)

Environmental Report			Environmental Impact Statement
Section	**Page**	**Environmental Report Statement**	**Sections**
		the soils. Any such irregularity will be removed. This will allow the water to run off the site as sheet flow. • No solid materials including demolition materials will be discharged to waters of the United States (U.S.), unless authorized under an approved permit. The erosion and sediment control measures and other protection measures will be maintained in effective operating condition. Maintenance will be performed on an "as needed" basis and as specified by state and local permits. Specific maintenance requirements include, but are not limited to: • Routine removal of sediment and other debris collected behind silt fences or hay bales; • Routine cleaning of sediment from detention ponds; and • Based on visual inspection, replacement of gravel and sediment from entrances/exits.	
4.6.3.5.1	4.6-5	The fueling stations will have temporary secondary containment around the fuel tanks. For specifics, see Section 4.6.3.5.8.	4.8
4.6.3.5.5.1	4.6-6	In general, excavated soils and stockpiles will be managed; management techniques are described below. • Stockpiles of excavated soils will be placed on plastic sheeting or other suitable material, if required, near the excavation areas. • If practical, stockpiles will be provided with liner, cover, and perimeter berm in order to prevent rupture, release or infiltration of liquids, and to prevent the re-suspension dispersion of dust. If it is not possible to cover stockpiles, it may be necessary to install a temporary sprinkler system to inhibit dust dispersion. • Polyethylene sheeting or other suitable material will be used for liners and covers. • A perimeter berm, typically hay bales placed beneath the liner, will be constructed to allow for collection of any free liquids draining from the stockpile. • Accumulated free liquids will be pumped, treated, and removed, as required. • Covers and perimeter berms will be secured in place when not in use and at the end of the workday, or will be secured as necessary in order to prevent wind dispersion or runoff from major precipitation events.	4.1, 4.4.1
4.6.3.5.6	4.6-6	Sediment and the generation of dust will be minimized using the methods noted in Section 4.6.3.3, thereby minimizing the amount that is tracked off site by vehicles.	4.4.1

Table K-1. (contd)

Environmental Report			Environmental Impact Statement Sections
Section	Page	Environmental Report Statement	
4.6.3.5.8.1	4.6-7	Fuel and waste tanks located on soil will be bermed with a perimeter dike of native material, or placed inside an open tank capable of containing its' maximum capacity, in case of rupture. When practical, areas inside the dike will be covered with an oil resistant membrane to minimize soil contamination in the event of a spill. Fuel and waste tanks located on concrete or steel foundations will be bermed with appropriate materials suitable for the application. These materials will allow for the containment of the full capacity of the tank while minimizing contamination of the surrounding area. Construction projects requiring fuel or waste tanks will maintain a sufficient number of spill kits to contain minor spills and leaks.	4.8
4.6.3.6 5.10.3.5	4.6-9 5.10-5	Traffic and traffic control impacts may include, but are not limited to: • Working adjacent to or in active roadways (day/night); • Traffic control zones; • Traffic control device installation and removal; • Flagging; • Inspection and maintenance of traffic control devices; • Equipment; and • General roadway traffic control zone safety. Regulatory guidance 29 CFR 1926 contains requirements for traffic control signs, signals, and barricades. Some state OSHA and DOT plans may have requirements that are more stringent. However, local, state, and federal requirements will be adhered to regarding traffic control on and off site from construction activities.	4.5, 5.5.1, 7.6
4.6.3.7	4.6-10	The construction will be confined to the EGC ESP Site and the existing transmission corridor. Proper mitigation and management methods implemented during construction will limit the potential water quantity and quality impacts to the surface water (e.g., Clinton Lake, stream crossings, and intermittent drainage ways) and adjacent groundwater.	4.4.2
4.6.3.7.1 4.2.1	4.6-11 4.2-2	Construction erosion control measures and comprehensive SWPPP are required under the Illinois Environmental Protection Act, the Illinois Pollution Control Rules, and the federal CWA. Where necessary, special erosion control measures will be implemented to minimize impacts to the lake and lake users and CPS operations. Typical stormwater control elements of a SWPPP are discussed in Section 4.6.3.4. A NOI will be filed with the federal and state agencies to receive authorization for land disturbance under the general stormwater permit. A SWPPP will also be prepared in accordance with the requirements of the general permit. A NOT will be filed with the IEPA upon completion of construction and stabilization of the disturbed areas.	4.5.1.4, 7.6

Table K-1. (contd)

Environmental Report			Environmental Impact Statement Sections
Section	Page	Environmental Report Statement	
4.6.3.7.1.2	4.6-11	Construction erosion control measures will be applied during the phases of site development to contain eroded soil on the construction site and remove sediment from stormwater prior to leaving the site. Design measures will be incorporated to avoid concentrated flow that has a high potential to transport sediment. Visual inspections of construction erosion control measures will be incorporated into the construction project to monitor the effectiveness of the control measures and to aid in determining if other mitigation measures are necessary. Mitigation measures will be incorporated into the requirements of the construction contracts and the SWPPP. Beyond the construction activity, stormwater management practices will be incorporated into the site design to minimize the long-term delivery of sediment to the lake.	4.1, 4.3, 4.4.1
4.6.3.7.1.3 4.2.1.3	4.6-12 4.2-6	The dewatering effluent obtained from the station excavation will be pumped and eventually discharged to an adjacent drainage way and into Clinton Lake. Measures will be implemented, such as sedimentation or filtration, so that erosion or siltation caused by the dewatering will be negligible. Existing sediment basin facilities will be considered or new facilities constructed to accommodate dewatering flows. Where possible, dewatering flows will be diverted to the south or to the discharge side of Clinton Lake in order to avoid impacts to the CPS intake and cooling system. A limited amount of silt deposition in the drainage ways and Clinton Lake will be unavoidable; however, the impacts from these activities will be confined to the construction period and will be monitored and controlled using best management practices for sediment control. Proper safeguards will be implemented to prevent long-term effects on downstream habitats resulting from the construction activities.	4.3.1, 4.3.2
4.6.3.7.2.1 4.2.2.2	4.6-13 4.2-7	The limited amount of additional sediment in stormwater related to construction activities will be first controlled by sight specific practices identified in the SWPPP. During construction of the new EGC ESP intake structure, the CPS intake structure will be protected to prevent suspended sediment from entering the cooling system. Special construction techniques, such as watertight sheet piling with dewatering of submerged areas to expose the construction zone, will be implemented where necessary to prevent migration of suspended solids. Water collected from dewatering operations will be settled or filtered before water is allowed to return to the lake. Where appropriate, stormwater runoff and treated dewatering water will be diverted to the discharge side of the lake to reduce CPS impacts.	4.3.3

Table K-1. (contd)

Environmental Report			Environmental Impact Statement Sections
Section	Page	Environmental Report Statement	
5.1.1.2	5.1-2	Quantification of impacts associated with salt drift will be reassessed, as appropriate, once the facility's cooling system configuration and design parameters have been determined. This analysis will be conducted at or before a later licensing stage.	5.1
5.2.1.1.1 5.10.3.7.1.1	5.2-2 5.10.7	The dam that forms Clinton Lake is operated to provide a minimum downstream release of 5 cfs from Clinton Lake to Salt Creek. This flow rate will not change under the operation of the EGC ESP Facility.	2.6, 5.3, 7.3
5.2.2.2.1	5.2-7	The EGC ESP Facility operation will comply with federal laws related to hydrology and water quality.	5.3.3
5.2.2.2.2	5.2-8	The combined discharge of the two plants will be within with the limits of the NPDES permit for the CPS.	5.4.2.2
5.2.2.3	5.2-8	As discussed above, it is anticipated that surface water (namely Clinton Lake) will be used to meet the operational water requirements of the EGC ESP Facility, and that groundwater will not be used as a source of water. In addition, based on the proposed design of the plant, no permanent groundwater dewatering system will be implemented.	5.3.2
5.3	5.3.-1	As described in Section 3.3, either mechanical draft or natural draft hyperbolic type cooling towers will be used for normal non-safety plant cooling and for safety-related cooling. The makeup water for the normal (non-safety) plant operations will be taken up through a new intake structure located approximately 65 feet south of the CPS intake structure on the northern basin of Clinton Lake. The intake will include a screening system similar in function to the CPS intake, but for a significantly smaller flow rate. Makeup water for the safety-related cooling towers will be supplied from the same intake structure, which will draw water from the bottom of the submerged impoundment within Clinton Lake (i.e., the UHS). The cooling tower(s) blowdown will be discharged to the CPS discharge flume that flows to the southern basin of Clinton Lake.	5.3
5.3.1.1.1	5.3-2	Design of the intake structure will include features that maintain an even distribution of intake flows. Where necessary, the intake area will be protected to prevent local areas of erosion.	2.7, 4.4, 5.4, 7.5, 8.5, 8.6
5.3.1.1.3 5.10.3.9.2.2	5.3-2 5.10-11	In addition, the piping system will need to be kept clean of aquatic organisms such as algae and shellfish. Standard practices that have been used by the utility industry include scraping, backwash with the heated cooling water and chemical treatment including certain biocides, anti-corrosion, and anti-scaling chemicals. These chemicals will ultimately be discharged to Clinton Lake through the thermal discharge piping, as described in Section 3.6.1. If a chemical addition is required to protect the new cooling system, this same approach may be used in the intake piping. It is anticipated that there will be a minor change in the quality of the water discharged. The selection of chemicals will be done in order to minimize the impacts on	2.7, 4.4, 5.4, 7.5

Table K-1. (contd)

Environmental Report			Environmental Impact Statement Sections
Section	Page	Environmental Report Statement	
		water quality. It is assumed that the discharges will be comparable to those associated with the CPS as approved under their NPDES permit.	
5.3.2.1.2 5.10.3.9.2.2	5.3-4 5.10-11	The chemicals used will be subject to review and approval for use by the IEPA, and releases will be in compliance with water quality standards and an approved NPDES permit. The total residual chemical concentrations in the discharges to Clinton Lake will be subject to limits that will be established by the IEPA.	2.7, 4.4, 5.4, 7.4, 7.5, 8.5, 8.6
5.3.3.3.2.1	5.3-9	It is not anticipated that the proposed heat dissipation system will have any adverse impacts on the terrestrial environment within the Clinton Lake State Recreation Area. The proposed system will not inhibit access to or use of the terrestrial system surrounding Clinton Lake.	5.1
5.3.4.2	5.3-11	...the operation of the EGC ESP Facility will result in significant heat dissipation to the atmosphere. Depending on the type of cooling system(s) used to dissipate this heat, the rejected heat will be manifested in the form of thermal and/or vapor plumes on and around the site. Quantification of these ambient impacts will necessarily require a more in depth assessment once the facility's cooling system configuration and design parameters have been determined. This analysis will be conducted at or before a later licensing stage.	5.2.1
5.4.1.3	5.4-3	Contained sources of radiation at the EGC ESP Facility will be shielded as was done at the CPS. It is assumed that the direct radiation from any of the EGC ESP Facility designs remains bounded by the CPS direct and skyshine dose from the turbine building.	5.9, 7.8, 8.11.8
5.5.1.2.1	5.5-1	Drains from radioactive sources or potentially radioactive sources will not be connected to the chemical waste drain system. Chemical waste discharges will be collected in a tank for sampling and pH adjustment before being discharged as neutralized wastes to Clinton Lake. The chemical wastes will be routed to the discharge flume of the CPS, which flows to Clinton Lake.	5.9, 7.8
5.5.1.2.2	5.5-2	Sanitary system wastes that are anticipated to be discharged to Clinton Lake during actual station operations include discharges from the potable and sanitary water treatment system. It is anticipated that the sanitary system effluents will receive tertiary treatment consisting of presettling, filtration, and chlorination prior to release to the environment via the circulating water discharge flume. The normal and maximum amount of sanitary discharges to Clinton Lake based on PPE data for the composite reactor (see SSAR Table 1.4-1) is presented in Chapter 3. These discharges will comply with the approved NPDES permit for the EGC ESP Facility.	5.8, 7.7
5.5.1.3 10.2.1.6 10.3.2	5.5-3 10.2-2 10.3-2	Air emissions will be in compliance with the limits that will be established and imposed by state and local regulations.	5.2.2

Table K-1. (contd)

Environmental Report			Environmental Impact Statement Sections
Section	Page	Environmental Report Statement	
5.5.2	5.5-4	However, if mixed waste is generated, the volume may be reduced or eliminated by one or more of the following basic types of treatment prior to disposal: decay, stabilization, neutralization, filtration, and chemical or thermal destruction by an off-site vendor. If required, programs will be implemented and mixed waste storage facilities constructed to store mixed waste for decay or for storage prior to shipment to an approved off-site treatment or disposal area. It is not the Applicant's intention to dispose of mixed waste on site.	5.8, 7.7
5.8	5.8-1	The operation workforce will consist of up to 580 people (see SSAR Table 1.4-1).	5.5.1, 7.6
5.8.1	5.8-1	The physical impacts are defined as noise, air, and aesthetic disturbances. Physical impacts will be controlled as specified by applicable regulations and will not significantly impact the site, vicinity, or region.	5.5.1, 7.6
5.8.1.1	5.8-1	The two largest cities within the vicinity include DeWitt, with a population of 188, and Weldon, with a population of 440 (U.S. Census Bureau 2001). These two cities are small rural communities that include small businesses, houses, and farm buildings. These communities will not experience any physical impact from station operation.	5.5.1, 7.6
5.8.2.7	5.8-5	Also, since private security guards will be used, dependence on local police forces will not be required.	5.5.3.6
5.8.3	5.8-6	Noise and air pollution will be controlled by following any federal, state, and local regulation.	5.5
5.9	5.9-1	According to the USNRC, decommissioning of a nuclear power plant has certain environmental consequences. The impacts on the proposed site will be discussed in detail at the COL stage.	6.3
5.9	5.9-1	As decommissioning plans are developed, efforts will be made to minimize or mitigate any adverse impacts from decommissioning.	6.3
5.10.3.3	5.10-3	The following goals and criteria will be applied, as applicable: • Erosion and sedimentation controls will be implemented in order to retain sediment on site to the greatest extent practicable. • In accordance with the manufacturer's specifications and good engineering practices, control measures will be selected, installed, and maintained. If periodic inspections or other information indicate that a particular erosion control measure is ineffective, the control measure will be modified or replaced as necessary. • If possible and if required, off-site accumulations of sediment will be removed in the event that sediment escapes the construction site in order to minimize the off-site impacts. • Sediment from sediment traps or sedimentation ponds will be routinely removed when design capacity, as a general rule, has been reduced by approximately 50 percent. This will limit the potential for trap or pond failure.	4.4.1, 5.8, 7.7

Table K-1. (contd)

Environmental Report			Environmental Impact Statement Sections
Section	Page	Environmental Report Statement	
		• Housekeeping practices will be implemented that prevent litter, debris, and chemicals exposed to stormwater from becoming a pollutant source for stormwater discharges. • Erosion and sediment runoff will be controlled through the use of structural and/or stabilization practices. Structural control practices may include the use of straw bales, silt fences, earth dikes, drainage swales, sediment traps, and sediment basins. Sediment traps and basins will be designed to accommodate the large potential load from the deep excavation dewatering operations. Stabilization practices may include temporary seeding, permanent seeding, mulching, geotextiles, sod stabilization, vegetative buffer strips, protection of trees, and preservation of mature vegetation.	
5.10.3.4.1	5.10-3	The fueling stations, as appropriate, will have secondary containment structures installed around the fuel tanks with a leak detection system to alert personnel in the event a tank leaks fuel to the secondary containment.	5.8, 7.7
5.10.3.4.2	5.10-4	Regular vehicle maintenance will be performed in an area designated for that purpose. Any spills will be cleaned up promptly. Precautions will be taken to prevent the release of pollutants to the environment from vehicle maintenance. Precautions will include the use of drip pans, mats, and other similar methods. No vehicle washwater will be allowed to run off the EGC ESP Site or enter local, state, or federal waters.	4.4.1
5.10.3.4.3 5.2.1.3	5.10-4 5.2-6	To prevent the mobilization of contaminants in stormwater runoff from entering and/or leaving excavated areas, the following controls on erosion and sedimentation controls will be implemented, as applicable and as found appropriate to control the material. • Stockpiles of excavated soils will be placed on plastic sheeting near the excavation areas. • Stockpiles will be provided with liner, cover, and perimeter berm to prevent rupture and release or infiltration of liquids. • Polyethylene sheeting will be used for liners and covers. • A perimeter berm, typically hay bales placed beneath the liner, will be constructed to allow for collection of any free liquids draining from the stockpile. • Accumulated free liquids will be pumped or otherwise removed to a sanctioned area or container. • Covers and perimeter berms will be secured in place when not in use and at the end of the workday, or as necessary to prevent wind dispersion or runoff from major precipitation events.	5.8, 7.7

Table K-1. (contd)

Environmental Report			Environmental Impact Statement Sections
Section	Page	Environmental Report Statement	
5.10.3.4.4	5.10-4	The following material handling and housekeeping practices described below will be implemented during EGC ESP Facility operations, as applicable and as found appropriate. • Auxiliary fuel tanks will have secondary containment. The area will be kept free of trash and spilled fuel. • Garbage receptacles will be equipped with covers. This includes such receptacles that contain materials that may be carried by the wind or contain water-soluble materials, (e.g., paint). • Empty storage containers including drums and bags will be stored inside a designated storage building or area. • Containers will be kept closed except as necessary to add or remove material. • Containers will be stored in such a manner to prevent corrosion that could result from contact between the container and ground surface, and in a release of material. • The containers will be appropriately labeled to show the name, type of substance, health hazards, and other appropriate information, if applicable. • MSDSs for chemical substances used or stored on site will be available for review and use.	5.8, 7.7
5.10.3.7.3 5.2.1.3	5.10-8 5.2-6	It is anticipated that surface water (namely Clinton Lake) will be used to meet the operational water requirements of the EGC ESP Facility; groundwater will not be used as a source of water. In addition, based on the planned design of the EGC ESP Facility, no permanent groundwater dewatering system will be implemented.	5.3.2
5.10.3.8.1.1 5.10.3.7.1.1	5.10-9 5.10.7	The 5-cfs minimum discharge from Clinton Lake to Salt Creek will be maintained in accordance with the CPS NPDES requirements.	2.7, 4.4, 5.4, 7.4, 8.5, 8.6
5.10.3.8.2.1 5.2.2.2.1	5.10-9 5.2-7	The EGC ESP Facility will be designed and operated to be compatible with the operation of the CPS and its NPDES permit.	5.3.3
5.10.3.8.2.1	5.10-9	The EGC ESP Facility operation will comply with federal laws related to hydrology and water quality.	5.3.3
5.10.3.8.3	5.10-10	In addition, based on the proposed design of the plant, no permanent groundwater dewatering system will be implemented. Thus, there are no anticipated groundwater use impacts resulting from the operation of the EGC ESP Facility.	5.3.3
5.10.3.9.4.1	5.10-13	Monitoring will be performed, as appropriate and if required, for the presence of thermophilic organisms, and the potential health risk will be evaluated during preapplication monitoring.	5.8, 7.7
5.10.3.9.4.1	5.10-14	If wet cooling is selected, the cooling tower water will be treated with biocides to prevent the growth of dangerous organisms. Monitoring programs will be established to test for the presence of thermophilic microorganisms once the EGC ESP Facility is operational, both to protect on-site workers and the public.	5.8, 7.7

Table K-1. (contd)

Environmental Report			Environmental Impact Statement Sections
Section	Page	Environmental Report Statement	
5.10.3.10.3	5.10-15	The EMP will utilize 10 CFR 50, Appendix B, compliant quality programs and processes to: Provide that personnel are trained and qualified to perform radiological monitoring; Create and approve procedures for sample collection, packaging, shipment, and receipt of samples for analysis, and prepare and analyze samples at the lab; Document lab processes such as maintenance, storage, and use of radioactivity reference standards, and document the cal bration and checks of radiation, radioactivity measurement systems, and sample tracking and control; Document the processes and procedures of the monitoring program; Conduct periodic audits of analysis laboratory functions and their facilities; Maintain records of sample collection, shipment, and receipt. Lab activity records will also be maintained including sample description, receipt, lab identification, coding, sample preparation and radiochemical processing, data reduction, and verification.	5.9, 7.8
5.10.3.10.3	5.10-15	In addition, the following activities will be performed: • Perform duplicate analysis of the samples (excluding TLDs) to check laboratory precision; • Routinely count quality indicator and control samples; and • Participation in inter-comparison programs, such as the Environmental Resource Associates (ERA) cross-check program. The analytical results provided by the laboratory will be reviewed monthly to validate that the required minimum sensitivities have been achieved and the correct analyses have been performed.	5.9, 7.8
6.1.1.2	6.1-2	Additional preapplication monitoring will be conducted to verify and update the baseline conditions at the time of the COL application. The proposed preapplication monitoring will include the collection of monthly temperature measurements from general locations described below and presented in Figure 6.1-1. • Locations Coincident with CPS Monitoring Locations – Site 16 is located upstream from the discharge canal. Data from this site will be used to characterize thermal conditions upstream of the discharge flume. – Site 2 is located offshore from the cooling water discharge flume. Data from this site will be used to characterize lake conditions at the point of thermal discharge to the lake. – Sites 8 and 13 are located along the path of the cooling loop between the discharge of water into the lake and the CPS intake. The data from these sites will be used to characterize conditions along the cooling loop.	2.7, 4.4, 5.4, 7.4, 8.5, 8.6

Table K-1. (contd)

Environmental Report			Environmental Impact Statement Sections
Section	Page	Environmental Report Statement	
		– Site 4 is located near the CPS screen house. The data from this location will be used to characterize lake conditions at the intake.	
6.1.1.2	6.1-3	At each site, the temperature measurements will be collected at the surface and 0.5-m (1.5-ft) depth intervals to the bottom using a "YSI Multiprobe or Multiparameter Instrument" (or equivalent meter). The depth of the water column will also be recorded. If thermal stratification (temperature gradient of at least 1°C [about 35°F] per 3-ft depth interval) is present, the water column will be segmented into epilimnion, metalimnion, and hypolimnion. The temperature measurements at each site will be taken at consistent depths and at a time of day (morning) that minimizes the effect of diurnal solar warming.	2.7, 4.4, 5.4, 7.4, 8.5, 8.6
6.1.2	6.1-3	The Preoperational Monitoring Program will consist of continuing the preapplication monitoring until the EGC ESP Facility is operational. The results of the preapplication sampling will be evaluated in order to determine if the scope and the frequency of thermal monitoring need to be modified to establish the baseline for water temperature in Clinton Lake and Salt Creek.	2.7, 4.4, 5.4, 7.4, 8.5, 8.6
6.1.2	6.1-3	Modifications to the Preoperational Monitoring Program will consider the following objectives: - Determine the average, extent, and surface area of the limiting excess temperature isotherm if one has been established by the IEPA; -Determine the temperature at positions that are appropriate in order to define the extent of existing mixing zones from the discharge flume; and -Establish time-temperature relationships at monitoring stations.	2.7, 4.4, 5.4, 7.4, 8.5, 8.6
6.2	6.2-1	The proposed radiological environmental monitoring program (REMP) for the EGC ESP Facility will be designed to monitor the radiological environment during the preconstruction and construction phases from active CPS Facility operations as well as the radiological environment surrounding the EGC ESP Facility during active facility operations.	2.5, 5.9, 7.8, 8.11.8
6.2.1	6.2-1	The proposed REMP will be implemented in accordance with the 10 CFR 20.1501 and Criterion 64 of 10 CFR 50, Appendix A.	2.5, 5.9, 7.8, 8.11.8
6.2.1	6.2-2	The scope of the program will include the monitoring of six environmental elements: • Direct radiation; • Atmospheric; • Aquatic; • Terrestrial environments; • Groundwater; and • Surface water.	2.5, 5.9, 7.8, 8.11.8

Table K-1. (contd)

Environmental Report			Environmental Impact Statement Sections
Section	Page	Environmental Report Statement	
6.2.2	6.2-3	Analyses performed on environmental samples collected will include the following: • Gross alpha and beta analysis; • Gamma spectroscopy analysis; • Tritium analysis; • Strontium analysis; and • Gamma dose (TLD only).	2.5, 5.9, 7.8, 8.11.8
6.2.2.1	6.2-3	TLDs will be used to measure the ambient gamma radiation levels at many locations surrounding the EGC ESP Facility.	2.5, 5.9, 7.8, 8.11.8
6.5	6.5-1	Furthermore, in an effort not to duplicate monitoring efforts, the Applicant will coordinate its Ecological Monitoring Programs with existing Ecological Monitoring Programs and efforts being performed by the CPS, IDNR, IEPA, and other applicable groups or agencies.	2.7.2.3
6.5	6.5-1	Site preparation and construction monitoring, preoperational monitoring, and operational monitoring programs will be provided at the COL phase, in accordance with the schedule provided in NUREG-1555.	2.7, 4.4, 5.4, 7.4, 8.5, 8.6
6.5.2.1	6.5-4	The program proposed in the CPS ER included fish sampling at five sampling locations that were identified in the preliminary baseline assessment. The CPS ER proposed that sampling be continued at these locations on a quarterly basis so that fishery resources are sampled during each season of the year (CPS 1973). Additionally, new locations within Clinton Lake will be monitored, associated with the proposed intake structure and discharge from the EGC ESP Facility, to evaluate effects on fishery resources during operation.	2.7, 4.4, 5.4, 7.4, 8.5, 8.6
6.5.2.2.1.3	6.5-5	As previously discussed, specific monitoring programs used to identify impacts to fishery resources resulting from operation of the EGC ESP Facility will be recommended once the final design has been confirmed. Representatives from EGC will coordinate their efforts with the IDNR to design a monitoring program that does not duplicate any of the IDNR's ongoing data collection/sampling efforts. In addition, the proposed program will provide the ability to monitor species of commercial and recreational value within the vicinity.	2.7, 5.4, 7.5
7.1	7-1	Analysis of severe accidents and mitigation of those accidents will be deferred until the COL stage.	5.10.2
8.1	8-1	[Need for Power] Therefore, this evaluation will be provided at the time an application for a construction permit or COL is submitted, in accordance with the applicable regulations (USNRC 1999).	Chapter 8

Table K-2. Key Statements Made in the Environmental Report Not Directly
Considered in the NRC Staff's Environmental Analysis

Environmental Report		
Section	Page	Environmental Report Statement
2.6	2.6-1	The potential effects of seismic loads, such as liquefaction and soil structure interaction, will be considered during design.
2.6	2.6-1	New cooling water detention ponds could be required, based on the final reactor selection. Although these ponds would have the potential to serve as a source of groundwater infiltration, the cooling water ponds will be lined to preclude such occurrences.
3.0	3-1	The EGC ESP Facility will be essentially independent of the CPS. With the exception of using the CPS UHS as a source of makeup water, no CPS safety-related systems or equipment will be shared or cross-connected. Raw water for cooling water makeup and other facility services will be provided from a new intake structure located on Clinton Lake adjacent to the CPS intake structure. Facility discharges will use the CPS discharge flume as a discharge path to Clinton Lake. Some structures, such as a warehouse, training buildings, and parking lots, may be shared. Some support facilities, such as domestic water supply and sewage treatment, may also be shared.
3.1.3	3.1-3	Raw water for cooling water makeup and other facility services will be provided from a new intake structure located on Clinton Lake adjacent to the CPS intake structure. Cooling tower blowdown and other facility discharges will use the CPS discharge flume as a discharge path to Clinton Lake.
3.1.3	3.1-3	The existing switchyard will be expanded to accommodate the output of the new facility and to provide the necessary off-site power. The switchyard area intended for the planned CPS Unit 2 will be utilized for this purpose. Existing transmission right-of-way will be used. Detailed information regarding this subject area is presented in Section 4.1.2.
3.3	3.3-1	Wastewater discharges from the proposed facility will be in strict compliance with an approved NPDES permit issued by the IEPA. This permit will make certain that discharges are controlled from systems (such as flumes, sewage treatment facilities, radwaste treatment systems, activated carbon treatment systems, water treatment waste systems, facility service water, stormwater runoff, etc.) to Clinton Lake. The effect on water quality in Clinton Lake due to the operation of the proposed facility will be carefully monitored in full compliance with the NPDES permit.
3.4.2.2	3.4-3	The CPS discharge flume will have to be modified to accommodate discharges from the EGC ESP Facility. The only modification to the discharge flume will be to connect discharge pipes from the EGC ESP Facility to the discharge flume. Discharge pipe connections will be in the portion of the existing flume discharge structure that was originally provided for the circulating water discharge from the cancelled CPS Unit 2.

Table K-2. (contd)

Environmental Report		
Section	Page	Environmental Report Statement
3.4.2.4	3.4-4	The UHS system will pump water from the safety-related (essential service water) cooling tower basins through the components cooled by the system. The water will then be returned to the cooling towers for heat rejection to the atmosphere. Normal makeup water for the UHS cooling tower basins will be supplied from Clinton Lake. Emergency makeup water will be supplied from the submerged pond below Clinton Lake in the event that Clinton Lake dam fails. Pumps for the normal and emergency UHS makeup water will be located in a new intake structure, the same one used for the NHS cooling towers, and positioned approximately 65 feet south of the CPS intake structure. Detailed design information regarding the new intake structure is not presently available but will be provided at the COL phase. Blowdown, from the discharge of the UHS system pumps, will be used to control the concentration of impurities in the water due to evaporation in the cooling tower.
3.4.2.5	3.4-4	Temperature monitoring instrumentation will be provided in the blowdown discharge pipe to monitor the discharge temperature.
3.5	3.5-1	Radioactive waste management and effluent control systems will be designed to minimize releases from active reactor operations to values as low as reasonably achievable (ALARA).
3.5.1	3.5-1	The release of radioactive liquid effluents from the plant will be controlled in such a manner as to not exceed the average annual effluent concentration limits (ECLs) specified in 10 CFR 20. The proposed EGC ESP Facility will be operated such that releases of radioactive liquid effluent to Clinton Lake are expected to be negligible.
3.5.2	3.5-2	The release of radioactive gaseous effluents from the plant will be controlled and monitored so that the regulatory limits specified in 10 CFR 20 and 10 CFR 50, Appendix I, are maintained.
3.5.3	3.5-3	In addition, the solid waste management system will provide storage of operations waste prior to processing or shipment. The system will be designed to collect and store radioactive wastes in a manner that will maintain radiation exposures ALARA and perform the following objectives: • Collect, hold for decay, monitor, package, and temporarily store the wet and dry solid radioactive wastes produced by the plant during operation and maintenance prior to • Provide a means for segregating trash by radioactivity level and temporarily store the • Minimize exposure to solid radioactive waste materials that could conceivably be hazardous to either operating personnel or the public, in accordance with 10 CFR 20 and 10 CFR 50, Appendix I. • Minimize the volume of solidified waste requiring shipment off site. • Take due account (through equipment selection, arrangement, remote handling, and shielding) of the necessity to keep radiation exposure of in-station personnel ALARA.
3.5.3	3.5-3	The waste will be packaged and shipped in accordance with the applicable regulatory requirements.

Table K-2. (contd)

Environmental Report		
Section	Page	Environmental Report Statement
4.2.1.1	4.2-2	The impacts to Salt Creek will be reduced by lake watershed stormwater management practices and the buffering effect of the lake on the rate and volume of runoff as well as water quality. The dam operating procedures will be reviewed and revised as necessary during the construction phase, to accommodate changes in the watershed hydrology and monitoring improvements to support the minimum 5 cfs discharge. These changes will be mitigated by incorporating construction erosion practices as required by federal and state law and stormwater best management practices following construction.
4.2.1.2	4.2-3	Construction erosion control measures will be applied during the phases of site development to contain eroded soil on the construction site and remove sediment from stormwater prior to leaving the site. Design measures will be incorporated to avoid concentrated flow that has a high potential to transport sediment. Visual inspections of construction erosion control measures will be incorporated into the construction project to monitor the effectiveness of the control measures and to aid in determining if other mitigation measures are necessary. Mitigation measures will be incorporated into the requirements of the construction contracts and the stormwater pollution prevention plans (SWPPP). Beyond the construction activity, stormwater management practices will be incorporated into the site design to minimize the long-term delivery of sediment to the lake.
4.2.1.2.2	4.2-5	A notice of intent (NOI) will be filed with the federal and state agencies to receive authorization for land disturbance under the General Stormwater Permit. A SWPPP will also be prepared in accordance with the requirements of the general permit. A notice of termination (NOT) will be filed with the IEPA upon completion of construction and stabilization of the disturbed areas.
4.2.1.2.3	4.2-5	These spoil areas will be maintained during construction in order to minimize water and wind erosion. Spoil areas will be kept graded, reasonably flat, and compacted by normal construction traffic. Spoil areas will be surrounded by a silt fence or a vegetated buffer strip, which will be maintained in order to minimize erosion. If necessary, water will be sprayed on the bare soil to minimize wind erosion during dry periods. If stockpiles are in place for more than a specified period of time, they will be vegetated in order to prevent erosion.
4.2.1.3	4.2-6	The excavation activities will be designed to minimize the amount of water to be handled as well as potential slope stability problems that may be caused by caving and dewatering of these unconsolidated materials.
4.2.1.3	4.2-6	Measures will be implemented, such as sedimentation or filtration, to ensure that erosion or siltation caused by the dewatering will be negligible. Proper safeguards will be implemented to prevent long-term effects on downstream habitats resulting from the construction activities.
4.2.2.3	4.2-8	Impacts from construction dewatering on the shallow wells will be evaluated during the preapplication monitoring (conducted at time of the COL application) for the EGC ESP Facility (see Section 6.3.1).

Table K-2. (contd)

Environmental Report Section	Page	Environmental Report Statement
4.6.3.5.7	4.6-7	The following material handling and storage practices will be implemented during construction activities, as applicable. • Materials on the construction site will be stored in areas designated for that purpose. Suitable measures will be taken in storage areas to reduce the likelihood of a discharge, such as straw bale barriers around the storage area. • Equipment not in use will be stored in a designated area. • Used oil tanks will be emptied frequently as necessary to avert overflow. The area will be kept free of trash and spilled oil. Tanks containing waste will have secondary containment. • Garbage receptacles will be equipped with covers. This includes such receptacles that contain materials that may be carried by the wind, or water soluble materials (e.g., paint). • Storage containers, including drums and bags, will be stored away from traffic to prevent accidental spills. • Containers will be kept closed except to add or remove material as necessary. • Containers will be stored in such a manner as to prevent corrosion that could result from contact between the container and ground surface, resulting in a release of material. • Containers will be appropriately labeled to show the name, type of substance, health hazards, and other appropriate information. • Material safety data sheets (MSDSs) for substances used or stored on the construction site will be available for review and use. • Hazardous substances such as used oil, anti-freeze, spent solvents, discarded paint cans, etc. will be controlled, stored and disposed of in accordance with the applicable MSDS.
4.6.3.5.8	4.6-7	During construction, the project specific waste management and health and safety plans will contain spill prevention, control, and response procedures that address site and activity specific conditions. These plans will be maintained on site. The general procedures for addressing spill prevention, control, and response are provided below, and will be implemented for on-site construction activities.
4.6.3.5.8.2	4.6-8	Fueling operations and vehicle maintenance will be performed at designated facilities, when practical. Spill sumps will be constructed around fuel and oil tanks. Drip pans will be used underneath oil barrels and other fluids that are used during construction activities. Spills of toxic or hazardous materials will be reported promptly to on-site authority (i.e., general contractor representative or site health and safety personnel) or their designee. The procedure, described below, will be followed for the clean up of small spills, as applicable. • Upon detection of any spill, personal safety is the first priority. The area of the spill and the nature of the spilled material will be evaluated in order to determine if remedial actions could result in additional health hazards, escalation of the spill, or station damage that may escalate the problem. If such conditions exist, a guard will be posted near the area (if practical), and the on-site authority or their designee will be promptly notified. • Identify the source of the spill (if possible), and then stop the flow of pollutants if it can be done in a safe manner as described above. • Record pertinent facts and information about the spill including type of pollutant, location, apparent source, estimated volume, and time of discovery.

Table K-2. (contd)

Environmental Report		
Section	**Page**	**Environmental Report Statement**
		• Spread absorbent materials on the area to soak up as much of the liquid as possible and prevent infiltration into the soil, and transfer the used materials to an appropriate container.
		• As soon as possible, the contaminated soil and absorbent material will be excavated and transported to a designated site for collection of such material.
		• If prompt transfer of the contaminated soil is not practical, the contaminated soil will be excavated and placed on polyethylene sheeting or other suitable material of sufficient thickness, and form a small berm to prevent breakout or infiltration.
		• If the general contractor responds to the spill, notify the site health and safety representative of the spill and provide in writing the amount of material, type of contaminant, and the source (location of the spill).
4.6.3.5.8.2	4.6-8	The procedure, described below, will be followed for the clean up of medium to large spills, as applicable.
		• Upon detection of any spill, personal safety will be the first priority. The area of the spill and the nature of the spilled material will be evaluated in order to determine if remedial actions could result in additional health hazards, escalation of the spill, or facility damage that may escalate the problem. If such conditions exist, a guard will be posted near the area (if practical). In addition, the on-site health and safety personnel or their designee, and other parties will be promptly notified. The responsible on-site authority will, in turn, notify appropriate agencies (e.g., National Response Center).
		• Identify the source of the spill (if possible) and stop the flow of pollutants if it can be done in a safe manner as described above.
		• Record pertinent facts and information about the spill including type of pollutant, location, apparent source, estimated volume, and time of discovery.
		• Promptly dispatch appropriate equipment (e.g., front-end loader) to the spill and construct a berm or berms downstream of it in order to minimize the spread.
		• Mobilize additional resources as necessary to address the spill.
		• Commence spill cleanup when the lateral spread has been contained and the notifications have been made.
		• Bail or pump free liquid into the appropriate container.
		• When the liquid has been bailed to the soil layer, apply absorbent materials to the surface, and transfer it to the appropriate container.
		• The remaining contaminant soils and absorbent material will be excavated and transferred to a temporary contaminant stockpile underlaid with polyethylene sheeting or other suitable material of sufficient thickness. The edges will be bermed to provide a dam to prevent inflow of water or leakage of the liquid.
		• Contaminated soil and absorbent material will be disposed, as appropriate.
4.6.3.5.8.3	4.6-9	The National Response Center will be contacted when a release containing a hazardous substance or oil in an amount equal to or in excess of a reportable quantity occurs during a 24-hr period, established under either 40 CFR 110, 40 CFR 117, or 40 CFR 302.
4.6.3.7.1.3	4.6-12	The excavation activities will be designed to minimize the amount of water to be handled as well as potential slope stability problems that may be caused by caving and dewatering of these unconsolidated materials.

Table K-2. (contd)

Environmental Report		
Section	Page	Environmental Report Statement
5.2.1.3	5.2-6	It is anticipated that surface water (namely Clinton Lake) will be used to meet the operational water requirements of the EGC ESP Facility; groundwater will not be used as a source of water. In addition, based on the planned design of the EGC ESP Facility, no permanent groundwater dewatering system will be implemented.
5.2.2.2.1	5.2-7	The EGC ESP Facility will be designed and operated to be compatible with the operation of the CPS and their respective NPDES permits.
5.2.2.2.1	5.2-7	Dam operation practices will be reviewed and revised where appropriate in conjunction with the CPS to maintain minimum flows in Salt Creek downstream of the dam and conserve water in the lake impoundment for power plant operation and recreational purposes.
5.3.1.1.3	5.3-2	The intake screens will be kept clean by mechanical means. The screens will be washed or scraped to remove algae, dead fish, trash, and debris that may have been drawn in. Captured material will be removed and disposed of onshore at an approved landfill site. There will be no direct discharge of these materials except for water to Clinton Lake.
5.3.3.2.1	5.3-8	Impacts to terrestrial ecosystems associated with salt drift will be assessed once the facility's cooling system configuration and design parameters have been determined. This analysis will be conducted before or during a later licensing stage.
5.3.3.4	5.3-10	The volume of the UHS is measured annually to track the progress of sedimentation. These annual measurements will be continued to confirm the available volume of the impoundment.
5.3.4.1	5.3-11	Additionally, the EGC ESP Facility thermal discharges will comply with the approved CPS NPDES permit, and therefore, operations will not increase the risk of the presence of *Naegleria fowleri* in Clinton Lake.
5.5.1.1	5.5-1	Solid nonradioactive and non-hazardous waste may include office waste, aluminum cans, laboratory waste, glass, metals, paper, etc., and will be collected from several on-site locations and deposited in dumpsters located throughout the site. Segregation and recycling of waste will be practiced to the greatest extent practical. The material will either be disposed of onsite or the Applicant will contract with an outside vendor who will perform weekly collections and disposal at area landfills. If collected and disposed of off site, it is not expected that the amount of solid waste generated will significantly contribute to the total amount of household waste disposed of weekly by area residents.
5.5.1.2.1	5.5-2	Other small volumes of wastewater, which may be released from other station sources, are described in the SSAR for the EGC ESP Facility. These will be discharged from sources such as the service water and auxiliary cooling systems, water treatment, laboratory and sampling wastes, floor drains, and stormwater runoff. These waste streams will be discharged as separate point sources or will be combined with the cooling water discharges.

Table K-2. (contd)

Environmental Report		
Section	Page	Environmental Report Statement
5.5.1.2.3	5.5-3	A SWPPP will be written, if deemed appropriate, that will meet the requirements of a permit for stormwater discharges from the EGC ESP Facility. The plan will include aspects of stormwater pollution prevention common to areas of the EGC ESP Facility that have a potential to discharge stormwater to waters of the U.S. The aspects common to activities will include site description and assessment, erosion and sediment control, stormwater management, identification and control of potential sources of pollution, implementation, maintenance, inspection, and stabilization.
5.5.1.2.4	5.5-3	The nonradioactive liquid wastes will be checked for proper pH and the presence of radiological and hazardous constituents, discharged as a separate point source or combined with plant circulating water prior to discharge to Clinton Lake. These discharges comply with the approved NPDES permit for the EGC ESP Facility issued by the IEPA.
5.5.2	5.5-4	In the event of a spill, emergency procedures will be implemented to limit any on-site impacts. Emergency response personnel will be properly trained and will be routinely provided with a facility inventory, which will include types, volumes, locations, hazards, control measures, and precautionary measures to be taken in the event of a spill.
5.5.2	5.5-4	If generated on site, mixed waste will be assessed based on the following regulatory guidance. Mixed waste (low level radioactive and hazardous waste) is waste that satisfies the definition of low level radioactive waste in the Low-Level Radioactive Waste Policy Amendments Act of 1985 (LLRWPAA) and contains hazardous waste that either: 1) is listed as a hazardous waste in 40 CFR 261(d); or 2) causes the waste to exhibit any of the hazardous waste characteristics identified in 40 CFR 261(c).
5.5.2 5.10.3.11.2	5.5-5 5.10-16	The EGC ESP Facility personnel will place primary importance on source reduction efforts to prevent pollution, and eliminate or reduce the generation of mixed waste. Potential pollutants and wastes that cannot be eliminated or minimized will be evaluated for recycling. Treatment to reduce the quantity, toxicity, or mobility of the mixed waste before storage or disposal will be considered only when prevention or recycling is not poss ble or practical. Environmentally safe disposal will be the last option (USNRC 1999).
5.5.2 5.10.3.11.2	5.5-5 5.10-16	A Pollution Prevention and Waste Minimization Program (PPWMP) will be developed, if deemed appropriate, and implemented before initial reactor operations.
5.5.2.1.1	5.5-5	Inventory management or control techniques will be used to reduce the poss bility of generating mixed waste resulting from excess or out-of-date chemicals and hazardous substances. Where necessary, techniques will be implemented to reduce inventory size of hazardous chemicals, size of containers, and amount of chemicals, while increasing inventory turnover.
5.5.2.1.1	5.5-6	A chemical management system, if required, will be established, prior to initial operation, and acquisition of new chemical supplies will be documented in a controlled process that addresses, as appropriate, the following: • Need for the chemical; • Availability of non-hazardous or less hazardous substitutes or alternatives; and • Amount of chemical required and the on-site inventory of the chemical.

Table K-2. (contd)

Environmental Report		
Section	Page	Environmental Report Statement
5.5.2.1.1	5.5-6	Excess chemicals will be managed in accordance with the station's chemical management procedures. Excess chemicals that are deemed usable will be handled through an excess chemical program. Material control operations will be revised or expanded to reduce raw material and finished product loss, waste material, and damage during handling, production, and storage. The inventory management procedures will be periodically assessed and updated, as appropriate, using criteria that include the following considerations: • If existing inventory management techniques are in accordance with existing pollution prevention and waste minimization guidelines, and regulatory guidelines; • How existing inventory management procedures can be applied more effectively; • Whether new techniques will be added to or substituted for current procedures; • If the review and evaluation approval procedures for the purchase of materials will be revised; • If additional employee training in the principles of inventory management is needed; • How specifications for the review and revision of procurement limit the purchase of environmentally sound products; and • How to increase the purchase of recycled products.
5.5.2.1.2	5.5-6	Equipment maintenance programs will be periodically reviewed to determine whether improvements in corrective and preventive maintenance can reduce equipment failures that generate mixed waste. The methods for maintenance cost tracking and preventive maintenance scheduling and monitoring will be examined. Maintenance procedures will be reviewed in order to determine which are contributing to the production of waste in the form of process materials, scrap, and cleanup residue. In addition, the need for revising operational procedures, modifying equipment, and source segregation and recovery will be determined.
5.5.2.1.3	5.5-6	Recycling of the waste types will be considered. Opportunities for reclamation and reuse of waste materials will be explored whenever feasible. Decontamination of tools, equipment, and materials for reuse or recycle will be used whenever possible to minimize the amount of waste for disposal.
5.5.2.1.4	5.5-7	When radiological or hazardous waste is generated, proper handling, containerization, and separation techniques will be employed, as applicable.
5.5.2.1.6	5.5-7	Prejob planning will be completed to determine what materials and equipment are needed to perform the anticipated work.
5.5.2.1.7	5.5-7	A tracking system will be developed, if required, to identify waste generation data and PPWMP opportunities.
5.5.2.1.8	5.5-8	A PPWMP will be developed and implemented, as required, that incorporates the following: • A waste minimization plan that will be routinely reviewed, revised, and implemented during the phases of the EGC ESP Facility construction and operation; • Educate employees of general environmental activities and hazards at the EGC ESP Facility and pollution prevention program and waste minimization requirements, goals, and accomplishments;

Appendix K

Table K-2. (contd)

Environmental Report		
Section	Page	Environmental Report Statement
		• Inform employees of specific environmental issues; • Train employees on their responsibilities in pollution prevention and waste minimization; • Recognize employees for efforts to improve environmental conditions through pollution prevention and waste minimization; and • Encourage employees to participate in pollution prevention and waste minimization.
5.5.2.1.9	5.5-8	The EGC ESP Facility will implement procurement practices that comply with regulatory guidance, and other requirements for the purchase of products with recovered materials. This includes the elimination of the purchase of ozone depleting substances and the minimization of the purchase of hazardous substances.
5.5.2.1.10	5.5-8	Policies and procedures will be developed, as applicable, to reflect a focus on integrating PPWMP objectives into EGC ESP Facility activities. The Environmental, Health, and Safety departments will review new procedures for EGC ESP Facility activities. The procedures will determine whether the elimination or revision of procedures can contribute to the reduction of waste (hazardous, radiological, or mixed). This will include incorporating PPWMP into the appropriate on-site work procedures. Changes to procurement procedures to require affirmative procurement of IEPA-designated recycled products, and reduction of procurement of ozone-depleting substances will also be completed.
5.5.2.2.1	5.5-9	The EGC ESP Facility Environmental Health and Safety management will implement and enforce the following guides if it is necessary to store mixed wastes on site: • Use the area only for storage of mixed waste and not for storing unrelated materials or equipment, or for other functions; • Follow proper storage protocols for different kinds of mixed waste; • Label the containers properly and in accordance with regulatory requirements; • Follow the container label requirements; • Post applicable material safety data sheets, emergency spill response procedures, and have a spill kit in the area; • Install fire detection and suppression equipment (if required), alternate water supply, telephone, and alarm at the area; • Make an emergency shower/eyewash station immediately available, where it is tested weekly and functioning; • Fence and lock the gate to the accumulation area or long-term storage area when authorized personnel are not present; • Post "MIXED HAZARDOUS WASTE AREA" and "DANGER—UNAUTHORIZED PERSONNEL—KEEP OUT" signs at the entrance; • Provide secondary containment for liquid mixed hazardous waste; • Conduct weekly inspections; and • Post "NO SMOKING OR OPEN FLAME" signs.
5.5.2.2.1	5.5-9	The EGC ESP Facility management will also develop and implement contingency plans, emergency preparedness, and prevention procedures that will be utilized in the event of a mixed waste spill. The EGC ESP Facility personnel who are designated to handle mixed waste or whose job function it is to provide emergency response to mixed waste spills will receive appropriate training in order to perform their work properly and safely.

Table K-2. (contd)

Environmental Report

Section	Page	Environmental Report Statement
5.5.2.2.1	5.5-9	If mixed waste is generated and shipped for treatment and disposal rather than stored, EGC ESP Facility management will identify potential disposal facilities considering the following selection criteria: • The desired method of treatment or disposal (e.g., incineration vs. land disposal); • The disposal facility's permit (e.g., can they accept polychlorinated biphenyls (PCBs), hazardous waste, or radioactive waste); • The disposal facility's turnaround time on approvals; • The form of waste, (e.g., is it soil, debris, semi-solid, or liquid); • The mass or volume of waste; and • The cost of transportation and disposal.
5.5.2.2.1	5.5-10	The EGC ESP Facility management will also identify one disposal facility as the primary facility, and a second facility will be identified as an alternate in the event that laboratory testing or other observations prove the waste to be different than initially determined.
5.5.2.2.2	5.5-10	If stored at the facility, the USEPA mandates that waste storage containers must be inspected on a weekly basis, and certain aboveground portions of waste storage tanks must be inspected on a daily basis. The purpose of these inspections is to detect leakage from, or deterioration of, containers (40 CFR 264). The USNRC recommends that waste in storage be inspected on at least a quarterly basis (10 CFR 20). The methods used for these inspections may include direct visual monitoring or the use of remote monitoring devices for detecting leakage or deterioration. The remote methods would reduce exposures due to direct visual inspections. Additionally, measures will be provided to promptly locate and segregate or remediate leaking containers.
5.8.1.2	5.8-2	Equipment manufacturers will be required to guarantee that specifications on allowable octave bands will be met.
5.8.1.2	5.8-2	noise control devices will be used when necessary.
5.8.1.3	5.8-2	Depending on the reactor technology selected, air pollution control devices may be needed and will be used to meet applicable regulations.
5.8.2.7	5.8-5	The EGC ESP Site will use their own on-site water and septic facilities.
5.10.3.1	5.10-2	Procedures and a Hearing Conservation Program will be developed for any employees exposed to excessive noise, which is defined as an 8-hr exposure of 85 dB or more.
5.10.3.10.3	5.10-15	To establish confidence and credibility that any radiological environmental monitoring data collected and reported are accurate and precise, monitoring activities will be incorporated into the construction phase quality assurance program established pursuant to 10 CFR 50, Appendix B, in concurrence with COL activities.
5.10.3.11.1.1	5.10-16	Solid nonradioactive and non-hazardous waste may include office waste, aluminum cans, laboratory waste, glass, metals, paper, etc., and will be collected from several on-site locations and deposited in dumpsters located throughout the site.
5.10.3.11.1.2	5.10-16	The nonradioactive liquid wastes will be combined with plant circulating water and checked for proper pH and the presence of radiological and hazardous constituents prior to discharge to Clinton Lake. These discharges will comply with an approved NPDES permit for the EGC ESP Facility issued by the IEPA.
5.10.3.11.1.3	5.10-16	The nonradioactive air emissions will be in compliance with the limits that will be established and imposed by the IEPA. These limits will be protective of the air quality in and around the EGC ESP Facility.

Table K-2. (contd)

Environmental Report		
Section	Page	Environmental Report Statement
5.10.3.11.2	5.10-16	The EGC ESP Facility personnel will place primary importance on source reduction efforts to prevent pollution and eliminate or reduce the generation of mixed waste. Potential pollutants and wastes that cannot be eliminated or minimized will be evaluated for recycling.
5.10.3.14.1.1	5.10-20	Any equipment that exceeds the noise abatement criteria will use noise control devices.
6.1.1.1	6.1-2	Although the existing thermal database is sufficient to describe the thermal conditions in Salt Creek, additional preapplication monitoring will be conducted to verify and update the baseline conditions at the time of the COL application. In addition to continued collection and evaluation of data collected at these locations, the proposed preapplication water quality monitoring will include monthly temperature measurements at a location downstream of the Clinton Lake Dam (Site E-3 on Figure 6.1-1). At each site, temperature measurements will be collected at the surface and 1.5-ft depth intervals to the bottom using a "YSI Multiprobe or Multiparameter Instrument" (or equivalent meter). The depth of the water column will also be recorded.
6.1.3	6.1-4	The Operational Thermal Monitoring Program will be implemented in order to establish changes in water temperature resulting from facility operation. The specific operational monitoring requirements will be developed in consultation with IEPA, relative to NPDES permit requirements and the monitoring requirements for the CPS at that time.
6.2.2.1	6.2-4	Monitoring stations will be placed in the facility proximity and approximately 5 mi from the proposed reactor in locations representing the 16 meteorological compass sectors. Other locations will be chosen to measure the radiation levels at places of special interest, such as nearby residences, meeting places, and population centers.
6.2.2.2	6.2-4	The inhalation and ingestion of radionuclides in the air is a direct exposure pathway to man. A network of ten active air samplers will be used to monitor this pathway.
6.2.2.2	6.2-4	The air sampling equipment will be maintained and calibrated by facility personnel using reference standards that are traceable back to the National Institute of Standards and Technology (NIST).
6.2.2.2	6.2-4	Air samples will be collected every week and analyzed for gross beta and Iodine-131 activities. Quarterly, the air particulate filters collected throughout this period will be combined and counted for gamma isotopic activity.
6.2.2.3	6.2-4	Aquatic monitoring will provide for the collection of fish and shoreline sediments to detect the presence of any radioisotopes related to the operation of the EGC ESP Facility. These samples will be analyzed for naturally occurring and manmade radioactive materials.
6.2.2.3.1	6.2-5	Various samples of fish will be collected from Clinton Lake and Lake Shelbyville... These samples will be collected semi-annually and analyzed by gamma spectroscopy.
6.2.2.3.2	6.2-5	Samples of shoreline sediments will be collected at Clinton Lake and Lake Shelbyville...Samples will be collected semi-annually and analyzed for gross beta, gross alpha, Strontium-90, and gamma isotopic activities.
6.2.2.4	6.2-5	In addition to direct radiation, radionuclides that are present in our atmosphere expose receptors when they are deposited on plants and soil, and subsequently consumed. To monitor this food pathway, samples of green leafy vegetables, grass, and milk will be analyzed. ... These samples will be analyzed by gamma spectroscopy.

Table K-2. (contd)

Environmental Report		
Section	Page	Environmental Report Statement
6.2.2.4.1	6.2-5	Milk samples will be collected from a dairy located about 14-mi west southwest of the facility (twice a month during May through October, and once a month during November through April). These samples will be analyzed for Iodine-131, Strontium-90, and gamma isotopic activities.
6.2.2.4.2	6.2-5	Grass samples will be collected at three indicator locations and at one control location. These samples will be collected twice a month during May through October, and once a month during November through April (when available). Grass samples will be analyzed for gamma isotopic activity including Iodine-131.
6.2.2.4.3	6.2-6	Broadleaf vegetable samples will be obtained from three indicator locations and at one control location. The indicator locations will be in the meteorological sectors with the highest potential for surface deposition. The control location will be a meteorological sector and distance approximately 13-mi downwind, which is considered to be unaffected by unit operations. Samples will be collected once a month during the growing season (June through September) and will be analyzed for gross beta and gamma isotopic activities including Iodine-131.
6.2.2.5	6.2-6	Water monitoring (e.g., the collection of drinking water, surface water, and groundwater [well water] samples) will be used to detect the presence of any radioisotopes relative to the operation of the EGC ESP Facility. ... Samples taken will be analyzed for naturally occurring and manmade radioactive isotopes.
6.2.2.5.1	6.2-6	A composite water sampler will be located at the service building for the EGC ESP Facility. ... This monthly composite sample will then be analyzed for gross alpha, gross beta, and gamma isotopic activities. ... This quarterly composite sample will then be analyzed for Tritium.
6.2.2.5.2	6.2-6	Composite water samplers will be installed at three locations to sample surface water from Clinton Lake. ... This water sample will be collected on a monthly basis. ...Tritium analyses will be performed quarterly from the monthly composites from the water composite sample locations.
6.2.2.5.3	6.2-7	Every quarter, both the treated and untreated well water samples will be collected from the well serving the Village of DeWitt and from a well serving the Illinois Department of Conservation at the Mascoutin State Recreational Area. Samples will be analyzed for Iodine-131, gross alpha, gross beta, Tritium, and gamma isotopic activities. See Table 6.2-2 for location of sample points.
6.2.3	6.2-7	To establish confidence and credibility that the data collected and reported are accurate and precise, EMP activities will be incorporated into the construction phase Quality Assurance Program established pursuant to 10 CFR 50, Appendix B, in pursuance of COL activities, The EMP will utilize quality programs and processes to: • Personnel will be trained and qualified to perform radiological monitoring. • Procedures for sample collection, packaging, shipment, and receipt of samples for analysis will be created and approved, and samples at the lab will be prepared and analyzed.

Table K-2. (contd)

Environmental Report		
Section	**Page**	**Environmental Report Statement**
		• Lab processes will be documented, such as maintenance, storage, and use of radioactivity reference standards; calibration and checks of radiation radioactivity measurement systems and sample tracking and control.
		• The processes and procedures of the monitoring program will be documented.
		• Periodic audits of analysis laboratory functions and their facilities will be conducted.
		• Records of sample collection, shipment and receipt will be maintained. Records will also be maintained of lab activities including sample description, receipt, lab identification, coding, sample preparation and radiochemical processing, data reduction, and verification. In addition, the following activities will be performed:
		• Duplicate analysis of the samples (excluding TLDs) will be performed to check laboratory precision.
		• Quality indicator and control samples will be routinely counted.
		• Inter-comparison programs will be participated in, such as the ERA cross-check program.
		• The analytical results provided by the laboratory will be reviewed monthly to validate that the required minimum sensitivities have been achieved, and that the correct analyses have been performed.
6.3.1.1	6.3-2	Although the hydrologic data collected provide a sufficient database to describe hydrologic conditions in Salt Creek, additional preapplication monitoring will be conducted in order to verify and update the baseline conditions at the time of the COL application. The proposed preapplication monitoring will include the following:
		• The continued collection and evaluation of mean daily flow in Salt Creek downstream of the dam at the Rowell gauging station; and
		• Monthly stream flow will be measured at Site E-3, concurrent with thermal and chemical monitoring (see Figure 6.1-1). Measurements will be made using a "Marsh McBirney Flowmeter" (or equivalent instrument) at a depth of 3-ft below the surface.
6.3.1.2	6.3-2	Although the existing database is sufficient to describe the conditions in Clinton Lake as presented in Section 2.3.1.2, additional preapplication monitoring will be conducted in order to verify and update the baseline conditions at the time of the COL application.
		The proposed preapplication monitoring for Clinton Lake will include the collection of the following data:
		• Mean daily stage of Clinton Lake;
		• Mean daily flow being discharged from Clinton Lake (namely through the dam);
		• Monthly current velocity, concurrent with thermal and chemical monitoring, measured at a depth of 3 ft from the surface using a "Marsh McBirney Flowmeter" (or equivalent instrument) (see Figure 6.1-1 for locations); and
		• Depth of water column at regular intervals along transects across the impoundment used to estimate the current volume of Clinton Lake.

Table K-2. (contd)

Environmental Report		
Section	**Page**	**Environmental Report Statement**
6.3.1.3	6.3-4	The proposed preapplication monitoring for the EGC ESP Facility will be implemented at the time of the COL application and is described below.
6.3.2.3	6.3-5	• Location and survey of previously installed CPS piezometers that have not been identified
6.3.3.3	6.3-6	as destroyed by construction activities.
		• Location and identification of existing private wells within 5 mi of the site.
		• Installation of additional shallow water table piezometers and deep piezometers (screened in discontinuous sand layer) spaced at suitable lateral intervals away from the EGC ESP Facility, between the EGC ESP Facility and the CPS Facility. In addition, piezometers located near Clinton Lake to help define the lateral continuity of sand layers and will be used during the pumping test.
		• Monitoring of water levels in the piezometers on a monthly basis to verify the hydrostatic loading on the power plant foundation, flow directions, and to estimate the amount of water that may need to be controlled during the excavation activities.
		• Installation of a 12-in. test well and performance of a long-term pumping test to help evaluate the potential impacts that may be caused from the dewatering activities and the amount of water that may need to be controlled during the excavation activities.
		• Installation of points to monitor for settlement or ground movement.
		The specific number, depths, and locations of the piezometers and the test well will be determined as the engineering design of the facility is better defined. The data collected will be used to define the baseline conditions at the time of the COL application and groundwater-related design elevations. In addition, the information will be used to identify additional locations that will be monitored during the construction of the EGC ESP Facility. The specific procedures of the operational monitoring requirements of Salt Creek are anticipated to be similar to the Preapplication and Preoperational Monitoring programs. The program may be modified based on data collected and consultations with IEPA and the CPS. The data will be evaluated in order to monitor for changes in the discharge from Clinton Lake to Salt Creek.
6.3.2.1	6.3-5	The construction-related impacts to Salt Creek are considered minimal, provided that the proper controls are implemented to minimize impacts to Clinton Lake. The proposed construction monitoring of Salt Creek will include continuing the Preapplication Monitoring Program.
6.3.2.2	6.3-5	A major element of the construction monitoring will be to monitor the amount of sediment deposited in Clinton Lake as a result of the construction activities. The proposed construction monitoring will include continuing the Preapplication Monitoring Program. In addition, the amount of sediment deposited at the stormwater outfalls will be monitored to determine if a sufficient thickness of sediment has accumulated in order to require removal upon completion of the construction.
6.3.2.3	6.3-5	Water levels from the piezometers installed for the Preapplication Monitoring Program will be measured at least daily during the active construction period in order to monitor lateral depression in the groundwater surface caused by dewatering. In addition, settlement points will be monitored to protect existing structures from settlement or ground movement during the excavation activities. These points will be monitored daily, at a minimum, and critical points may be monitored continuously.

Table K-2. (contd)

Environmental Report		
Section	Page	Environmental Report Statement
6.3.4	6.3-6	The Operational Hydrological Monitoring Program will be designed to establish the impacts from the operation of the EGC ESP Facility and detect any unexpected impacts from facility operation. Based on the monitoring data for the CPS, the Operational Hydrological Monitoring Program is anticipated to extend over a five-year period or until conditions appear to have stabilized based on the trend analysis. Modifications to the monitoring program (e.g., changes in monitoring locations or collection procedures) will be assessed regularly over the duration of the monitoring program.
6.3.4.1	6.3-6	The specific procedures of the operational monitoring requirements of Salt Creek are anticipated to be similar to the Preapplication and Preoperational Monitoring programs. The program may be modified based on data collected and consultations with IEPA and the CPS. The data will be evaluated in order to monitor for changes in the discharge from Clinton Lake to Salt Creek.
6.3.4.2	6.3-7	The data from this monitoring program will be evaluated in order to determine changes in the cooling system flows, water levels in Clinton Lake, and discharges from Clinton Lake to Salt Creek.
6.3.4.3	6.3-7	A limited Operational Hydrological Monitoring Program will be implemented in order to establish the impacts to the groundwater system from the operation of the EGC ESP Facility and detect any unexpected impacts from facility operation. ...The monitoring will consist of extending preoperational monitoring for an additional five-year period or until conditions appear to have stabilized based on the trend analysis of groundwater and surface water conditions. The need for modifications to the monitoring program (e.g., changes in monitoring locations or frequency of collection) will be assessed regularly over the duration of the monitoring program.
6.5.1.1	6.5-1	A Terrestrial Monitoring Program was established for the CPS to monitor, on a low-level basis, the wildlife and vegetation communities in the vicinity of the site. ... A similar program will be implemented for the EGC ESP Facility. This monitoring program will document changes in plant and animal species composition over time, and will build on the database gathered during the CPS preliminary baseline environmental assessment and monitoring. In addition, monitoring of terrestrial resources along the proposed transmission right-of-way will be implemented as appropriate.
6.5.1.1.1	6.5-2	Sampling methodologies for the five communities will continue with the generally accepted techniques of quadrant, quarter, and transect sampling.
6.5.1.1.2	6.5-2	The results of these surveys [bird] will be reviewed, as necessary, to document avian communities in the vicinity.
6.5.1.1.2	6.5-2	Monitoring surveys of waterfowl at Clinton Lake and other waterbodies within the vicinity will be performed, as appropriate, in order to confirm that changes in composition, abundance, or distrbution are not occurring as a result of operation of the EGC ESP Facility.
6.5.1.1.3	6.5-2	The CPS ER proposed that monitoring programs for small mammal populations be conducted during May and November at five locations within the vicinity (CPS 1973). Trap-lines were set to help determine the composition and abundance of small mammal populations, and roadside counts were performed in order to determine the presence of cottontail rabbits in the vicinity (CPS 1973 and CPS 1982). It is anticipated that the continuation of this program will be adequate to identify any adverse effects that the EGC ESP Facility may have on small mammal populations in the vicinity. During monitoring efforts, records will also be kept of mammal sightings or signs of presence including tracks

Table K-2. (contd)

Environmental Report		
Section	**Page**	**Environmental Report Statement**
		or scat.
6.6.1.1	6.6-1	Although the existing chemical database is sufficient to describe the chemical conditions in Salt Creek, additional preapplication monitoring will be conducted to verify and update the baseline conditions at the time of the COL application. In addition to continued collection and evaluation of data collected at the Rowell gauging station, the proposed preapplication water quality monitoring will include sampling at a location downstream of the Clinton Lake Dam (Site E-3 on Figure 6.1-1). Water samples will be collected monthly (at a minimum), concurrent with the thermal monitoring (see Section 6.1). Dissolved oxygen, specific conductance, and pH will be measured *in situ* from the water surface, and at 1.5-ft depth intervals at each site using a "YSI Multiprobe or Multiparameter Instrument" or equivalent meter. Water samples will be collected using non-metallic Van Dorn, Kemmerer, or Beta type bottles from 3-ft below the surface. The data gathered will be used to assess conditions in Salt Creek between the Clinton Lake Dam and the Rowell gauging station.
6.6.1.3	6.6-3	A similar limited Preapplication Monitoring Program will be implemented to define baseline groundwater quality conditions. Selected piezometers and public or private wells will be sampled on a quarterly basis. The specific number and locations of the piezometers/wells and the analytical parameters will be determined based on the groundwater flow patterns in and around the EGC ESP Facility, as determined by the measured water levels and consultation with IEPA. The results will be used to verify and update the baseline chemical conditions of the glacial drift aquifers underlying the EGC ESP Facility and in the vicinity of the site at the time of the COL application. The baseline conditions are established to monitor potential impacts from the construction and operation of the EGC ESP Facility.
6.6.1.3	6.6-4	In addition, water quality will be evaluated prior to and after the pumping test in order to monitor potential changes in water quality during the construction dewatering activities.
6.6.2	6.6-4	The chemical monitoring of surface water and groundwater will be conducted to provide data necessary to assess water quality changes that result from construction and operation of the EGC ESP Facility.
6.6.2.1	6.6-4	The data from the preapplication sampling of Salt Creek and Clinton Lake will be evaluated. This will determine if the scope and the frequency of chemical monitoring will need to be modified in order to establish the baseline for water quality in Salt Creek. In addition, the need for changes to the monitoring program (e.g., changes in monitoring locations, parameters, collection, or analytical procedures) will be assessed regularly over the duration of the monitoring program.
6.6.2.2	6.6-4	The results of the preapplication sampling will be evaluated, and will determine if the scope and the frequency of chemical monitoring will be to be modified in order to establish the baseline for water quality. In addition, the need for modifications to the monitoring program (e.g., changes in monitoring locations, parameters, collection, or analytical procedures) will be assessed regularly and over the duration of the monitoring program.
6.6.2.3	6.6-4	The chemical monitoring of groundwater will be conducted in order to provide data necessary to assess water quality changes that result from construction dewatering and operation of the EGC ESP Facility.

Table K-2. (contd)

Environmental Report		
Section	**Page**	**Environmental Report Statement**
6.6.2.3	6.6-4	The results of the preapplication sampling will be evaluated, and will determine if the scope and the frequency of chemical monitoring will be modified in order to establish the baseline for groundwater quality. In addition, the need for modifications to the monitoring program (e.g., changes in monitoring locations, parameters, collection, or analytical procedures) will be assessed regularly and over the duration of the monitoring program.
6.6.3	6.6-5	An Operational Monitoring Program will be implemented to identify changes in water quality that results from operation of the EGC ESP Facility. A consideration in the development of the Operational Monitoring Program is the ability to update the estimates of the effectiveness of various effluent treatment systems, and to provide real time warnings of any failures in the effluent treatment systems. The specific elements of the Operational Monitoring Program for the assessment of surface water quality will be developed in consultation with the IEPA, relative to NPDES permit requirements and with consideration of monitoring conducted for the CPS.
6.6.3.1	6.6-5	operational monitoring for Salt Creek ... The data will be evaluated by monitoring for water quality changes of the discharge from Clinton Lake to Salt Creek.
6.6.3.2	6.6-5	[Lakes and Impoundments Operational Monitoring] The data will be evaluated for chemical variability along the flow path and temporal trends. The results of the operational monitoring and previous sampling events will be evaluated to determine if the scope and the frequency of chemical monitoring will be modified. The need for modifications to the monitoring program (e.g., changes in monitoring locations, parameters, collection, or analytical procedures) will be assessed regularly and over the duration of the monitoring program.
6.6.3.3	6.6-5	The groundwater data from the preapplication and preoperational sampling events will be evaluated, and the scope and/or the frequency of chemical monitoring will be modified, as needed. The need for modifications to the monitoring program (e.g., changes in monitoring locations, parameters, collection, or analytical procedures) will be assessed regularly and over the duration of the monitoring program.
6.7.2	6.7-1	The programs that are listed in Table 6.7-1 will continue into the preoperational phase.
6.7.3	6.7-1	Operational monitoring is proposed to begin after construction is complete and the EGC ESP Facility is operating. ... The need for modifications (e.g., changes in monitoring locations, parameters, collection, or analytical procedures) will be assessed regularly, over the duration of the monitoring programs.
9.4	9.4-1	Based on the evaluations provided in this ER, the site will accommodate the operational and environmental requirements for any one of them. Therefore, alternative facility systems will be discussed at the COL stage, when the full spectrum of design alternatives will be available.
10.3.2	10.3-2	Radiological monitoring programs will be enacted to measure and reduce radiation levels emitted by the facility.

Table K-3. Key Statements Made in the Environmental Report Related to Actions and Activities of Others and the Impacts of Those Activities Considered in the NRC Staff's Environmental Analysis

Environmental Report			Environmental Impact Statement Sections
Section	**Page**	**Environmental Report Statement**	
2.2.2	2.2-3	The transmission corridor does not interfere with the county's land use plan since only existing right-of-way will be used for the transmission corridor.	2.2, 4.1.1, 5.1, 7.1
2.2.2	2.2-3	The transmission corridor will not conflict with any proposed zoning for the county.	2.2, 4.1.1, 5.1, 7.1
3.7	3.7-1	An RTO or the owner, both regulated by Federal Energy Regulatory Commission (FERC), will bear the ultimate responsibility for defining the nature and extent of system improvements, and the design and routing of connecting transmission and the impacts of such improvements. Therefore, the construction described in this section is based on the existing infrastructure, Illinois Power Company system design preferences, and best transmission practices. The guiding assumption for transmission route design is that the new construction will follow in parallel with some of the existing transmission serving the CPS, and that it is only required to reach the nearest substation providing connection to the greater area grid. Impacts to the grid will be addressed by the system owner after submission of an interconnect request.	3.3, 4.4.1, 5.4.1
3.7.1.1 5.1.2 5.6	3.7-2 5.1-2 5.6-1	The EGC ESP Facility will rely on an interconnection with Illinois Power Company, and anticipates that the configuration of the transmission system and corridor will be similar to the existing system.	4.4.1, 5.4.1
3.7.2 4.1.2	3.7-2 4.1-3	However, in order to accommodate the bounding case of an output of 2180 MWe, new lines will be required, as there is insufficient capacity on the existing system to carry the load, and the existing structures were not designed for additional circuits. Parallel lines are required in each direction because a single line can not carry the full output of both the EGC ESP Facility and CPS. Four new transmission lines will be required to connect the EGC ESP Facility to the existing transmission grid in southern Illinois. Two parallel, double circuit transmission lines will depart the station north to an interconnect point at the Brokaw substation near Bloomington, Illinois, approximately 15 mi from the site (see Figure 2.2-4). A second pair of parallel double circuit lines will depart the station south to an interconnect point on Illinois Power Company's Latham-Rising 345-kV line (Number 4571), approximately 9 mi from the site (see Figure 2.2-4).	2.2, 3.3, 4.1, 4.1.1, 4.4.1, 5.1.1, 5.4.1, 7.1, 8.5, 8.6

Table K-3. (contd)

Environmental Report			Environmental Impact Statement Sections
Section	Page	Environmental Report Statement	
3.7.2 5.6.3.3 5.10.3.12.3.2	3.7-3 5.6-6 5.10-18	Transmission system design, construction, and operation will comply with the relevant local, state, and industry standards including the National Electric Safety Code (NESC) and various ANSI/Institute of Electrical and Electronics Engineers (IEEE) standards. This includes ground clearances, electromagnetic fields (EMF), radio interference (RI), television interference (TVI), audible noise, aviation safety, and other factors as appropriate.	5.8.4
3.7.4 5.6.3.3 5.10.3.12.3.2	3.7-4 5.6-6 5.10-8	The EMF reduction measures will be incorporated into the line and station designs so that the EMF strengths will be minimized.	5.8.4
3.7.5 5.6.3.5 5.10.3.12.3.4	3.7-4 5.6-7 5.10-18	To minimize these induced ground currents and distribute ground fault currents, the tangent or inline structure will be grounded. The tangent structure will have an electrical connection between the shield wire and ground lead, which will be connected to ground rods. Ground resistance tests will be made at the tangent structure before the shield wire is electrically connected to the ground lead. Sufficient ground rods will be installed to reduce the resistance to 10 ohms or less under normal atmospheric conditions. Angle or corner structures will have a low voltage insulator installed between the shield wire and down guys to avoid possible anchor corrosion problems.	5.8.4
4.1.2	4.1-3	In general, construction of transmission corridor in off-site areas will have a minimal impact on land use due to the fact that it is assumed that only existing rights-of way will be used.	2.2, 3.3, 4.1, 4.4.1, 5.4.1, 7.1, 8.5, 8.6
4.1.2	4.1-3	The northern section will run north of the EGC ESP Facility and then turn west and run towards Bloomington, Illinois. The southern section will run southeast of the EGC ESP Facility, west past Clinton Lake and then turn south and run towards the southern boundary of DeWitt County.	2.2, 3.3
4.1.2.1.1	4.1-4	Where temporary access is required, short routes of nongraded overland access will be constructed for as long as access to the site is required, after which they will be reclaimed. Standard design techniques, such as installing water bars and dips to control erosion, will be employed along with minimizing construction during wet seasons.	4.1.2, 7.1
4.1.2.1.2	4.1-4	Any area disturbed by the storage operations, not already in use for substation operations or construction activities, will be restored consistent with existing and natural vegetation.	4.1.2, 7.1
4.1.2.1.4	4.1-6	Where necessary, culverts and fence openings will be installed to allow access to and along the right-of-way during clearing and construction activities. Except where requested by landowners, the culverts and fence openings will be removed following completion of construction activities.	4.1.2, 7.1

Table K-3. (contd)

Environmental Report			Environmental Impact Statement Sections
Section	Page	Environmental Report Statement	
4.1.2.1.5	4.1-6	The H-Frame structures will be direct buried in the ground except where site conditions dictate a concrete foundation. Foundation holes will typically be excavated with rubber tire or track mounted augers, which will leave a minimum footprint of disturbed ground. Following erection of the H-Frames into the foundation holes, the holes will be backfilled with the removed soil and compacted. Excess soil will be distributed evenly around the legs and graded to match the existing ground profile. The small amount of excess soil will not require off-site disposal. The poles, connecting hardware, insulators, and guys required for H-Frame construction will be delivered to the construction site from the storage yard on suitable rubber tire trucks and trailers. At the erection site, a rubber tire rough duty mobile crane will be used to move the sections during assembly and to install the completed H-Frames. During this operation an area approximately 100 ft-long by 40-ft wide will be required for component laydown, the preassembly of structures, and vehicle access at each H-Frame location.	4.1.2, 7.1
4.1.2.1.5	4.1-7	On completion of construction, the right-of-way will be restored as near as possible to its original condition. As the contractor completes the operations, the right-of-way will be backbladed with a bulldozer and the area will be graded. Customary practices for erosion prevention will then be used.	4.4.1
4.1.2.2.1	4.1-8	The transmission corridor will not cause long-term changes to special agricultural resources, such as prime or unique farmland, since the transmission corridor will be constructed in existing right-of-way. There are no known significant mineral resources (sand and gravel, coal oil, natural gas, and ores) within the transmission corridor (Masters et al. 1999). No construction activities for the transmission corridor will take place within a coastal zone (USGS 1990) or wild and scenic river (USFWS 2002). Clinton Lake is considered a 100-yr floodplain. There are also three other 100-yr floodplains within the transmission corridor (IDNR 1986). There are minor wetland areas within the vicinity (IDNR 1987). Careful consideration of these floodplains and wetlands will take place when constructing the transmission corridor. Transmission towers required for the proposed transmission system will be sited in upland areas within the existing utility corridor. Adverse impacts to watercourses, wetlands, and floodplains within the existing right-of-way will be avoided to the greatest extent possible.	2.7, 4.1.2, 4.4.1, 5.4.1, 7.4, 7.5, 8.5, 8.6
4.2	4.2-1	The construction will be confined to the station site and the existing transmission corridor. Proper mitigation and management methods implemented during construction will limit the potential water quantity and quality impacts to the surface water (e.g., Clinton Lake, stream crossings, and intermittent drainage ways) and adjacent groundwater.	4.4.2, 4.3

Table K-3. (contd)

Environmental Report			Environmental Impact Statement Sections
Section	Page	Environmental Report Statement	
4.3.1.2	4.3-1	As previously discussed, transmission system improvements will be required to support the EGC ESP Facility. These modifications will be located within or immediately adjacent to the existing substation at the CPS and along the existing transmission corridor. The proposed transmission line improvements will be sited within the existing utility rights-of-way to the greatest extent possible. Construction of the proposed transmission line improvements will temporarily habitats within the existing rights-of-way; however, the agricultural and open field areas will be allowed to revegetate to preconstruction conditions. There will be no significant loss of agricultural or open field habitats resulting from construction of the transmission systems. Where right-of-way expansion is required in forested lands, clearing will be required. Forested habitats do not make up a significant amount of the proposed utility corridor; therefore, significant impacts to forested lands are not anticipated.	4.4.1, 5.4.1
4.3.2.2	4.3-4	Construction of the proposed transmission corridor will temporarily impact watercourses existing along the proposed right-of-way. These temporary impacts will be short-term and temporary in nature, and there will be no net loss of resource area.	4.4.1, 5.4.1, 7.4, 7.5, 8.5, 8.6
4.6.3.7	4.6-10	The construction will be confined to the EGC ESP Site and the existing transmission corridor. Proper mitigation and management methods implemented during construction will limit the potential water quantity and quality impacts to the surface water (e.g., Clinton Lake, stream crossings, and intermittent drainage ways) and adjacent groundwater.	4.4.2
5.1.2	5.1-2	It has been assumed that operation and maintenance activities will be conducted in a similar manner to the existing transmission facilities because it is anticipated that the transmission corridor will, most likely, be within the existing right-of-way.	5.1, 5.4.1, 7.4, 8.4, 8.5, 8.6
5.1.2.1	5.1-2	A major portion, approximately 88 percent, of the transmission line right-of-way that will most likely serve the EGC ESP Facility will cross agricultural land. As part of the existing right-of-way agreements, it is assumed that farmers will continue to cultivate this land except for a small area around the H-Frame structure. Therefore, it is anticipated that existing access to the right-of-way is adequate and no permanent roads will be built on the right-of-way for either construction or maintenance. However, road construction may become necessary if the landowner requires it as a condition of the right-of-way or for access to a switching structure. A road will be constructed to the following general specifications: • Aligned to avoid impacts to wetland resource areas; • Grades will be minimized to eliminate erosion; • Grading, ditches, cut and fill areas, or other disturbed areas will be re-vegetated to prevent erosion;	2.2, 3.3, 2.7, 4.1, 4.4.1, 5.1, 5.4.1

Table K-3. (contd)

Environmental Report			Environmental Impact Statement Sections
Section	Page	Environmental Report Statement	Sections
		• Culverts will be installed where needed to prevent erosion and prevent flooding of the road; and • The surface of the road will be paved with crushed rock or natural gravelly material to withstand expected loads. Once constructed, these roads will be permitted to "grassover"for grazing, aesthetics, and minimal maintenance.	
5.1.2.2	5.1-3	Vegetation control will be performed in accordance with customary practices. With such a high percentage of the transmission right-of-way crossing productive agricultural land, there will be a minimal amount of vegetation control required. Where the transmission line crosses wooded areas, trees with the potential to impact the lines may be removed or pruned during construction. For maintenance purposes those tree species with the potential for resprouting may be controlled with an environmentally acceptable selective basal spray herbicide. It is not customary for trees to be allowed directly under the transmission lines for approximately 50 ft on either side of the centerline. Trees outside of the 50-ft limit may be maintained through periodic trimming in order to keep them out of the danger timber zone, see Figure 5.1-2. Where the transmission line crosses public roads, a screen of trees may be left to minimize visual impacts from the line. Any new access to the right-of-way, though not anticipated, may be constructed at oblique angles to the road in order to prevent line of sight down the right-of-way, see Figure 5.1-3. Routine inspections of the right-of-way for vegetation control monitoring will be conducted periodically. It is assumed that inspections will be conducted by aircraft in order to determine the need for roads and minimize associated impacts. Maintenance and repair inspections required by cause, such as storms that may down timber on or near the lines, will be conducted by air, road, or foot, as required by the circumstances.	5.1, 5.4.1
5.6	5.6-1	Trees and shrubs that obstruct access along the transmission line right-of-way or pose a safety concern to the lines and pole structures will be removed. The right-of-way will periodically be maintained to control vegetative growth using mechanical mowing (e.g., brush hogs) and selective use of herbicides to control noxious species such as vines that climb poles. It has been assumed that the transmission line will be operated and maintained in accordance with existing approved Illinois Power Company plans and procedures.	2.2.2, 2.7, 3.3, 4.1.2, 4.4, 5.1.2, 5.4, 7.4, 8.5, 8.6
5.6.1	5.6-1	Rights-of-way will be maintained in accordance with the transmission corridor owner or operators plans and procedures.	
5.6.1.1.2	5.6-2	Transmission towers and lines will be located in the vicinity of existing towers and lines; therefore, mortality to any state-listed species of concern (including a variety of birds species discussed in Section 2.4) is not anticipated to increase significantly over current levels.	4.4.1, 5.4.1

Table K-3. (contd)

Environmental Report			Environmental Impact Statement Sections
Section	Page	Environmental Report Statement	
5.6.1.2.3	5.6-3	Appropriate best management practices will be utilized so that adverse impacts to any environmentally sensitive areas potentially occurring along the proposed corridor are avoided during periodic maintenance activities.	5.1
5.6.2 5.10.3.12.2	5.6-4 5.10-17	Appropriate construction procedures and best management practices will be used to minimize disturbances to existing wetlands, floodplains, and other aquatic ecosystems located within or along the existing corridor, during operation and maintenance activities. In marsh and emergent growth, wetlands vegetation maintenance is typically not required. In shrub and forested wetland areas, mowing and trimming is periodically required to keep growth outside of the line areas and away from poles. Periodic maintenance will be performed in accordance with the transmission corridor owner or operators plans and procedures.	4.4.1, 5.4.1
5.6.2.3	5.6-5	Periodic maintenance activities will be performed in accordance with the transmission corridor owner or operators plans and procedures.	5.4.1
5.10.3.12.1	5.10-17	There will be no construction of new right-of-way or access roadways required for the proposed transmission system.	5.4.1
5.10.3.12.1	5.10-17	There may be temporary disturbances to agricultural activities during construction of the proposed transmission system, but following construction, the disturbed areas will be restored to preconstruction activities.	4.4.1
5.10.3.12.1.2 5.6.2 5.6.1.2.5	5.10-17 5.6-4 5.6-3	Towers required to support the proposed transmission system will be sited in upland areas to the greatest extent possble. Appropriate construction procedures and best management practices will be utilized to make certain that the adverse impacts to any environmentally sensitive areas or important habitats potentially occurring along the proposed corridor are avoided.	4.4.1, 5.4.1

NRC FORM 335 (9-2004) NRCMD 3.7	U.S. NUCLEAR REGULATORY COMMISSION	1. REPORT NUMBER (Assigned by NRC, Add Vol., Supp., Rev., and Addendum Numbers, if any.)
BIBLIOGRAPHIC DATA SHEET *(See instructions on the reverse)*		NUREG-1815

2. TITLE AND SUBTITLE	3. DATE REPORT PUBLISHED	
Environmental Impact Statement for an Early Site Permit (ESP) at the Exelon ESP Site, Final Report Volume 2, Appendices A through K	MONTH	YEAR
	July	2006
	4. FIN OR GRANT NUMBER	

5. AUTHOR(S)	6. TYPE OF REPORT
See Appendix A of Report	Technical
	7. PERIOD COVERED *(Inclusive Dates)*

8. PERFORMING ORGANIZATION - NAME AND ADDRESS *(If NRC, provide Division, Office or Region, U.S. Nuclear Regulatory Commission, and mailing address; if contractor, provide name and mailing address.)*

Division of New Reactor Licensing
Office of Nuclear Reactor Regulation
U. S. Nuclear Regulatory Commission
Washington, D.C. 20555-0001

9. SPONSORING ORGANIZATION - NAME AND ADDRESS *(If NRC, type "Same as above"; if contractor, provide NRC Division, Office or Region, U.S. Nuclear Regulatory Commission, and mailing address.)*

Same as above.

10. SUPPLEMENTARY NOTES
Docket No. 52-007

11. ABSTRACT *(200 words or less)*

This report has been prepared in response to an application submitted to the NRC by Exelon Generation Company, LLC, for an early site permit (ESP) for the Exelon ESP site located adjacent to the Clinton Power Station in Clinton, Illinois. The ESP does not authorize construction and operation of a nuclear power plant. However, the application does include a site redress plan that, if approved, would allow limited site preparation work.

The staff's recommendation to the Commission related to the environmental aspects of the proposed action is that the ESP should be issued. This recommendation is based on (1) the application, including the Environmental Report (ER), submitted by Exelon; (2) consultation with Federal, State, Tribal, and local agencies; (3) the staff's independent review; (4) the staff's consideration of comments related to the environmental review that were received; and (5) the assessments summarized in this EIS, including the potential mitigation measures identified in the ER and this EIS. In addition, in making its recommendation, the staff determined that there are no environmentally preferable or obviously superior sites. Finally, the staff has concluded that the site-preparation and construction activities allowed by 10 CFR 50.10(e)(1) requested by Exelon in its application would not result in any significant adverse environmental impact that cannot be redressed.

12. KEY WORDS/DESCRIPTORS *(List words or phrases that will assist researchers in locating the report.)*	13. AVAILABILITY STATEMENT
Clinton Early Site Permit ESP National Environmental Policy Act NEPA Exelon Environmental Impact Statement	unlimited
	14. SECURITY CLASSIFICATION
	(This Page) unclassified
	(This Report) unclassified
	15. NUMBER OF PAGES
	16. PRICE

NRC FORM 335 (9-2004)
PRINTED ON RECYCLED PAPER

www.ingramcontent.com/pod-product-compliance
Lightning Source LLC
Chambersburg PA
CBHW081428170526
45166CB00008B/2126